# Ergebnisse der Mathematik und ihrer Grenzgebiete 91

*A Series of Modern Surveys in Mathematics*

Robert C. Gunning

# Riemann Surfaces and Generalized Theta Functions

With 5 Figures

Springer-Verlag
Berlin Heidelberg New York 1976

Robert C. Gunning

Princeton University, Dept. of Mathematics,
Princeton, NJ 08540/U.S.A.

AMS Subject Classification (1970): primary 14 K 25,
secondary 14 H 40, 14 F 05, 30 A 46, 32 L 05

ISBN-13: 978-3-642-66384-0 e-ISBN-13: 978-3-642-66382-6
DOI: 10.1007/978-3-642-66382-6

Library of Congress Cataloging in Publication Data. Gunning, Robert Clifford, 1931—
Riemann surfaces and generalized theta functions. (Ergebnisse der Mathematik und ihrer Grenzgebiete; 91). Includes bibliographical references and index. 1. Riemann surfaces. 2. Functions, Theta. I. Title. II. Series. QA333.G83. 515'.73. 76-12579.


Softcover reprint of the hardcover 1st edition 1976

*To Wanda*

# Preface

The investigation of the relationships between compact Riemann surfaces (algebraic curves) and their associated complex tori (Jacobi varieties) has long been basic to the study both of Riemann surfaces and of complex tori. A Riemann surface is naturally imbedded as an analytic submanifold in its associated torus; and various spaces of linear equivalence classes of divisors on the surface (or equivalently spaces of analytic equivalence classes of complex line bundles over the surface), classified according to the dimensions of the associated linear series (or the dimensions of the spaces of analytic cross-sections), are naturally realized as analytic subvarieties of the associated torus.

One of the most fruitful of the classical approaches to this investigation has been by way of theta functions. The space of linear equivalence classes of positive divisors of order $g-1$ on a compact connected Riemann surface $M$ of genus $g$ is realized by an irreducible $(g-1)$-dimensional analytic subvariety, an irreducible hypersurface, of the associated $g$-dimensional complex torus $J(M)$; this hypersurface $W_{g-1} \subseteq J(M)$ is the image of the natural mapping $M^{g-1} \to J(M)$, and is birationally equivalent to the $(g-1)$-fold symmetric product $M^{g-1}/S_{g-1}$ of the Riemann surface $M$. There is a well developed theory of linear equivalence classes of divisors on a complex torus $J(M)$, or equivalently of complex analytic equivalence classes of complex line bundles over $J(M)$; and that theory leads to an explicit description of any hypersurface of $J(M)$ as the zero locus of a complex analytic function on the universal covering space $\mathbb{C}^g$ of $J(M)$ with specified periodicity properties. Thus the hypersurface $W_{g-1} \subseteq J(M)$ can be described as the zero locus of an explicitly given theta function on $\mathbb{C}^g$. The original Riemann surface, realized as an analytic submanifold $W_1 \subseteq J(M)$, is almost fully determined by the hypersurface $W_{g-1}$ alone, hence by the explicitly given theta function, albeit in a rather indirect and complicated manner; that is the essential content of Torelli's theorem. Many other properties of the subvarieties of special positive divisors in $J(M)$, hence of the original Riemann surface $M$, can also be determined merely from the hypersurface $W_{g-1}$ or its defining theta function; and that is the topic of many classical and modern treatises.

The object of this volume however is to demonstrate that the original Riemann surface, realized as the submanifold $W_1 \subseteq J(M)$, and the subvarieties of special positive divisors in $J(M)$ can be described more directly and simply in terms of certain natural generalizations of the classical theta functions. These generalized theta functions are vector-valued functions on $\mathbb{C}^g$ which represent cross-sections of some complex vector bundles over $J(M)$, as the classical theta functions are scalar-valued functions on $\mathbb{C}^g$ which represent cross-sections of some complex line bundles over $J(M)$. Paralleling the classical case there are two aspects to the theory of generalized theta functions: the first is the classification and explicit description of complex analytic vector bundles over $J(M)$ and of their complex analytic cross-sections; the second is the determination of the particular vector bundles, the cross-sections of which describe the subvarieties of special positive divisors in $J(M)$, and the analysis of the properties of these bundles which most closely reflect properties of the Riemann surface $M$. The first aspect really involves the study of the complex manifolds $J(M)$, and is still relatively little developed; the second aspect involves the study of the Riemann surfaces $M$, and is somewhat better developed. The present volume will be limited to the second aspect alone, so can be viewed primarily as a study of Riemann surfaces; the task of completing this study and making it more explicit by a corresponding treatment of the first aspect remains for another work.

In order to make this volume more nearly self contained two preliminary background chapters and an appendix have been included. The first chapter contains a general survey of complex manifolds and complex vector bundles, developed more or less from the beginning. Complex tori, as particular examples of complex manifolds, are examined in more detail. The representation of complex vector bundles over tori and Riemann surfaces by factors of automorphy is also examined in some detail, since it provides the most explicit and classically oriented approach to complex vector bundles as well as the approach which seemed most convenient to use in the present study. The second chapter contains a survey of some properties of compact Riemann surfaces, also developed more or less from the beginning. Needless to say no attempt has been made to include complete proofs of everything; but those results not proved have been limited to a few easily stated general theorems (the basic existence theorem for functions on Riemann surfaces, the weak form of Abel's theorem, and the Riemann-Roch theorem), explicit references for which are given. Also needless to say no attempt has been made to give a complete and balanced survey of the theory of Riemann surfaces; the topics chosen and the points of view adopted have been rather obviously influenced by the main object of this volume, the study of generalized theta functions.

The generalized theta functions are introduced and studied in some detail in the third chapter, which is the core of this volume. The principal application of the generalized theta functions in this chapter is to the explicit description of the subvarieties of special positive divisors in the Jacobi variety. It might be worth noting in passing that Jacobi varieties of Riemann surfaces can by characterized among all Abelian varieties as those admitting a complex vector bundle of rank $g$ having the properties of the vector bundle described by the theta factor of automorphy of rank $g$; but this is not a particularly useful characterization without a more explicit description of the theta factor of automorphy of rank $g$, and that is not within the scope of this volume. Another application of the generalized theta functions, to the description of Prym differentials and their periods, is given in the fourth chapter. The explicit determination of the periods of Prym differentials is a longstanding if rather implicitly treated problem. The Prym periods seem to be functions on the Teichmüller space rather than on the Torelli space, so cannot be expressed in terms of simple functions of the Jacobi variety such as the classical theta functions. However the generalized theta functions describe the Riemann surface itself, within its Jacobi variety, so do involve more than just the Torelli point representing the surface; and the Prym periods can be determined to a considerable extent merely in terms of the theta factors of automorphy restricted to the imbedded Riemann surface, as noted in §§ 13 and 14.

The appendix contains a discussion of some topics in the classical theory of theta functions, included both for completeness and as a model for a theory of generalized theta functions. The first two sections cover the classification and explicit description of factors of automorphy and their associated relatively automorphic functions, the first aspect of the classical theory of theta functions. The third section covers the determination of those factors of automorphy which describe the subvariety $W_{g-1} \subseteq J(M)$ and includes a discussion of some properties of theta functions more closely related to properties of the original Riemann surface, the second aspect of the classical theory of theta functions; it should be noted that much of this discussion does not really involve the explicit form of the theta factors of automorphy or the theta functions, although these forms have been used as is traditional to simplify the treatment somewhat.

In conclusion I should like here to express my thanks to the National Science Foundation for support while the research for this book and the writing of it were being undertaken, to many students and colleagues for their very helpful suggestions and comments, to Lorretta Landolt for the typing of the manuscript, and to the staff of Springer-Verlag for all their efforts.

Robert C. Gunning

Princeton, New Jersey
December, 1975

# Table of Contents

# Chapter I. Complex Manifolds and Vector Bundles

## § 1. Complex Manifolds and the Example of the Complex Tori

An $n$-dimensional topological manifold is a Hausdorff topological space $M$ such that every point $p \in M$ has an open neighborhood homeomorphic to an open subset of the $n$-dimensional number space $\mathbb{R}^n$. A *coordinate covering* $\{U_\alpha, z_\alpha\}$ of such a manifold $M$ consists of a covering of $M$ by open subsets $U_\alpha$ together with homeomorphisms $z_\alpha\colon U_\alpha \longrightarrow V_\alpha$ between the sets $U_\alpha$ and open subsets $V_\alpha \subseteq \mathbb{R}^n$; the sets $U_\alpha$ are called *coordinate neighborhoods* and the mappings $z_\alpha$ are called *coordinate mappings*. A topological manifold of course always admits coordinate coverings. Note that on each nonempty intersection $U_\alpha \cap U_\beta$ of coordinate neighborhoods there are thus two homeomorphisms into $\mathbb{R}^n$; the compositions

$$f_{\alpha\beta} = z_\alpha \circ z_\beta^{-1}\colon z_\beta(U_\alpha \cap U_\beta) \longrightarrow z_\alpha(U_\alpha \cap U_\beta)$$

are called the *coordinate transition functions* of the coordinate covering, and for any point $p \in U_\alpha \cap U_\beta$ the two coordinate mappings are related by $z_\alpha(p) = f_{\alpha\beta}(z_\beta(p))$. The manifold $M$ is completely determined by the sets $\{V_\alpha\}$ and the mappings $\{f_{\alpha\beta}\}$; for $M$ can be obtained from the disjoint union of all the sets $V_\alpha$ by identifying a point $z_\alpha \in V_\alpha$ and a point $z_\beta \in V_\beta$ whenever $z_\alpha = f_{\alpha\beta}(z_\beta)$.

Suppose that $\{U_\alpha, z_\alpha\}$ and $\{U'_\beta, z'_\beta\}$ are two coordinate coverings of the manifold $M$. The *union* of these coordinate coverings is the coordinate covering consisting of all the coordinate neighborhoods and coordinate mappings from the two given coordinate coverings. It is important to observe that the set of coordinate transition functions for the union of these two coordinate coverings is properly larger than the union of the sets of coordinate transition functions of the two separate coordinate coverings; for in addition to the coordinate transition functions corresponding to nonempty intersections $U_\alpha \cap U_\beta$ or $U'_\mu \cap U'_\nu$ there are those corresponding to nonempty intersections $U_\alpha \cap U'_\mu$.

Coordinate coverings with special properties can be used to describe additional structures on a topological manifold. If $M$ is a topological manifold of dimension $2n$ and if the vector space $\mathbb{R}^{2n}$ is identified with the complex vector space $\mathbb{C}^n$ then the coordinate mappings of a coordinate covering can be viewed as mappings into $\mathbb{C}^n$. A coordinate covering $\{U_\alpha, z_\alpha\}$ is then called a *complex analytic coordinate covering* if all the coordinate transition functions are complex analytic mappings. Two complex analytic coordinate coverings are called *equivalent* if their union is also a complex analytic coordinate covering. This is readily seen to be an equivalence relation in the traditional sense; indeed symmetry and reflexivity are quite trivial, and it is only necessary to verify transitivity. For that purpose consider complex analytic coordinate coverings $\{U'_\beta, z'_\beta\}$ equivalent to $\{U_\alpha, z_\alpha\}$ and $\{U''_\gamma, z''_\gamma\}$ equivalent to $\{U'_\beta, z'_\beta\}$, and consider a point $p \in U''_\gamma \cap U_\alpha$; there is then a coordinate neighborhood $U'_\beta$ containing $p$, and since in an open neighborhood of that point $f_{\alpha\gamma} = f_{\alpha\beta} \circ f_{\beta\gamma}$ and $f_{\alpha\beta}, f_{\beta\gamma}$ are complex analytic mappings, necessarily $f_{\alpha\gamma}$ is also a complex analytic mapping and consequently $\{U''_\gamma, z''_\gamma\}$ is equivalent to $\{U_\alpha, z_\alpha\}$ as desired. An equivalence class of complex analytic coordinate coverings of $M$ is called a *complex structure* on $M$; and a topological manifold with a fixed complex structure is called a *complex manifold.*

The only property of complex analytic mappings needed to show that equivalence is indeed an equivalence relation is that the composition of two complex analytic mappings is again a complex analytic mapping; consequently other structures can similarly be defined on topological manifolds, corresponding to any other classes of local homeomorphisms which are closed under composition. For example, a coordinate covering is called a $C^\infty$ coordinate covering if the coordinate transition functions are $C^\infty$ mappings; an equivalence class of $C^\infty$ coordinate coverings is called a $C^\infty$ structure, and a topological manifold with a fixed $C^\infty$ structure is called a $C^\infty$ manifold. A complex analytic coordinate covering is of course also a $C^\infty$ coordinate covering, and equivalent complex analytic coordinate coverings are also equivalent $C^\infty$ coordinate coverings; thus a complex structure on a topological manifold is associated to a unique $C^\infty$ structure on that manifold. In this sense a complex structure is finer than a $C^\infty$ structure, or equivalently, a complex manifold is a $C^\infty$ manifold with an additional structure.

A complex-valued function $f$ on an open subset $D$ of a complex manifold $M$ is called a *complex analytic function* if for all complex analytic coordinate coverings $\{U_\alpha, z_\alpha\}$ of $M$ the functions $f \circ z_\alpha^{-1}: z_\alpha(U_\alpha \cap D) \longrightarrow \mathbb{C}$ are complex analytic functions in the usual sense; it is clearly sufficient merely to impose this condition for any one complex analytic coordinate covering of $M$. Similarly a mapping $f: M \longrightarrow M'$ between two complex manifolds is called a *complex analytic mapping* if for all complex analytic coordinate coverings $\{U_\alpha, z_\alpha\}$ of $M$ and $\{U'_\beta, z'_\beta\}$ of $M'$ and all points $p \in U_\alpha$ with $f(p) \in U'_\beta$ the mappings $z_\beta \circ f \circ z_\alpha^{-1}$ are complex analytic mappings in the usual sense in some open neighborhoods of the points $z_\alpha(p) \in \mathbb{C}^n$; again it is clearly sufficient merely to impose this condition for any one coordinate covering of $M$ and any one coordinate covering of $M'$. Note that complex analytic functions or mappings are necessarily continuous, indeed are even $C^\infty$. Two complex analytic manifolds $M$ and $M'$ are called *isomorphic* if there is a complex analytic mapping $f: M \longrightarrow M'$ which is a homeomorphism between these two topological spaces; the inverse mapping $f^{-1}: M' \longrightarrow M$ is then of course also a complex analytic mapping. As a matter of practice, isomorphic complex manifolds will usually be identified with one another, so that a complex manifold will really be a class of isomorphic complex manifolds.

Two classes of complex manifolds will be of particular interest in this book. The first is the class of one-dimensional complex manifolds or *Riemann surfaces*;

a detailed discussion of the properties of these manifolds will be taken up later. The second is the class of complex tori; some elementary properties of these manifolds will be discussed here, to provide explicit examples.

A lattice subgroup $\mathscr{L}$ of a vector space $\mathbb{R}^n$ is an additive subgroup generated by a set of vectors which are linearly independent over the real numbers; the number of generators is called the rank of $\mathscr{L}$, and is of course at most $n$. After a change of coordinates in $\mathbb{R}^n$ the generators of $\mathscr{L}$ can be reduced to the first $r$ basis vectors of $\mathbb{R}^n$, where $r$ is the rank of $\mathscr{L}$; and consequently the quotient group $\mathbb{R}^n/\mathscr{L}$ can be identified with $(\mathbb{R}/\mathbb{Z})^r \times \mathbb{R}^{n-r}$, which is topologically the product of an $r$-dimensional torus (the product of $r$ circles) with an $(n-r)$-dimensional real vector space. In particular when $r=n$ the quotient group is a compact $n$-dimensional torus. A complex vector space $\mathbb{C}^n$ can also be viewed as a real vector space $\mathbb{R}^{2n}$, and by a lattice subgroup $\mathscr{L}$ of $\mathbb{C}^n$ is meant merely a lattice subgroup of the underlying real vector space $\mathbb{R}^{2n}$; thus the rank of $\mathscr{L}$ is at most $2n$, and is equal to $2n$ precisely when the quotient group $\mathbb{C}^n/\mathscr{L}$ is topologically a compact torus of dimension $2n$. The quotient space $\mathbb{C}^n/\mathscr{L}$ has a canonical complex structure, in which any open subset of $\mathbb{C}^n$ which is naturally mapped homeomorphically into $\mathbb{C}^n/\mathscr{L}$ can be taken as a coordinate neighborhood; the coordinate transition functions for this coordinate covering are indeed just translations. A complex manifold of the form $\mathbb{C}^n/\mathscr{L}$ where $\mathscr{L}$ has rank $2n$ is called a *complex torus.*

A complex linear mapping from $\mathbb{C}^m$ into $\mathbb{C}^n$ which maps a lattice subgroup $\mathscr{L}\subset\mathbb{C}^m$ into a lattice subgroup $\mathscr{L}'\subset\mathbb{C}^n$ induces a complex analytic mapping between the complex manifolds $\mathbb{C}^m/\mathscr{L}$ and $\mathbb{C}^n/\mathscr{L}'$. If $\mathbb{C}^m/\mathscr{L}$ is a complex torus (so that $\mathscr{L}$ has rank $2m$) then conversely any complex analytic mapping from $\mathbb{C}^m/\mathscr{L}$ into $\mathbb{C}^n/\mathscr{L}'$ is induced by a complex linear mapping from $\mathbb{C}^m$ into $\mathbb{C}^n$. To see this note that any complex analytic mapping $f\colon \mathbb{C}^m/\mathscr{L} \longrightarrow \mathbb{C}^n/\mathscr{L}'$ arises from a complex analytic mapping $F\colon \mathbb{C}^m \longrightarrow \mathbb{C}^n$, since $\mathbb{C}^m$, $\mathbb{C}^n$ are the universal covering spaces of $\mathbb{C}^m/\mathscr{L}$, $\mathbb{C}^n/\mathscr{L}'$ respectively, and that for any element $\lambda\in\mathscr{L}$ there will exist an element $\lambda'\in\mathscr{L}'$ such that $F(p+\lambda)=F(p)+\lambda'$ for all points $p\in\mathbb{C}^m$. Each coordinate function $F_i$ of the mapping $F$ then satisfies the corresponding relation $F_i(p+\lambda)=F_i(p)+\lambda'_i$, hence the first partial derivatives of that function are invariant under translation by any element of $\mathscr{L}$; since $\mathbb{C}^m/\mathscr{L}$ is compact these partial derivatives attain their maxima and must therefore be constant as a consequence of the maximum modulus theorem, and as a result the function $F_i$ is linear as desired. In particular therefore it is evident that two complex tori $\mathbb{C}^n/\mathscr{L}$ and $\mathbb{C}^n/\mathscr{L}'$ are isomorphic precisely when there is a nonsingular linear mapping $F\colon \mathbb{C}^n \longrightarrow \mathbb{C}^n$ such that $F(\mathscr{L})=\mathscr{L}'$.

The generators of the lattice subgroup defining a complex torus are linearly independent over the real numbers but not of course over the complex numbers; two lattice subgroups of rank $2n$ in $\mathbb{C}^n$ cannot generally be transformed into one another by a nonsingular complex linear change of coordinates in $\mathbb{C}^n$, so that not all complex tori of complex dimension $n$ are isomorphic complex manifolds. To discuss the isomorphism classification of complex tori it is convenient to be somewhat more explicit. Elements of the complex vector space $\mathbb{C}^n$ will be viewed as column vectors of length $n$, or equivalently as $n\times 1$ matrices, with the usual matrix operations; and the transpose of a matrix $\lambda$ will be denoted by ${}^t\lambda$, so that if $\lambda \in \mathbb{C}^n$ then ${}^t\lambda$ is a row vector of length $n$, or equivalently a $1 \times n$ matrix. Choosing a set of generators $\lambda_1, \ldots, \lambda_{2n}$ for a lattice subgroup $\mathcal{L} \subset \mathbb{C}^n$ of rank $2n$, where each $\lambda_i \in \mathbb{C}^n$ is then viewed as a column vector of length $n$, the corresponding *period matrix* for the lattice $\mathcal{L}$ is the $n \times 2n$ complex matrix $\Lambda = (\lambda_1, \ldots, \lambda_{2n})$. The lattice $\mathcal{L}$ is then of course described by the period matrix $\Lambda$, in the sense that $\mathcal{L} = \Lambda \cdot \mathbb{Z}^{2n} = \{\Lambda u \mid u \in \mathbb{Z}^{2n}\}$ where $\mathbb{Z}^{2n}$ denotes the set of all column vectors of length $2n$ having integral entries.

**Theorem 1a.** *An $n \times 2n$ complex matrix $\Lambda$ is the period matrix for an n-dimensional complex torus precisely when it satisfies any one of the following equivalent conditions:*

(i) *if $\Lambda x = 0$ for $x \in \mathbb{R}^{2n}$ then $x = 0$;*
(ii) *if ${}^t z \Lambda \in \mathbb{R}^{2n}$ for $z \in \mathbb{C}^n$ then $z = 0$;*
(iii) *the $2n \times 2n$ matrix*

$$\begin{pmatrix} \Lambda \\ \overline{\Lambda} \end{pmatrix}$$

*is nonsingular.*

*Proof.* By definition the columns of an $n \times 2n$ matrix $\Lambda$ generate a lattice subgroup $\mathcal{L} \subset \mathbb{C}^n$ of rank $2n$ precisely when they are linearly independent over the real numbers, which is exactly condition (i). To see that conditions (i) and (iii) are equivalent note that the converse of condition (i) is that there exists a nonzero vector $x \in \mathbb{R}^{2n}$ such that $\Lambda x = 0$; writing $\Lambda = \Lambda' + i\Lambda''$ where $\Lambda', \Lambda''$ are $n \times 2n$ real matrices, the latter condition can be rewritten as $\Lambda' x = \Lambda'' x = 0$, or alternatively as $\begin{pmatrix} \Lambda' \\ \Lambda'' \end{pmatrix} x = 0$, and that is just the condition that the matrix $\begin{pmatrix} \Lambda' \\ \Lambda'' \end{pmatrix}$ is singular, which is evidently equivalent to the converse of condition (iii). To see that conditions (ii) and (iii) are equivalent note that the converse of condition (ii) is that there exists a nonzero vector $z \in \mathbb{C}^n$ such that ${}^t z \Lambda \in \mathbb{R}^{2n}$, hence such that ${}^t z \Lambda = {}^t \bar{z} \bar{\Lambda}$; but then $({}^t z, -{}^t \bar{z}) \begin{pmatrix} \Lambda \\ \overline{\Lambda} \end{pmatrix} = 0$, which implies the converse of condition (iii). On the other hand the converse of condition (iii) can be rewritten as the assertion that there exists a nonzero vector $x \in \mathbb{R}^{2n}$ such that ${}^t x \begin{pmatrix} \Lambda' \\ \Lambda'' \end{pmatrix} = 0$, where $\Lambda = \Lambda' + i\Lambda''$ as before, or setting ${}^t x = ({}^t b, {}^t a)$ where $a, b \in \mathbb{R}^n$, such that ${}^t b \Lambda' + {}^t a \Lambda'' = 0$; but then $z = a + ib \in \mathbb{C}^n$ is a nonzero vector such that $\operatorname{Im} {}^t z \Lambda = {}^t a \Lambda'' + {}^t b \Lambda' = 0$, which implies the converse of condition (ii). That suffices to conclude the proof of the theorem.

**Theorem 1b.** *The complex tori described by period matrices $\Lambda, \Lambda'$ are isomorphic complex manifolds precisely when $\Lambda' = M \Lambda N$ for some matrices $M \in GL(n, \mathbb{C})$, $N \in GL(2n, \mathbb{Z})$.*

*Proof.* Note first that two period matrices $\Lambda, \Lambda'$ describe the same lattice subgroup of $\mathbb{C}^n$ precisely when the columns of $\Lambda$ belong to the lattice $\Lambda' \mathbb{Z}^{2n}$ and the columns of $\Lambda'$ belong to the lattice $\Lambda\mathbb{Z}^{2n}$, or equivalently when $\Lambda=\Lambda'N'$ and $\Lambda'=\Lambda N$ for some $2n\times 2n$ integral matrices $N, N'$. Since then

$$\begin{pmatrix}\Lambda\\ \overline{\Lambda}\end{pmatrix}=\begin{pmatrix}\Lambda'\\ \overline{\Lambda'}\end{pmatrix}N'=\begin{pmatrix}\Lambda\\ \overline{\Lambda}\end{pmatrix}NN'$$

while $\begin{pmatrix}\Lambda\\ \overline{\Lambda}\end{pmatrix}$ is nonsingular by Theorem 1a, the matrix $N$ must be an invertible integral matrix; that is to say, $N\in GL(2n,\mathbb{Z})$ where as usual $GL(m,R)$ denotes the group of invertible $m\times m$ matrices over the ring $R$, and $N'=N^{-1}$. On the other hand, it was noted earlier that the complex tori described by lattice subgroups $\mathscr{L},\mathscr{L}'$ are isomorphic complex manifolds precisely when there is a nonsingular linear transformation $F:\mathbb{C}^n\longrightarrow\mathbb{C}^n$ such that $\mathscr{L}'=F(\mathscr{L})$; and if the transformation $F$ is represented by a matrix $M\in GL(n,\mathbb{C})$ this is just the condition that $\Lambda'=C\Lambda$ where $\Lambda',\Lambda$ are some period matrices describing the lattice subgroups $\mathscr{L}',\mathscr{L}$ respectively. The desired theorem is an immediate consequence of these two observations.

**Theorem 1c.** *Any complex torus of complex dimension $n$ can be described by a period matrix of the form $(I,\Lambda_0)$ where $I$ is the $n\times n$ identity matrix and $\Lambda_0$ is an $n\times n$ complex matrix with nonsingular imaginary part. The complex tori described by period matrices $(I,\Lambda_0)$ and $(I,\Lambda_0')$ are isomorphic complex manifolds precisely when $\Lambda_0'=(A+\Lambda_0C)^{-1}(B+\Lambda_0D)$ where $A,B,C,D$ are $n\times n$ integer matrices such that $\begin{pmatrix}A & B\\ C & D\end{pmatrix}\in GL(2n,\mathbb{Z})$.*

*Proof.* It is clear from Theorem 1a that the period matrix $\Lambda$ of any complex torus of dimension $n$ is of rank $n$. Upon interchanging the columns of $\Lambda$ by replacing $\Lambda$ by $\Lambda N$ for suitable matrix $N\in GL(2n,\mathbb{Z})$ it can be assumed, recalling Theorem 1b, that the first $n$ columns of $\Lambda$ are linearly independent complex vectors; thus $\Lambda=(\Lambda_1,\Lambda_2)$ where $\Lambda_1,\Lambda_2$ are $n\times n$ complex matrices with $\Lambda_1\in GL(n,\mathbb{C})$, and replacing $\Lambda$ by $\Lambda_1^{-1}\Lambda$ yields a period matrix of the form $(I,\Lambda_0)$ describing the same complex torus, again recalling Theorem 1b. It is evident from Theorem 1a (iii) that a matrix of the form $(I,\Lambda_0)$ is the period matrix of a complex torus precisely when the imaginary part of the matrix $\Lambda_0$ is nonsingular. It further follows from Theorem 1b that period matrices $(I,\Lambda_0)$ and $(I,\Lambda_0')$ describe isomorphic complex tori precisely when $(I,\Lambda_0')=M(I,\Lambda_0)N$ for some matrices $M\in GL(n,\mathbb{C})$, $N\in GL(2n,\mathbb{Z})$. Writing $N=\begin{pmatrix}A & B\\ C & D\end{pmatrix}$ where $A,B,C,D$ are $n\times n$ integral matrices, the latter condition reduces to the two equations $I=M(A+\Lambda_0C)$ and $\Lambda_0'=M(B+\Lambda_0D)$; consequently $M=(A+\Lambda_0C)^{-1}$ and $\Lambda_0'=(A+\Lambda_0C)^{-1}(B+\Lambda_0D)$ as desired. That serves to conclude the proof.

As a particular consequence of Theorem 1c, one-dimensional complex tori are described by period matrices of the form $(1,\lambda)$ where $\lambda$ is a complex number with nonzero imaginary part; and period matrices $(1,\lambda)$ and $(1,\lambda')$ describe isomorphic complex tori precisely when $\lambda'=(a+\lambda c)^{-1}(b+\lambda d)$ where $\begin{pmatrix} a & b \\ c & d \end{pmatrix} \in GL(2,\mathbb{Z})$. This is of course a very familiar result. The equivalence classes of such complex numbers $\lambda$ under this action of the group $GL(2,\mathbb{Z})$ are called the moduli of the associated complex tori, since they describe the isomorphism classes of these tori. Thus the isomorphism classes of one-dimensional complex tori can be put in one-to-one correspondence with the quotient space of the set of complex numbers with nonzero imaginary part under this action of the group $GL(2,\mathbb{Z})$, or equivalently with the quotient space of the set of complex numbers with positive imaginary part under this action of the subgroup $SL(2,\mathbb{Z})$ consisting of those matrices in $GL(2,\mathbb{Z})$ having determinant 1; and it is again quite familiar that this quotient space also has the structure of a one-dimensional complex manifold isomorphic to the full complex plane. The situation in the higher-dimensional case is somewhat more complicated, since the group $GL(2n,\mathbb{Z})$ does not act properly discontinuously on the set of all $n\times n$ matrices with nonsingular imaginary part when $n>1$, so the set of isomorphism classes of all complex tori of dimension $n$ cannot be given the natural structure of a complex manifold (or even of a complex variety) when $n>1$.

## § 2. Complex Analytic Vector Bundles and the Example of Line Bundles of Divisors

A *complex analytic family of vector spaces* over a complex manifold $M$ is a complex manifold $\xi$ together with a complex analytic mapping $\pi: \xi \longrightarrow M$ and a finite-dimensional complex vector space structure on the subset $\pi^{-1}(p)\subseteq\xi$ for each point $p\in M$. The mapping $\pi$ is called the *projection*; and the subset $\pi^{-1}(p)$, also denoted by $\xi_p$, is called the *fibre* over the point $p$. Note that for any open subset $U\subseteq M$ the inverse image $\pi^{-1}(U)\subseteq\xi$ has the natural structure of a complex analytic family of vector spaces over $U$; this is called the *restriction* of the family $\xi$ to the subset $U$ and is denoted by $\xi|U$. The simplest example of a complex analytic family of vector spaces over $M$ is a product manifold $\xi=M\times\mathbb{C}^n$ with the natural projection mapping $\pi: M\times\mathbb{C}^n\longrightarrow M$ and the given vector space structure on each fibre $p\times\mathbb{C}^n$; this is called a *trivial* family of vector spaces over $M$. A *homomorphism* from one complex analytic family of vector spaces $\pi': \xi'\longrightarrow M$ to another $\pi'': \xi''\longrightarrow M$ is a complex analytic mapping $\phi: \xi'\longrightarrow\xi''$ such that $\pi''\phi=\pi'$ and the restriction $\phi|\xi'_p: \xi'_p\longrightarrow\xi''_p$ is a complex linear mapping for each point $p\in M$. An *isomorphism* is a homomorphism for which the mapping $\phi: \xi'\longrightarrow\xi''$ is a topological homeomorphism, hence for which the inverse mapping $\phi^{-1}: \xi''\longrightarrow\xi'$ is also a homomorphism; and two complex analytic families of vector spaces are called *isomorphic* if there is an isomorphism from one to another. Again as a matter of practice isomorphic families will usually be identified with one another, so that a family of vector spaces will really be a class of isomorphic families. A *complex analytic vector bundle* over the complex manifold $M$ is a complex analytic family of vector spaces $\pi: \xi \longrightarrow M$ which is locally trivial in the sense that each point $p \in M$ has an open neighborhood $U$ such that the restriction $\xi|U$ is isomorphic to a trivial family of complex vector spaces over $U$. Note that for a complex analytic vector bundle $\pi: \xi \longrightarrow M$ the dimension of the complex vector space $\xi_p$ is constant as a function of $p \in M$ on each connected component of $M$; that constant is called the *rank* of the complex analytic vector bundle over that component. A complex analytic vector bundle of rank one is also called a *complex analytic line bundle.*

Of course the corresponding concepts can be introduced in the categories of differentiable manifolds and differentiable mappings or of topological spaces and continuous mappings as in the category of complex analytic manifolds and complex analytic mappings; these are the concepts of differentiable complex vector bundles or of continuous complex vector bundles, the latter of which are usually simply called complex vector bundles. As in the case of manifolds a complex analytic vector bundle is associated to a unique underlying differentiable complex vector bundle, and a differentiable complex vector bundle is in turn associated to a unique underlying continuous complex vector bundle. Two complex analytic bundles are called topologically isomorphic if their underlying continuous complex vector bundles are isomorphic.

If $\pi: \xi \longrightarrow M$ is a complex analytic vector bundle of rank $r$ over the complex manifold $M$ then the complex manifold $\xi$ admits special coordinate coverings which reflect this additional structure. There are coordinate coverings $\{U_\alpha, z_\alpha\}$ of the complex manifold $M$ for which each restriction $\xi|U_\alpha$ is isomorphic to a trivial complex analytic vector bundle of rank $r$ over $U_\alpha$, hence for which there are complex analytic homeomorphisms $\xi_\alpha: \pi^{-1}(U_\alpha) \longrightarrow U_\alpha \times \mathbb{C}^r$ such that for any intersection $\pi^{-1}(U_\alpha) \cap \pi^{-1}(U_\beta) = \pi^{-1}(U_\alpha \cap U_\beta)$ the composite mapping

$$\xi_\alpha \circ \xi_\beta^{-1}: (U_\alpha \cap U_\beta) \times \mathbb{C}^r \longrightarrow (U_\alpha \cap U_\beta) \times \mathbb{C}^r$$

is a complex analytic homeomorphism of the form $\xi_\alpha \circ \xi_\beta^{-1}(p,t) = (p, \xi_{\alpha\beta}(p,t))$ where $\xi_{\alpha\beta}(p,t)$ is complex linear in $t \in \mathbb{C}^r$; thus representing the points of $\mathbb{C}^r$ by column vectors of length $r$ as before, there is a complex analytic mapping $\xi_{\alpha\beta}: U_\alpha \cap U_\beta \longrightarrow GL(r, \mathbb{C})$ such that $\xi_{\alpha\beta}(p,t) = \xi_{\alpha\beta}(p) \cdot t$. The composition of the homeomorphisms $\xi_\alpha: \pi^{-1}(U_\alpha) \longrightarrow U_\alpha \times \mathbb{C}^r$ and $z_\alpha: U_\alpha \longrightarrow z_\alpha(U_\alpha) \subseteq \mathbb{C}^n$ is then a homeomorphism $\tilde{\xi}_\alpha: \pi^{-1}(U_\alpha) \longrightarrow z_\alpha(U_\alpha) \times \mathbb{C}^r \subseteq \mathbb{C}^n \times \mathbb{C}^r$, where $M$ is an $n$-dimensional complex manifold; and the sets $\pi^{-1}(U_\alpha)$ together with the homeomorphisms $\tilde{\xi}_\alpha$ form a coordinate covering of the complex manifold $\xi$ with coordinate transition functions of the form

$$(z_\alpha, t_\alpha) = \tilde{\xi}_\alpha \circ \tilde{\xi}_\beta^{-1}(z_\beta, t_\beta) = (f_{\alpha\beta}(z_\beta), \xi_{\alpha\beta}(z_\beta) t_\beta)$$

where $f_{\alpha\beta}$ are the coordinate transition functions for the given coordinate covering of the complex manifold $M$ and $\xi_{\alpha\beta}(z_\beta(p)) = \xi_{\alpha\beta}(p)$ for any point $p \in U_\alpha \cap U_\beta$. A coordinate covering of this form is called a *complex analytic coordinate bundle* for the complex analytic vector bundle $\pi: \xi \longrightarrow M$; the coordinates $t_\alpha \in \mathbb{C}^r$ are called the *fibre coordinates*, and the mappings $\xi_{\alpha\beta}: U_\alpha \cap U_\beta \longrightarrow GL(r, \mathbb{C})$ are called the *coordinate transformations* for the coordinate bundle. The complex manifold $\xi$ is as usual completely determined by the sets $z_\alpha(U_\alpha)\times\mathbb{C}^r\subseteq\mathbb{C}^n\times\mathbb{C}^r$ and the mappings $f_{\alpha\beta}$, $\xi_{\alpha\beta}$; for $\xi$ can be obtained from the disjoint union of the sets $z_\alpha(U_\alpha)\times\mathbb{C}^r$ by identifying a point $(z_\alpha,t_\alpha)\in z_\alpha(U_\alpha)\times\mathbb{C}^r$ and a point $(z_\beta,t_\beta)\in z_\beta(U_\beta)\times\mathbb{C}^r$ whenever $z_\alpha=f_{\alpha\beta}(z_\beta)$ and $t_\alpha=\xi_{\alpha\beta}(z_\beta)t_\beta$.

Note that the description of a complex analytic coordinate bundle for a complex analytic vector bundle $\pi:\xi\longrightarrow M$ really involves merely a coordinate covering $\{U_\alpha,z_\alpha\}$ of the manifold $M$ and the coordinate transformations $\{\xi_{\alpha\beta}\}$; and these coordinate transformations evidently satisfy the consistency conditions $\xi_{\alpha\beta}(p)\,\xi_{\beta\gamma}(p)=\xi_{\alpha\gamma}(p)$ whenever $p\in U_\alpha\cap U_\beta\cap U_\gamma$. Conversely given any coordinate covering $\{U_\alpha,z_\alpha\}$ of a complex manifold $M$ and any complex analytic mappings $\xi_{\alpha\beta}\colon U_\alpha\cap U_\beta\longrightarrow GL(r,\mathbb{C})$ such that $\xi_{\alpha\beta}(p)\,\xi_{\beta\gamma}(p)=\xi_{\alpha\gamma}(p)$ whenever $p\in U_\alpha\cap U_\beta\cap U_\gamma$ there is a complex analytic vector bundle $\pi:\xi\longrightarrow M$ with a complex analytic coordinate bundle having the coordinate transformations $\{\xi_{\alpha\beta}\}$. Indeed $\xi$ is obtained by factoring the disjoint union of the sets $U_\alpha\times\mathbb{C}^r$ by the equivalence relation that $(p_\alpha,t_\alpha)\sim(p_\beta,t_\beta)$ whenever $p_\alpha=p_\beta\in U_\alpha\cap U_\beta$ and $t_\alpha=\xi_{\alpha\beta}(p_\beta)t_\beta$; that this is an equivalence relation is an immediate consequence of the consistency conditions for the coordinate transformations, and it is a straightforward matter to verify that the result is a complex manifold. The projection is of course the mapping which assigns to the equivalence class containing a point $(p_\alpha,t_\alpha)\in U_\alpha\times\mathbb{C}^r$ the point $p_\alpha\in M$. Similar constructions can clearly be carried out for differentiable or continuous complex vector bundles.

A *complex analytic cross-section* of a complex analytic vector bundle $\pi:\xi\longrightarrow M$ is a complex analytic mapping $f\colon M\longrightarrow\xi$ such that $\pi\circ f(p)=p$ for every point $p\in M$. If the vector bundle is described by coordinate transformations $\{\xi_{\alpha\beta}\}$ in terms of a coordinate covering $\{U_\alpha,z_\alpha\}$ of the complex manifold $M$ then a complex analytic cross-section is evidently described by a collection of complex analytic mappings $f_\alpha\colon U_\alpha\longrightarrow\mathbb{C}^r$ such that $f_\alpha(p)=\xi_{\alpha\beta}(p)f_\beta(p)$ whenever $p\in U_\alpha\cap U_\beta$. The set of all complex analytic cross-sections of the bundle $\pi:\xi\longrightarrow M$ clearly form a complex vector space, which will be denoted by $\Gamma(\xi)$; and the dimension of this vector space will be denoted by $\gamma(\xi)=\dim_{\mathbb{C}}\Gamma(\xi)$. The conventional dropping of the projection mapping in the notation for a vector bundle is a convenient abbreviation that will frequently be used.

Suppose that $\pi'\colon\xi'\longrightarrow M$ and $\pi''\colon\xi''\longrightarrow M$ are two complex analytic vector bundles of ranks $r'$ and $r''$ over the same complex manifold $M$ and that they are described by the coordinate transformations $\{\xi'_{\alpha\beta}\}$ and $\{\xi''_{\alpha\beta}\}$ in terms of the same coordinate covering $\{U_\alpha,z_\alpha\}$ of $M$. If $\phi\colon\xi'\longrightarrow\xi''$ is a homomorphism of complex analytic vector bundles then the restriction of $\phi$ to a coordinate neighborhood $U_\alpha\times\mathbb{C}^{r'}$ for the bundle $\xi'$ is evidently a complex analytic mapping from that coordinate neighborhood to the corresponding coordinate neighborhood $U_\alpha\times\mathbb{C}^{r''}$ for the bundle $\xi''$ and is of the form $\phi(p_\alpha,t'_\alpha)=(p_\alpha,\phi_\alpha(p_\alpha)t'_\alpha)$ where $\phi_\alpha$ is a complex analytic mapping from $U_\alpha$ to the space $\mathbb{C}^{r''\times r'}$ of $r''\times r'$ complex matrices; and if $p\in U_\alpha\cap U_\beta$ then the various fibre coordinates over $p$ are related by $t''_\alpha=\phi_\alpha(p)t'_\alpha=\phi_\alpha(p)\,\xi'_{\alpha\beta}(p)t'_\beta$ and $t''_\alpha=\xi''_{\alpha\beta}(p)t''_\beta=\xi''_{\alpha\beta}(p)\,\phi_\beta(p)t'_\beta$, so that $\phi_\alpha(p)\,\xi'_{\alpha\beta}(p)=\xi''_{\alpha\beta}(p)\,\phi_\beta(p)$. On the other hand if $\phi_\alpha\colon U_\alpha\longrightarrow\mathbb{C}^{r''\times r'}$ are any complex analytic mappings such that $\phi_\alpha(p)\,\xi'_{\alpha\beta}(p)=\xi''_{\alpha\beta}(p)\,\phi_\beta(p)$ whenever $p\in U_\alpha\cap U_\beta$ then the complex analytic mappings $\tilde\phi_\alpha\colon U_\alpha\times\mathbb{C}^{r'}\longrightarrow U_\alpha\times\mathbb{C}^{r''}$ defined by $\tilde\phi_\alpha(p_\alpha,t'_\alpha)=(p_\alpha,\phi_\alpha(p_\alpha)t'_\alpha)$ are readily seen to be compatible with the coordinate transition functions of the manifolds $\xi'$ and $\xi''$ and to determine a homomorphism $\phi: \xi' \longrightarrow \xi''$ of complex analytic vector bundles. In particular if $r'=r''=r$ then the complex analytic vector bundles $\pi': \xi' \longrightarrow M$ and $\pi'': \xi'' \longrightarrow M$ are isomorphic precisely when there are complex analytic mappings $\phi_\alpha: U_\alpha \longrightarrow GL(r, \mathbb{C})$ such that $\xi''_{\alpha\beta}(p) = \phi_\alpha(p)\xi'_{\alpha\beta}(p)\phi_\beta(p)^{-1}$ whenever $p \in U_\alpha \cap U_\beta$.

A useful observation at this point is that if $\phi: D \longrightarrow \mathbb{C}^{r'' \times r'}$ is a complex analytic matrix-valued function in an open subset $D \subseteq \mathbb{C}^n$ and is of constant rank $r$ throughout $D$ then for each point $p_0 \in D$ there are an open neighborhood $U$ of $p_0$ in $D$ and complex analytic mappings $\theta': U \longrightarrow GL(r', \mathbb{C})$ and $\theta'': U \longrightarrow GL(r'', \mathbb{C})$ such that

$$\theta''(p)\phi(p)\theta'(p) = \begin{pmatrix} I_r & 0 \\ 0 & 0 \end{pmatrix} \quad \text{for all} \quad p \in U$$

where $I_r$ is the $r \times r$ identity matrix and 0 is a zero matrix of an appropriate size. To see this note that there is a matrix $C' \in GL(r', \mathbb{C})$ such that the first $r$ columns of $\phi(p_0)C'$ are linearly independent vectors; the first $r$ columns of $\phi(p)C'$ are then also linearly independent vectors for all points $p$ in some open neighborhood $U'$ of the point $p_0$ in $D$, and the remaining $r'-r$ columns of $\phi(p)C'$ are unique linear combinations of the first $r$ columns with coefficients which are necessarily complex analytic functions in $U'$. Thus writing $\phi C' = (\phi_1, \phi_2)$ where $\phi_1$ consists of the first $r$ columns of $\phi C'$ it follows that $\phi_2 = \phi_1 \psi$ for a uniquely determined complex analytic mapping $\psi: U' \longrightarrow \mathbb{C}^{r \times (r'-r)}$; and consequently

$$\phi C' \begin{pmatrix} I_r & -\psi \\ 0 & I_{r'-r} \end{pmatrix} = (\phi_1, 0)$$

so that $\phi\theta' = (\phi_1, 0)$ for some complex analytic mapping $\theta': U' \longrightarrow GL(r', \mathbb{C})$. A similar argument applied to the rows of the mapping $\phi_1: U' \longrightarrow \mathbb{C}^{r' \times r}$ then shows that in a subneighborhood $U \subseteq U'$ of the point $p_0$ there is a complex analytic mapping $\theta'': U \longrightarrow GL(r'', \mathbb{C})$ such that $\theta''\phi\theta'$ has the desired form.

Now suppose again that $\pi': \xi' \longrightarrow M$ and $\pi'': \xi'' \longrightarrow M$ are two complex analytic vector bundles of ranks $r'$ and $r''$ over the same complex manifold $M$, and let $\phi: \xi' \longrightarrow \xi''$ be a homomorphism of complex analytic vector bundles with the property that the restrictions $\phi | \xi'_p: \xi'_p \longrightarrow \xi''_p$ are linear transformations of constant rank $r$ as $p$ varies through $M$. Then as a consequence of the preceding observation it is clear that for a sufficiently fine coordinate covering $\{U_\alpha, z_\alpha\}$ of $M$, in terms of which the bundles are described by coordinate transformations $\{\xi'_{\alpha\beta}\}$ and $\{\xi''_{\alpha\beta}\}$ and the homomorphism $\phi$ by local complex analytic mappings $\{\phi_\alpha\}$, there exist complex analytic mappings $\theta'_\alpha: U_\alpha \longrightarrow GL(r', \mathbb{C})$ and $\theta''_\alpha: U_\alpha \longrightarrow GL(r'', \mathbb{C})$ for which

$$\theta''_\alpha(p)\phi_\alpha(p)\theta'_\alpha(p)^{-1} = \begin{pmatrix} I_r & 0 \\ 0 & 0 \end{pmatrix} \quad \text{whenever} \quad p \in U_\alpha.$$

The mappings $\{\theta'_\alpha\}$ and $\{\theta''_\alpha\}$ can be viewed as describing isomorphisms of complex analytic vector bundles $\theta': \xi' \longrightarrow \tilde{\xi}'$ and $\theta'': \xi'' \longrightarrow \tilde{\xi}''$ where the image bundles are described by the coordinate transformations $\tilde{\xi}'_{\alpha\beta}(p)=\theta'_\alpha(p)\,\xi'_{\alpha\beta}(p)\,\theta'_\beta(p)^{-1}$ and $\tilde{\xi}''_{\alpha\beta}(p)=\theta''_\alpha(p)\,\xi''_{\alpha\beta}(p)\,\theta''_\beta(p)^{-1}$, and the composite bundle homomorphism $\tilde{\phi}=\theta''\circ\phi\circ(\theta')^{-1}:\tilde{\xi}'\longrightarrow\tilde{\xi}''$ is then described by mappings $\{\tilde{\phi}_\alpha\}$ where

$$\tilde{\phi}_\alpha(p)=\begin{pmatrix} I_r & 0\\ 0 & 0\end{pmatrix}\quad \text{whenever}\quad p\in U_\alpha;$$

or alternatively the mappings $\{\theta'_\alpha\}$ and $\{\theta''_\alpha\}$ can be viewed merely as changes of coordinates for the bundles $\xi'$ and $\xi''$ which reduce the coordinate transformations to the forms $\{\tilde{\xi}'_{\alpha\beta}\}$ and $\{\tilde{\xi}''_{\alpha\beta}\}$ and the description of the homomorphism $\phi$ to the simpler form $\{\tilde{\phi}_\alpha\}$. In either case since $\tilde{\phi}_\alpha(p)\,\tilde{\xi}'_{\alpha\beta}(p)=\tilde{\xi}''_{\alpha\beta}(p)\,\tilde{\phi}_\beta(p)$ whenever $p\in U_\alpha\cap U_\beta$ it follows that

$$\tilde{\xi}'_{\alpha\beta}(p)=\begin{pmatrix}\xi'_{11\alpha\beta}(p) & 0\\ \xi'_{21\alpha\beta}(p) & \xi'_{22\alpha\beta}(p)\end{pmatrix},\qquad \tilde{\xi}''_{\alpha\beta}(p)=\begin{pmatrix}\xi''_{11\alpha\beta}(p) & \xi''_{12\alpha\beta}(p)\\ 0 & \xi''_{22\alpha\beta}(p)\end{pmatrix},$$

where $\xi'_{11\alpha\beta}(p)\in GL(r,\mathbb{C})$, $\xi'_{22\alpha\beta}(p)\in GL(r'-r,\mathbb{C})$, $\xi''_{11\alpha\beta}(p)\in GL(r,\mathbb{C})$, and $\xi''_{22\alpha\beta}(p)\in GL(r''-r,\mathbb{C})$. Now the subspace of $\mathbb{C}^{r'}$ consisting of those vectors the first $r$ components of which are zero can be viewed as a subspace of the fibre $\tilde{\xi}'_p$, since that subspace is evidently preserved by all the coordinate transformations $\{\tilde{\xi}'_{\alpha\beta}\}$. The union of these subspaces can evidently be given the structure of a complex analytic vector bundle over $M$, to form a *subbundle* of $\tilde{\xi}'$ described by the coordinate transformations $\{\xi'_{22\alpha\beta}\}$; and the union of the quotient spaces of $\tilde{\xi}'_p$ modulo these subspaces can also evidently be given the structure of a complex analytic vector bundle over $M$, to form a *quotient bundle* of $\tilde{\xi}'$ described by the coordinate transformations $\{\xi'_{11\alpha\beta}\}$. The subbundle $\xi'_2\subseteq\tilde{\xi}'$ is of rank $r'-r$ and the quotient bundle $\xi'_1=\tilde{\xi}'_2/\xi'_2$ is of rank $r$. By similarly considering the subspace of $\mathbb{C}^{r''}$ consisting of those vectors the last $r''-r$ components of which are zero there arise the subbundle $\xi''_1\subseteq\tilde{\xi}''$ of rank $r$ described by the coordinate transformations $\{\xi''_{11\alpha\beta}\}$ and the quotient bundle $\xi''_2=\tilde{\xi}''/\xi''_1$ of rank $r''-r$ described by the coordinate transformations $\{\xi''_{22\alpha\beta}\}$. Moreover the subbundle $\xi'_2\subseteq\tilde{\xi}'$ is precisely the *kernel* of the homomorphism $\tilde{\phi}$, in the sense that its fibres are the kernels of the restrictions of $\tilde{\phi}$ to the fibres of $\tilde{\xi}'$, and the subbundle $\xi''_1\subseteq\tilde{\xi}''$ is precisely the *image* of the homomorphism $\tilde{\phi}$, in the corresponding sense; and the homomorphism $\tilde{\phi}$ establishes an isomorphism between the bundles $\xi'_1$ and $\xi''_1$. A convenient notation for this situation is that

$$0\longrightarrow\xi'_2\overset{i}{\longrightarrow}\tilde{\xi}'\overset{\tilde{\phi}}{\longrightarrow}\xi''_1\longrightarrow 0$$

is an *exact sequence* of complex analytic vector bundles over $M$, where $i:\xi'_2\longrightarrow\tilde{\xi}'$ is the inclusion mapping viewed as a homomorphism of complex analytic vector bundles.

Algebraic operations other than homomorphisms can also be applied to complex analytic vector bundles. The *direct sum* of two complex analytic vector bundles $\pi':\xi'\longrightarrow M$ and $\pi'':\xi''\longrightarrow M$ is the complex analytic vector bundle $\pi:\xi'\oplus\xi''\longrightarrow M$ having as fibre over any point $p\in M$ the direct sum $\xi'_p\oplus\xi''_p$, defined by the coordinate transformations $\{\xi'_{\alpha\beta}\oplus\xi''_{\alpha\beta}\}$; the original bundles $\xi'$ and $\xi''$ are both subbundles of $\xi'\oplus\xi''$. Similarly the *tensor product* of the bundles $\pi':\xi'\longrightarrow M$ and $\pi'':\xi''\longrightarrow M$ is the complex analytic vector bundle $\pi:\xi'\otimes\xi''\longrightarrow M$ having as fibre over any point $p\in M$ the tensor product $\xi'_p\otimes_{\mathbb{C}}\xi''_p$, defined by the coordinate transformations $\{\xi'_{\alpha\beta}\otimes\xi''_{\alpha\beta}\}$. If the bundle $\xi'$ is of rank $r'=1$ the tensor products $\xi'_{\alpha\beta}\otimes\xi''_{\alpha\beta}$ can be identified with the ordinary products $\xi'_{\alpha\beta}\xi''_{\alpha\beta}$ of the scalars $\xi'_{\alpha\beta}$ and the matrices $\xi''_{\alpha\beta}$; and if the bundle $\xi''$ is also of rank $r''=1$ the products $\xi'_{\alpha\beta}\otimes\xi''_{\alpha\beta}=\xi'_{\alpha\beta}\xi''_{\alpha\beta}$ are also scalars hence describe another bundle of rank 1. The operation of tensor product can thus be used to impose the structure of an abelian group on the set of all complex analytic line bundles over $M$; the inverse of the line bundle defined by the coordinate transformations $\{\xi_{\alpha\beta}\}$ is the line bundle defined by the coordinate transformations $\{\xi_{\alpha\beta}^{-1}\}$.

Complex analytic line bundles occur quite naturally in the study of complex manifolds, as in the following example. Considering first the case of a Riemann surface $M$, a *divisor* on $M$ is a mapping $\mathfrak{d}: M\longrightarrow\mathbb{Z}$ such that $\mathfrak{d}(p)\neq 0$ only for a discrete set of points in $M$. It is notationally convenient to write $\mathfrak{d}=\sum_i v_i\cdot p_i$ where $\{p_i\}$ is a discrete set of points in $M$, $\mathfrak{d}(p_i)=v_i\in\mathbb{Z}$, and $\mathfrak{d}(p)=0$ unless $p\in\{p_i\}$. The *order* of a divisor $\mathfrak{d}$ which is nonzero for only finitely many points of $M$ is the integer $|\mathfrak{d}|=\sum_{p\in M}\mathfrak{d}(p)$, so that $|v_1\cdot p_1+\cdots+v_r\cdot p_r|=v_1+\cdots+v_r$. The set of all divisors on $M$ clearly form an abelian group under pointwise addition of the values of the functions. The set of divisors can also be given a partial ordering by setting $\mathfrak{d}_1\geq\mathfrak{d}_2$ provided that $\mathfrak{d}_1(p)\geq\mathfrak{d}_2(p)$ for all points $p\in M$, and this ordering is evidently compatible with the group structure. A divisor such that $\mathfrak{d}\geq 0$, that is such that $\mathfrak{d}(p)\geq 0$ for all points $p\in M$, is called a *positive divisor*. To every meromorphic function $f$ on the Riemann surface $M$ there can be associated a divisor $\mathfrak{d}(f)$, called the *divisor of the function* $f$, by assigning to each point $p\in M$ the order of the function $f$ at that point; thus if $f$ has a zero of order $v_1$ at a point $p_1\in M$ then $\mathfrak{d}(f)(p_1)=v_1$, and if $f$ has a pole of order $v_2$ at a point $p_2\in M$ then $\mathfrak{d}(f)(p_2)=-v_2$. It is of course not the case that every divisor on $M$ is necessarily the divisor of some meromorphic function on $M$; for instance on a compact Riemann surface no nontrivial positive divisor can be the divisor of a meromorphic function since a complex analytic function on such a surface must be constant. However on any coordinate neighborhood of $M$ which contains but finitely many points at which the divisor is nontrivial there obviously exists a meromorphic function having the given divisor on that neighborhood. Therefore for any divisor $\mathfrak{d}$ on $M$ there exist a coordinate covering $\{U_\alpha, z_\alpha\}$ of $M$ and meromorphic functions $f_\alpha$ on each coordinate neighborhood $U_\alpha$ such that $\mathfrak{d}(f_\alpha)=\mathfrak{d}\,|\,U_\alpha$. On each intersection $U_\alpha\cap U_\beta$ the functions $f_\alpha$ and $f_\beta$ have the same divisor, so that the quotient $\zeta_{\alpha\beta}=f_\alpha/f_\beta$ is complex analytic and nonvanishing; and evidently $\zeta_{\alpha\beta}(p)\zeta_{\beta\gamma}(p)=\zeta_{\alpha\gamma}(p)$ whenever $p\in U_\alpha\cap U_\beta\cap U_\gamma$. The collection of functions $\{\zeta_{\alpha\beta}\}$ can therefore be taken to be the coordinate transformations describing a complex analytic line bundle over $M$; that line bundle is called the *line bundle of the divisor* $\mathfrak{d}$ and is denoted by $\zeta_{\mathfrak{d}}$. The functions $\{f_\alpha\}$ are of course not uniquely determined; however if $\{f'_\alpha\}$ are any other functions having the same divisor then $f'_\alpha=f_\alpha h_\alpha$ where $h_\alpha$ is complex analytic and nonvanishing and $\zeta'_{\alpha\beta}=f'_\alpha/f'_\beta=\zeta_{\alpha\beta}h_\alpha/h_\beta$, so that the coordinate transformations $\{\zeta_{\alpha\beta}\}$ and $\{\zeta'_{\alpha\beta}\}$ describe isomorphic line bundles. The line bundle $\zeta_{\mathfrak{d}}$ is therefore uniquely determined by the divisor $\mathfrak{d}$ up to isomorphism. The line bundle $\zeta_p$ associated to a divisor of the form $\mathfrak{d}=1\cdot p$ is called the *point bundle* associated to the point $p\in M$.

Note that in the preceding construction the functions $\{f_\alpha\}$ can be viewed as describing a meromorphic cross-section of the line bundle $\zeta_\mathfrak{d}$, since $f_\alpha(p)=\zeta_{\alpha\beta}(p)f_\beta(p)$ whenever $p\in U_\alpha\cap U_\beta$. On the other hand if $\zeta$ is any complex analytic line bundle on $M$ described by coordinate transformations $\{\zeta_{\alpha\beta}\}$ in terms of a coordinate covering $\{U_\alpha, z_\alpha\}$ and if $\zeta$ admits a nontrivial meromorphic cross-section $f$ described by local meromorphic functions $f_\alpha$ in the various coordinate neighborhoods $U_\alpha$ then the functions $f_\alpha$ and $f_\beta$ evidently have the same divisor in $U_\alpha\cap U_\beta$ so that the notion of the divisor $\mathfrak{d}(f)$ of the meromorphic section $f$ is evidently well defined; and clearly $\zeta=\zeta_{\mathfrak{d}(f)}$. Therefore the complex analytic line bundles on $M$ associated to divisors on $M$ are precisely those line bundles which admit nontrivial meromorphic cross-sections. The basic **existence theorem** on Riemann surfaces is the assertion that *every complex analytic line bundle does indeed admit nontrivial meromorphic sections*, or equivalently that *every complex analytic line bundle is the line bundle of some divisor*. (For the case of compact Riemann surfaces the existence theorem is proved in just this form as Theorem 12 in [10]; for the case of noncompact Riemann surfaces the existence theorem, in the stronger form that every complex analytic line bundle admits nontrivial complex analytic sections, follows from Cartan's Theorem A [22] since such surfaces are necessarily Stein manifolds as proved in [2].)

The correspondence which associates to any divisor $\mathfrak{d}$ on $M$ the line bundle $\zeta_\mathfrak{d}$ is easily seen to be a homomorphism from the group of divisors on $M$ to the group of complex analytic line bundles over $M$. The basic existence theorem, which is just the assertion that this homomorphism is surjective, has a particularly useful consequence for compact Riemann surfaces. Every complex analytic line bundle $\zeta$ is the line bundle of some divisor $\mathfrak{d}$, and on a compact Riemann surface that divisor is nontrivial only for finitely many points so that $\mathfrak{d}=\nu_1\cdot p_1+\cdots+\nu_r\cdot p_r$ and $\zeta=\zeta_{p_1}^{\nu_1}\ldots\zeta_{p_r}^{\nu_r}$; thus on a compact Riemann surface every complex analytic line bundle can be expressed in this form as some combination of point bundles. The homomorphism from divisors to line bundles is not generally injective. If a divisor $\mathfrak{d}$ is in the kernel of this homomorphism then in terms of a suitable coordinate covering $\{U_\alpha, z_\alpha\}$ of $M$ there are local meromorphic functions $\{f_\alpha\}$ defining that divisor, the associated line bundle is described by coordinate transformations $\zeta_{\alpha\beta}=f_\alpha/f_\beta$, and the triviality of this line bundle is exhibited by nonvanishing complex analytic functions $\{h_\alpha\}$ in the various coordinate neighborhoods such that $\zeta_{\alpha\beta}=h_\alpha/h_\beta$ in $U_\alpha\cap U_\beta$; but then clearly $f_\alpha/h_\alpha=f_\beta/h_\beta$ in $U_\alpha\cap U_\beta$, so that the local meromorphic functions $\{f_\alpha/h_\alpha\}$ comprise a global meromorphic function $f$ such that $\mathfrak{d}=\mathfrak{d}(f)$. Conversely if $\mathfrak{d}=\mathfrak{d}(f)$ is the divisor of a global meromorphic function $f$ on $M$ then for any coordinate covering $\{U_\alpha, z_\alpha\}$ of $M$ the restrictions $f_\alpha=f|U_\alpha$ can be taken as the local meromorphic functions describing that divisor; the coordinate transformations describing the associated line bundle are $\zeta_{\alpha\beta}=f_\alpha/f_\beta=1$, hence the line bundle $\zeta_\mathfrak{d}$ is trivial. Thus the kernel of the homomorphism from divisors to line bundles consists precisely of the divisors of global meromorphic functions; and consequently $\zeta_{\mathfrak{d}'}=\zeta_{\mathfrak{d}''}$ for two divisors $\mathfrak{d}'$ and $\mathfrak{d}''$ precisely when there is a meromorphic function $f$ such that $\mathfrak{d}(f)=\mathfrak{d}'-\mathfrak{d}''$. Traditionally two divisors $\mathfrak{d}'$ and $\mathfrak{d}''$ are called *linearly equivalent* if there is a meromorphic function $f$ such that $\mathfrak{d}(f)=\mathfrak{d}'-\mathfrak{d}''$, and the group of linear equivalence classes of divisors on $M$ is called the *divisor class group* of $M$; the homomorphism from divisors to line bundles thus exhibits an isomorphism between the divisor class group of $M$ and the group of complex analytic line bundles over $M$.

Consider a complex analytic line bundle $\zeta_{\mathfrak{d}}$ over $M$ described by coordinate transformations $\zeta_{\alpha\beta}=f_\alpha/f_\beta$ where $\mathfrak{d}(f_\alpha)=\mathfrak{d}|U_\alpha$. If $h\in\Gamma(\zeta_{\mathfrak{d}})$ is a complex analytic cross-section, described by local functions $h_\alpha$ which are complex analytic in the various coordinate neighborhoods $U_\alpha$ and which satisfy $h_\alpha=\zeta_{\alpha\beta}h_\beta$ in the intersections $U_\alpha\cap U_\beta$, then the quotients $\tilde{h}_\alpha=h_\alpha/f_\alpha$ are meromorphic functions in the coordinate neighborhoods $U_\alpha$ and $\tilde{h}_\alpha=\tilde{h}_\beta$ in $U_\alpha\cap U_\beta$; there is therefore a well defined meromorphic function $\tilde{h}$ on $M$ such that $\tilde{h}|U_\alpha=\tilde{h}_\alpha$, and the correspondence which associates to $h\in\Gamma(\zeta_{\mathfrak{d}})$ the meromorphic function $\tilde{h}$ is clearly a linear mapping. On each coordinate neighborhood $U_\alpha$ note that $\mathfrak{d}(\tilde{h}|U_\alpha)+\mathfrak{d}|U_\alpha=\mathfrak{d}(\tilde{h}|U_\alpha)+\mathfrak{d}(f_\alpha)=\mathfrak{d}((\tilde{h}|U_\alpha)f_\alpha)=\mathfrak{d}(h_\alpha)\geq 0$ and consequently $\mathfrak{d}(\tilde{h})+\mathfrak{d}\geq 0$. Conversely if $\tilde{h}$ is any meromorphic function on $M$ such that $\mathfrak{d}(\tilde{h})+\mathfrak{d}\geq 0$ then the functions $h_\alpha=(\tilde{h}|U_\alpha)f_\alpha$ are complex analytic in the various coordinate neighborhoods $U_\alpha$ and satisfy $h_\alpha=\zeta_{\alpha\beta}h_\beta$ in $U_\alpha\cap U_\beta$, hence they describe a complex analytic cross-section $h\in\Gamma(\zeta_{\mathfrak{d}})$ to which the given meromorphic function $\tilde{h}$ is canonically associated. Thus the complex vector space $\Gamma(\zeta_{\mathfrak{d}})$ is isomorphic to the space of those meromorphic functions $\tilde{h}$ on the Riemann surface $M$ such that $\mathfrak{d}(\tilde{h})+\mathfrak{d}\geq 0$.

Precisely the same consideration of the line bundles associated to divisors can be carried out for an $n$-dimensional complex manifold $M$ if in the general case divisors are defined as equivalence classes of nontrivial local meromorphic functions, where two meromorphic functions $f$ and $g$ on an open subset $U$ of $M$ are considered equivalent if the quotient $f/g$ is regular analytic and nowhere vanishing in $U$. The zero locus or pole locus of a meromorphic function on an $n$-dimensional complex manifold is an $(n-1)$-dimensional complex analytic subvariety; and the notion of the multiplicity of a zero or pole is well defined and constant on each irreducible component of the zero or pole locus. This leads to the interpretation of divisors as formal expressions $\mathfrak{d}=\sum_i \nu_i\cdot V_i$, where $V_i$ are irreducible $(n-1)$-dimensional complex analytic subvarieties of $M$ such that $\bigcup_i V_i$ is also a complex analytic subvariety and where $\nu_i\in\mathbb{Z}$; when $n=1$ the components $V_i$ are points and $\bigcup_i V_i$ is a complex analytic subvariety precisely when the set of points $\{V_i\}$ is a discrete subset of $M$, so that this reduces to the notion of a divisor on a Riemann surface as considered here.

## § 3. Factors of Automorphy and Complex Analytic Vector Bundles

The universal covering space $\tilde{M}$ of a complex manifold $M$ has a natural complex structure induced by the covering projection $\rho: \tilde{M}\longrightarrow M$; for if $\{U_\alpha, z_\alpha\}$ is a coordinate covering of $M$ in which each set $U_\alpha$ is simply connected then any connected component of $\rho^{-1}(U_\alpha)$ is homeomorphic to $U_\alpha$ under the projection $\rho$, so that the mapping $z_\alpha \circ \rho$ can be used as a coordinate mapping, and the coordinate transition functions reduce to those for the covering $\{U_\alpha, z_\alpha\}$ of $M$ and are hence complex analytic. The covering translations form a group $\Gamma$ of complex analytic homeomorphisms from $\tilde{M}$ to itself such that the quotient space $\tilde{M}/\Gamma$ can be identified with $M$; for each covering translation is represented by the identity mappings in the induced coordinate covering of $\tilde{M}$, hence is a complex analytic mapping, and the remaining assertions are general properties of covering spaces. Any complex analytic vector bundle $\pi: \xi \longrightarrow M$ over $M$ induces a complex analytic vector bundle $\tilde{\pi}: \tilde{\xi} \longrightarrow \tilde{M}$ over the universal covering space $\tilde{M}$ of $M$ as well. Indeed if $\xi$ is described by coordinate transformations $\{\xi_{\alpha\beta}\}$ in terms of the above coordinate covering $\{U_\alpha, z_\alpha\}$ of $M$ then the mappings $\{\xi_{\alpha\beta}\circ\rho\}$ in the induced coordinate covering of $\tilde{M}$ clearly satisfy the consistency conditions required to describe a complex analytic vector bundle over $M$. The identity mappings in the induced coordinate coverings of $\tilde{\xi}$ and $\xi$ evidently comprise a complex analytic mapping $\tilde{\rho}: \tilde{\xi} \longrightarrow \xi$ such that $\rho\tilde{\pi} = \pi\tilde{\rho}$ and that for each point $\tilde{p} \in \tilde{M}$ the restriction $\tilde{\rho} \,|\, \tilde{\xi}_{\tilde{p}}: \tilde{\xi}_{\tilde{p}} \longrightarrow \xi_{\rho(\tilde{p})}$ is a linear mapping. Furthermore any covering translation $T \in \Gamma$ clearly extends to a bundle homomorphism $\tilde{T}: \tilde{\xi} \longrightarrow \tilde{\xi}$, also represented by the identity mappings in the induced coordinate coverings, such that $\tilde{\rho}\tilde{T} = \tilde{\rho}$; the set of all these homomorphisms form a group $\tilde{\Gamma}$ of complex analytic homeomorphisms from $\tilde{\xi}$ to itself, and the mapping $\tilde{\rho}: \tilde{\xi} \longrightarrow \xi$ identifies the quotient space $\tilde{\xi}/\tilde{\Gamma}$ with $\xi$.

A particularly interesting situation is that in which the complex analytic vector bundle $\pi: \xi \longrightarrow M$ induces a trivial bundle $\tilde{\pi}: \tilde{\xi} \longrightarrow \tilde{M}$ over the universal covering space $\tilde{M}$ of $M$. Upon making the identification $\tilde{\xi} = \tilde{M} \times \mathbb{C}^r$ each bundle homomorphism $\tilde{T} \in \tilde{\Gamma}$ corresponding to a covering translation $T \in \Gamma$ must take the form $\tilde{T}(\tilde{p}, t) = (T\tilde{p}, \xi(T, \tilde{p})t)$, where $\xi: \Gamma \times \tilde{M} \longrightarrow GL(r, \mathbb{C})$ is a complex analytic mapping from $\tilde{M}$ into $GL(r, G)$ for each fixed $T \in \Gamma$. The group property implies that for any two elements $S, T \in \Gamma$

$$\begin{aligned}
\tilde{S}\tilde{T}(\tilde{p}, t) &= (ST\tilde{p}, \xi(ST, \tilde{p})t) \\
&= \tilde{S}(T\tilde{p}, \xi(T, \tilde{p})t) \\
&= (ST\tilde{p}, \xi(S, T\tilde{p})\xi(T, \tilde{p})t),
\end{aligned}$$

and consequently $\xi(ST, \tilde{p}) = \xi(S, T\tilde{p})\xi(T, \tilde{p})$. On the other hand given any mapping $\xi: \Gamma \times \tilde{M} \longrightarrow GL(r, \mathbb{C})$ which satisfies these group consistency relations and which is a complex analytic mapping from $\tilde{M}$ into $GL(r, \mathbb{C})$ for each fixed $T \in \Gamma$, it is possible to exhibit $\Gamma$ as a group of complex analytic homeomorphisms from $\tilde{M} \times \mathbb{C}^r$ to itself by setting $T(\tilde{p}, t) = (T\tilde{p}, \xi(T, \tilde{p})t)$; these mappings are of course bundle homomorphisms, when the product $\tilde{M} \times \mathbb{C}^r$ is viewed as a trivial complex analytic vector bundle $\tilde{\xi}$ over $\tilde{M}$, and the quotient $\tilde{\xi}/\Gamma$ is a complex analytic vector bundle $\xi$ over $M$ which induces the bundle $\tilde{\xi}$ over $\tilde{M}$ under the covering projection $\rho: \tilde{M} \longrightarrow M$. Note that any cross-section $f: M \longrightarrow \xi$ induces a cross-section $\tilde{f}: \tilde{M} \longrightarrow \tilde{\xi} = \tilde{M} \times \mathbb{C}^r$ which commutes with the action of the group $\Gamma$; the cross-section $\tilde{f}$ is of course described by a mapping $\tilde{f}: \tilde{M} \longrightarrow \mathbb{C}^r$, and the mapping must satisfy $\tilde{f}(T\tilde{p}) = \xi(T, \tilde{p})\tilde{f}(\tilde{p})$ for each $T \in \Gamma$. Conversely every such mapping does describe a cross-section of $\xi$. Furthermore if

$\pi' : \xi' \longrightarrow M$ and $\pi'' : \xi'' \longrightarrow M$ are two complex analytic vector bundles over $M$ which induce trivial bundles over $\tilde{M}$ and which are described as quotients of $\tilde{M} \times \mathbb{C}^{r'}$ and $\tilde{M} \times \mathbb{C}^{r''}$ under the actions of the covering translation group $\Gamma$ of the form $T(\tilde{p}, t') = (T\tilde{p}, \xi'(T, \tilde{p})t')$ and $T(\tilde{p}, t'') = (T\tilde{p}, \xi''(T, \tilde{p})t')$ for any $T \in \Gamma$, then any bundle homomorphism $\phi: \xi' \longrightarrow \xi''$ induces a bundle homomorphism $\tilde{\phi}: \tilde{M} \times \mathbb{C}^{r'} \longrightarrow \tilde{M} \times \mathbb{C}^{r''}$ which commutes with the actions of the group $\Gamma$; thus $\tilde{\phi}$ must be of the form $\tilde{\phi}(\tilde{p}, t') = (\tilde{p}, \tilde{\phi}(\tilde{p})t')$ for some complex analytic mapping $\tilde{\phi}: \tilde{M} \longrightarrow \mathbb{C}^{r'' \times r'}$ from $\tilde{M}$ into the space of $r'' \times r'$ complex matrices, and $\tilde{\phi} T(p, t') = T\tilde{\phi}(\tilde{p}, t')$ for each $T \in \Gamma$ so that $(T\tilde{p}, \tilde{\phi}(T\tilde{p})\xi'(T, \tilde{p})t') = (T\tilde{p}, \xi''(T, \tilde{p})\tilde{\phi}(\tilde{p})t')$ or equivalently $\tilde{\phi}(T\tilde{p})\xi'(T, \tilde{p}) = \xi''(T, \tilde{p})\tilde{\phi}(\tilde{p})$ for each $T \in \Gamma$. Conversely of course any complex analytic mapping $\tilde{\phi}: \tilde{M} \longrightarrow \mathbb{C}^{r'' \times r'}$ satisfying these conditions determines a bundle homomorphism $\tilde{\phi}: \tilde{M} \times \mathbb{C}^{r'} \longrightarrow \tilde{M} \times \mathbb{C}^{r''}$ which commutes with the actions of $\Gamma$ and hence in turn determines a bundle homomorphism $\phi: \xi' \longrightarrow \xi''$. In particular the bundles $\xi'$ and $\xi''$ are isomorphic precisely when they are of the same rank $r' = r'' = r$ and there exists a complex analytic mapping $\tilde{\phi}: \tilde{M} \longrightarrow GL(r, \mathbb{C})$ such that $\tilde{\phi}(T\tilde{p})\xi'(T, \tilde{p}) = \xi''(T, \tilde{p})\tilde{\phi}(\tilde{p})$ for each $T \in \Gamma$.

To establish the terminology suppose more generally that $\Gamma$ is a group of complex analytic automorphisms of a complex manifold $\tilde{M}$; thus to each element $T \in \Gamma$ there is associated a complex analytic homeomorphism $T: \tilde{M} \longrightarrow \tilde{M}$ so that products in the group $\Gamma$ correspond to compositions of these homeomorphisms. A *factor of automorphy* $\xi$ of rank $r$ for the action of the group $\Gamma$ on $\tilde{M}$ is a mapping $\xi: \Gamma \times \tilde{M} \longrightarrow GL(r, \mathbb{C})$ which is complex analytic on $\tilde{M}$ and satisfies $\xi(ST, \tilde{p}) = \xi(S, T\tilde{p})\xi(T, \tilde{p})$ for any elements $S, T \in \Gamma$ and any point $\tilde{p} \in \tilde{M}$. A *relatively automorphic function* for such a factor of automorphy is a mapping $f: \tilde{M} \longrightarrow \mathbb{C}^r$ such that $f(T\tilde{p}) = \xi(T, \tilde{p})f(\tilde{p})$ for any element $T \in \Gamma$ and any point $\tilde{p} \in \tilde{M}$; and these functions can of course be restricted to be continuous, $C^\infty$, complex analytic, and so on. Two factors of automorphy $\xi'$ and $\xi''$ of the same rank $r$ are *analytically equivalent* if there exists a complex analytic mapping $h: \tilde{M} \longrightarrow GL(r, \mathbb{C})$ such that $\xi''(T, \tilde{p}) = h(T\tilde{p})\xi'(T, \tilde{p})h(\tilde{p})^{-1}$ for any element $T \in \Gamma$ and any point $\tilde{p} \in \tilde{M}$, and are *continuously equivalent* if there exists a continuous mapping $h: \tilde{M} \longrightarrow GL(r, \mathbb{C})$ satisfying the same conditions. Thus if $\tilde{M}$ is the universal covering space of $M$ and $\Gamma$ is the group of covering translations, equivalence classes of factors of automorphy for the actions of the group $\Gamma$ on $\tilde{M}$ correspond to isomorphism classes of complex analytic vector bundles over $M$ which induce trivial bundles over $\tilde{M}$; and the complex analytic relatively automorphic functions correspond to the complex analytic cross-sections of the associated complex analytic vector bundle over $M$.

A factor of automorphy $\xi: \Gamma \times \tilde{M} \longrightarrow GL(r, \mathbb{C})$ is said to be *flat* if the mapping $\xi$ is constant on $\tilde{M}$; thus a flat factor of automorphy merely consists of a mapping $\xi: \Gamma \longrightarrow GL(r, \mathbb{C})$ satisfying $\xi(ST) = \xi(S)\xi(T)$, for any two elements $S, T \in \Gamma$, or equivalently consists of an element $\xi \in \mathrm{Hom}(\Gamma, GL(r, \mathbb{C}))$ of the set of all group homomorphisms from $\Gamma$ into $GL(r, \mathbb{C})$. If $\xi'$ and $\xi''$ are flat factors of automorphy there are two relevant notions of equivalence: as before they are analytically equivalent factors of automorphy if there exists a complex analytic mapping $h: \tilde{M} \longrightarrow GL(r, \mathbb{C})$ such that $\xi''(T) = h(T\tilde{p})\xi'(T)h(\tilde{p})^{-1}$ for any element $T \in \Gamma$ and any point $\tilde{p} \in \tilde{M}$; and they are *flatly equivalent* if they are analytically equivalent and the mapping $h$ exhibiting this is constant on $\tilde{M}$, or equivalently if the representations $\xi'$ and $\xi''$ are conjugate in the sense that $\xi''(T)=h\xi'(T)h^{-1}$ for some element $h \in GL(r, \mathbb{C})$. It is evident that flat equivalence is a stricter equivalence relation than analytic equivalence; that it is indeed generally a properly stricter equivalence relation will later become apparent. It is quite easy to see that in the correspondence between complex analytic vector bundles and factors of automorphy the flat factors of automorphy correspond precisely to those complex analytic vector bundles which can be represented by coordinate transformations which are constant mappings, and in particular that all such complex analytic vector bundles induce the trivial flat bundle on the universal covering space of any complex manifold; the details are easily supplied, or the reader can be referred to [11].

Factors of automorphy can be used to represent all complex analytic vector bundles over most Riemann surfaces and all complex tori; for *any complex analytic vector bundle over the universal covering space of a complex torus or of a Riemann surface other than the Riemann sphere is trivial.* In all these cases the universal covering space is a contractible Stein manifold, indeed is either $\mathbb{C}^n$ or the unit disc in $\mathbb{C}^1$; and it is a general result of Grauert that any topologically trivial complex analytic vector bundle over a Stein manifold is analytically trivial. For a proof the reader is referred either to the original paper of Grauert [8] or to the exposition of this result by H. Cartan [4]. The special case of line bundles is considerably easier; a proof can be found in [22] or in most general texts on functions of several complex variables.

# Chapter II. Riemann Surfaces

## § 4. Markings of Riemann Surfaces and Characteristic Classes of Factors of Automorphy

Turning aside from the rather general topics treated in the preceding part, consider now the special case of a compact connected Riemann surface $M$. Some familiarity with the topology of surfaces will be presupposed; so it can be taken as known that topologically $M$ is a sphere with $g$ handles, where the integer $g$ is called the genus of the surface. The surface $M$ can then be dissected into a contractible set by cutting along $2g$ paths which issue from a common base point $p_0 \in M$ and are otherwise disjoint, one pair of paths for each handle, as illustrated in Fig. 1; note that these paths are viewed as oriented paths, with the

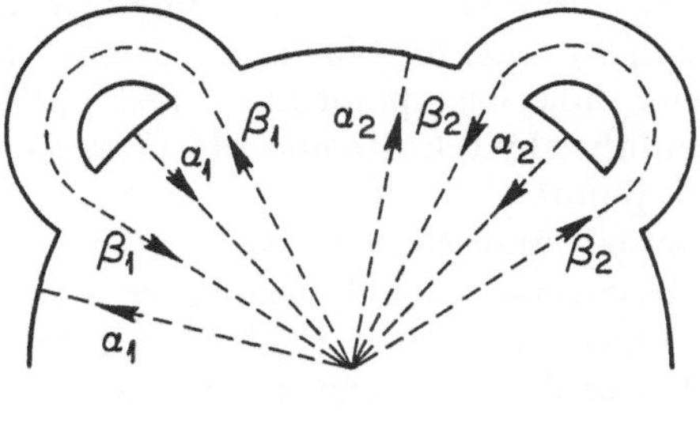


FIG. 1

orientations as indicated by the directions of the arrows. The complement of this set of paths is the interior of an oriented polygonal region $\Delta$ having $4g$ boundary arcs, one pair of boundary arcs corresponding to the two sides of each path of the dissection of $M$, as illustrated in Fig. 2; these boundary arcs inherit orientations from the orientations of the paths of the dissection, and the oriented boundary of $\Delta$ is then the chain

$$\partial\Delta = \sum_{i=1}^{g} (\alpha_i' + \beta_i' - \alpha_i'' - \beta_i'').$$

The choice of a base point $p_0 \in M$ and a canonical dissection by such oriented paths $\alpha_1, \dots, \alpha_g, \beta_1, \dots, \beta_g$ is called a *marking* of the surface $M$, and a surface with a particular marking is called a *marked surface*. A Riemann surface has a

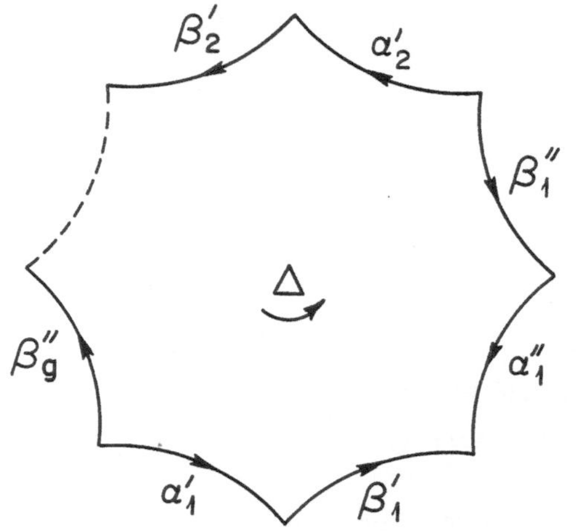


FIG. 2

canonical orientation, and markings of a Riemann surface will be restricted to those for which the orientation of $\Delta$ is the canonical orientation. There are of course a great many possible markings for any Riemann surface $M$. Any orientation preserving homeomorphism from $M$ to itself clearly transforms one marking of $M$ to another; and it is easy to see that conversely any two markings of $M$ can be transformed into one another by some orientation preserving homeomorphism, since the polygonal regions $\Delta$ are evidently homeomorphic. Two markings of $M$ with the same base point are called *equivalent* if they can be transformed into one another by a homeomorphism which is homotopic to the identity (with fixed base point).

If $\tilde{M}$ is the universal covering space of $M$ then the choice of a base point $z_0 \in \tilde{M}$ lying over the base point $p_0 \in M$ establishes a canonical isomorphism between the covering translation group $\Gamma$ of $\tilde{M}$ and the fundamental group $\pi_1(M, p_0)$ of $M$. That isomorphism associates to an element $T \in \Gamma$ the homotopy class of loops in $\pi_1(M, p_0)$ represented by the image in $M$ of any path from $z_0$ to $Tz_0$ in $\tilde{M}$. Note for this purpose that the product $\sigma \cdot \tau$ of two loops $\sigma$ and $\tau$ based at $p_0$ is the loop obtained by traversing first $\sigma$ and then $\tau$. Note further that the choice of a different base point in $\tilde{M}$ also lying over $p_0$ alters the isomorphism by an inner automorphism of the group $\pi_1(M, p_0)$. Now for any marking $(p_0; \alpha_1, \ldots, \alpha_g, \beta_1, \ldots, \beta_g)$ of $M$ and choice of base point $z_0 \in \tilde{M}$ let $\tilde{\alpha}_1, \ldots, \tilde{\alpha}_g, \tilde{\beta}_1, \ldots, \tilde{\beta}_g$ be the uniquely determined paths in $\tilde{M}$ beginning at $z_0$ and lying over the corresponding paths of the given marking; the path $\tilde{\alpha}_i$ runs from $z_0$ to $A_i z_0$, where $A_i \in \Gamma$ corresponds to the homotopy class of the loop $\alpha_i$ in $\pi_1(M, p_0)$ under this isomorphism, and the path $\tilde{\beta}_i$ runs from $z_0$ to $\beta_i z_0$, where $B_i \in \Gamma$ corresponds to $\beta_i$. The set $\Delta \subset M$ lifts to a number of disjoint open subsets of $\tilde{M}$, all of which are transformed homeomorphically to one another by the action of the covering translation group and all of which are homeomorphic to $\Delta$; referring back to Figs. 1 and 2 again it is easy to see that one of these liftings, which will be called the *canonical fundamental polygon* and will be denoted by $\tilde{\Delta}$, has the form indicated in Fig. 3. In that figure the elements $C_i \in \Gamma$ are the commutators $C_i = A_i B_i A_i^{-1} B_i^{-1}$; and the boundary of $\tilde{\Delta}$ is the closed path

$$\partial\tilde{\Delta} = \prod_{i=1}^{g} (C_1 \dots C_{i-1}\tilde{\alpha}_i)(C_1 \dots C_{i-1} A_i \tilde{\beta}_i)(C_1 \dots C_i B_i \tilde{\alpha}_i)^{-1}(C_1 \dots C_i \tilde{\beta}_i)^{-1},$$

where the simplified notation used to describe this product of paths is sufficiently unambiguous since the product $\sigma \cdot \tau$ of two paths is only defined when the end of $\sigma$ is the beginning of $\tau$. It is clear from Fig. 3 that the elements $C_i \in \Gamma$ are sub-

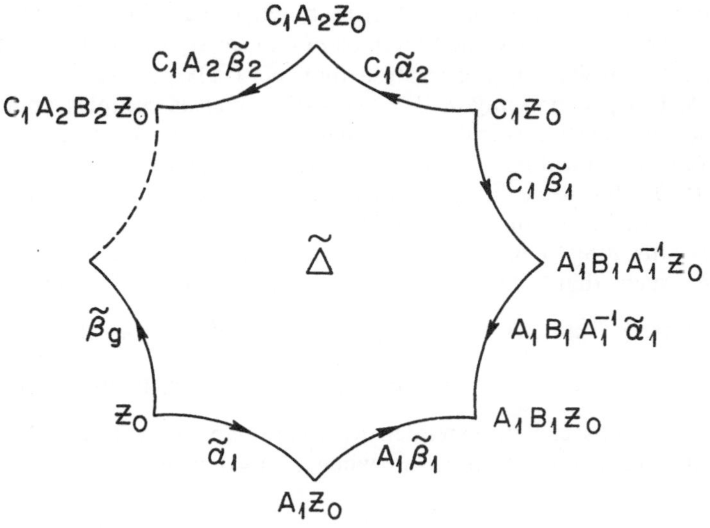


FIG. 3

ject to the relation $C_1 C_2 \dots C_g = 1$; and it is well known that the group $\Gamma$ is generated by the elements $A_1, \dots, A_g, B_1, \dots, B_g$ subject to this single relation. Correspondingly of course the fundamental group $\pi_1(M, p_0)$ is generated by the elements $\alpha_1, \dots, \alpha_g, \beta_1, \dots, \beta_g$ subject to the single relation $\gamma_1 \dots \gamma_g = 1$ where $\gamma_i = \alpha_i \beta_i \alpha_i^{-1} \beta_i^{-1}$.

Fixing a marking $(p_0; \alpha_1, \dots, \alpha_g, \beta_1, \dots, \beta_g)$ for the Riemann surface $M$ and a base point $z_0 \in \tilde{M}$, consider next a factor of automorphy $\xi$ of rank one for the action of the covering translation group $\Gamma$ on $\tilde{M}$. This factor of automorphy is completely determined by the functions $\xi(A_i, z)$ and $\xi(B_i, z)$ for the canonical generators $A_1, \dots, A_g, B_1, \dots, B_g$ of $\Gamma$ associated to the given marking; for any element $T \in \Gamma$ can be expressed as a word in these generators and the factor relation then determines $\xi(T, z)$ explicitly in terms of the functions $\xi(A_i, z)$ and $\xi(B_i, z)$, so that for example $\xi(A_i^{-1}, z) = \xi(A_i, A_i^{-1} z)^{-1}$ and $\xi(C_i, z) = \xi(A_i B_i A_i^{-1} B_i^{-1}, z) = \xi(A_i, B_i A_i^{-1} B_i^{-1} z) \cdot \xi(B_i, A_i^{-1} B_i^{-1} z) \cdot \xi(A_i, A_i^{-1} B_i^{-1} z)^{-1} \cdot \xi(B_i, B_i^{-1} z)^{-1}$. Furthermore the functions $\xi(A_i, z)$ and $\xi(B_i, z)$ can be arbitrary complex analytic nowhere vanishing functions on $\tilde{M}$ subject only to the condition $\prod_{i=1}^{g} \xi(C_i, C_{i+1} \dots C_g z) = 1$

corresponding to the relation $C_1 \dots C_g = 1$ among the generators of $\Gamma$, that is to say, to the condition that

$$\prod_{i=1}^{g} \frac{\xi(A_i, B_i A_i^{-1} B_i^{-1} C_{i+1} \dots C_g z)\xi(B_i, A_i^{-1} B_i^{-1} C_{i+1} \dots C_g z)}{\xi(A_i, A_i^{-1} B_i^{-1} C_{i+1} \dots C_g z)\xi(B_i, B_i^{-1} C_{i+1} \dots C_g z)} = 1 .$$

Since each function $\xi(T, z)$ is complex analytic and nowhere vanishing on the simply connected Riemann surface $\tilde{M}$ it can be written $\xi(T, z) = \exp 2\pi i \sigma(T, z)$ for some complex analytic function $\sigma(T, z)$ on $\tilde{M}$, and the function $\sigma(T, z)$ is unique up to an additive integer; this expression has the advantage of ease in handling the restriction that the function $\xi(T, z)$ be nowhere vanishing. It would be particularly convenient to have the functions $\sigma(T, z)$ satisfy the additive analogue of the factor relation, namely the relation that $\sigma(ST, z) = \sigma(S, Tz) + \sigma(T, z)$ for any elements $S, T \in \Gamma$; a mapping $\sigma: \Gamma \times \tilde{M} \longrightarrow \mathbb{C}$ which is complex analytic on $\tilde{M}$ and satisfies this relation is called a *summand of automorphy*. In general of course the functions $\sigma(T, z)$ will merely satisfy a relation of the form $\sigma(ST, z) = \sigma(S, Tz) + \sigma(T, z) + \sigma(S, T)$ for some integer $\sigma(S, T)$; so the question is whether the functions $\sigma(T, z)$ can be modified by adding an integer to each in order to reduce the expressions $\sigma(S, T)$ to zero, or equivalently whether there is a mapping $\sigma: \Gamma \longrightarrow \mathbb{Z}$ such that $\sigma(ST) = \sigma(S) + \sigma(T) + \sigma(S, T)$ for all elements $S, T \in \Gamma$. Choose arbitrary branches of the logarithms to define $\sigma(A_i, z) = \frac{1}{2\pi i} \log \xi(A_i, z)$ and $\sigma(B_i, z) = \frac{1}{2\pi i} \log \xi(B_i, z)$; and note as in the case of factors of automorphy that these functions can be extended to a summand of automorphy $\sigma$ for the action of the group $\Gamma$ on $\tilde{M}$ precisely when $c(\xi) = 0$ where

$$\begin{aligned} c(\xi) = \textstyle\sum_{i=1}^{g} [&\sigma(A_i, B_i A_i^{-1} B_i^{-1} C_{i+1} \dots C_g z) + \sigma(B_i, A_i^{-1} B_i^{-1} C_{i+1} \dots C_g z) \\ &- \sigma(A_i, A_i^{-1} B_i^{-1} C_{i+1} \dots C_g z) - \sigma(B_i, B_i^{-1} C_{i+1} \dots C_g z)] . \end{aligned}$$

It is clear that this expression is always an integer, since $\exp 2\pi i\, c(\xi) = 1$, and that this integer is independent of the choices of the branches of the logarithms. The integer $c(\xi)$ is called the *characteristic class* of the factor of automorphy $\xi$; and the preceding observations can be summarized in the assertion that *there exists a summand of automorphy $\sigma$ such that $\xi(T, z) = \exp 2\pi i \sigma(T, z)$ precisely when $c(\xi) = 0$.* Note also that the set of factors of automorphy of rank 1 form a group under multiplication, and that the mapping which associates to each factor of automorphy its characteristic class is a homomorphism from this group into the additive group of the integers $\mathbb{Z}$.

It is evident from the preceding observation that whether the characteristic class of a factor of automorphy vanishes or not is really independent of the choice of a marking of the surface. Actually the numerical value of the characteristic class is also independent of the choice of a marking. That is most easily demonstrated by deriving another interpretation of that characteristic class as follows.

**Theorem 2a.** *If $f$ is a meromorphic relatively automorphic function for a factor of automorphy $\xi$ of rank 1 associated to the action of the covering translation group on the universal covering space of a compact Riemann surface $M$ then the characteristic class of $\xi$ is equal to the order of the divisor of $f$ viewed as a divisor on $M$.*

*Proof.* It is clear that the divisor $\mathfrak{d}(f)$ of the function $f$ is invariant under the action of the covering translation group $\Gamma$, hence can be viewed as a divisor on $M$; and that the order of that divisor is given by

$$|\mathfrak{d}(f)| = \frac{1}{2\pi i}\int_{\partial\tilde{\Delta}} df(z)/f(z),$$

where $\tilde{\Delta}$ is the canonical fundamental polygon associated to the marking of $M$, if the divisor $\mathfrak{d}(f)$ vanishes along all the paths of that marking. It is obvious that this last condition can be assumed to hold. As before set $\sigma(A_i, z) = \frac{1}{2\pi i}\log\xi(A_i, z)$ and $\sigma(B_i, z) = \frac{1}{2\pi i}\log\xi(B_i, z)$ for some choices of branches of the logarithm; and extend these functions to a summand of automorphy for the free group on the letters $A_1, \ldots, A_g$, $B_1, \ldots, B_g$ in the obvious manner, so that for instance $\sigma(C_i, z) = \sigma(A_i, B_i A_i^{-1} B_i^{-1} z) + \sigma(B_i, A_i^{-1} B_i^{-1} z) - \sigma(A_i, A_i^{-1} B_i^{-1} z) - \sigma(B_i, B_i^{-1} z)$ and $\sigma(C_1 \ldots C_g, z) = c(\xi)$. Now recalling the explicit form of the canonical fundamental polygon $\tilde{\Delta}$ observe that

$$\begin{aligned}
&\frac{1}{2\pi i}\int_{\partial\tilde{\Delta}} df(z)/f(z)\\
&= \frac{1}{2\pi i}\sum_{i=1}^{g}\int_{C_1\ldots C_{i-1}\tilde{\alpha}_i + C_1\ldots C_{i-1}A_i\tilde{\beta}_i - C_1\ldots C_i B_i\tilde{\alpha}_i - C_1\ldots C_i\tilde{\beta}_i} df(z)/f(z)\\
&= \frac{1}{2\pi i}\sum_{i=1}^{g}\int_{\tilde{\alpha}_i}[df(C_1\ldots C_{i-1}z)/f(C_1\ldots C_{i-1}z) - df(C_1\ldots C_i B_i z)/f(C_1\ldots C_i B_i z)]\\
&\quad + \frac{1}{2\pi i}\sum_{i=1}^{g}\int_{\tilde{\beta}_i}[df(C_1\ldots C_{i-1}A_i z)/f(C_1\ldots C_{i-1}A_i z) - df(C_1\ldots C_i z)/f(C_1\ldots C_i z)]\\
&= \sum_{i=1}^{g}\int_{\tilde{\alpha}_i}[d\sigma(C_1\ldots C_{i-1}, z) - d\sigma(C_1\ldots C_i B_i, z)]\\
&\quad + \sum_{i=1}^{g}\int_{\tilde{\beta}_i}[d\sigma(C_1\ldots C_{i-1}A_i, z) - d\sigma(C_1\ldots C_i, z)]\\
&= \sum_{i=1}^{g}\int_{\tilde{\alpha}_i}[d\sigma(C_1\ldots C_{i-1}, z) - d\sigma(C_1\ldots C_i, B_i z) - d\sigma(B_i, z)]\\
&\quad + \sum_{i=1}^{g}\int_{\tilde{\beta}_i}[d\sigma(C_1\ldots C_{i-1}, A_i z) - d\sigma(C_1\ldots C_i, z) + d\sigma(A_i, z)]\\
&= \sum_{i=1}^{g}[\sigma(C_1\ldots C_{i-1}, A_i z_0) - \sigma(C_1\ldots C_i, B_i A_i z_0) - \sigma(B_i, A_i z_0)\\
&\qquad - \sigma(C_1\ldots C_{i-1}, z_0) + \sigma(C_1\ldots C_i, B_i z_0) + \sigma(B_i, z_0)\\
&\qquad + \sigma(C_1\ldots C_{i-1}, A_i B_i z_0) - \sigma(C_1\ldots C_i, B_i z_0) + \sigma(A_i, B_i z_0)\\
&\qquad - \sigma(C_1\ldots C_{i-1}, A_i z_0) + \sigma(C_1\ldots C_i, z_0) - \sigma(A_i, z_0)]\\
&= c(\xi) + \sum_{i=1}^{g}[\sigma(C_1\ldots C_{i-1}, A_i B_i z_0) - \sigma(C_1\ldots C_i, B_i A_i z_0) + \sigma(A_i, B_i z_0)\\
&\qquad - \sigma(B_i, A_i z_0) + \sigma(B_i, z_0) - \sigma(A_i, z_0)]
\end{aligned}$$

since $\sigma(C_1 \dots C_g, z_0) = c(\xi)$ and the remaining terms cancel. However noting that

$$\begin{aligned}\sigma(C_1 \dots C_i, B_i A_i z_0) &= \sigma(C_1 \dots C_{i-1}, A_i B_i z_0) + \sigma(C_i, B_i A_i z_0) \\ &= \sigma(C_1 \dots C_{i-1}, A_i B_i z_0) + \sigma(A_i, B_i z_0) + \sigma(B_i, z_0) \\ &\quad - \sigma(A_i, z_0) - \sigma(B_i, A_i z_0)\end{aligned}$$

it follows that

$$\frac{1}{2\pi i} \int_{\partial \tilde{\Delta}} df(z)/f(z) = c(\xi),$$

which suffices to conclude the proof of the theorem.

A few simple consequences of this theorem should be noted here. First any meromorphic function $f$ on the surface $M$ can be viewed as a meromorphic relatively automorphic function for the trivial factor of automorphy on $\tilde{M}$, and consequently $|\mathfrak{d}(f)| = 0$; hence linearly equivalent divisors on a compact Riemann surface must have the same order. Next since the factor of automorphy $\zeta_p$ associated to a point bundle admits a complex analytic relatively automorphic function $f$ such that $\mathfrak{d}(f) = 1 \cdot p$ if follows that $c(\zeta_p) = 1$; hence the characteristic class mapping is a surjective homomorphism from the multiplicative group of factors of automorphy for $\Gamma$ to the additive group $\mathbb{Z}$ of the integers. Finally since the basic existence theorem for Riemann surfaces implies that every factor of automorphy admits a nontrivial relatively automorphic function the result of Theorem 2a can be used to describe the characteristic class of any factor of automorphy. Now if two factors of automorphy are analytically equivalent any relatively automorphic function for one can be transformed into a relatively automorphic function for the other by multiplying by a nowhere vanishing complex analytic function; therefore analytically equivalent factors of automorphy have the same characteristic class. Somewhat more can be said as follows.

**Theorem 2b.** *Two factors of automorphy of rank* 1 *for the action of the covering translation group on the universal covering space of a compact Riemann surface are continuously equivalent precisely when they have the same characteristic class.*

*Proof.* It clearly suffices merely to prove that a factor of automorphy $\xi$ is continuously equivalent to the trivial factor of automorphy precisely when $c(\xi) = 0$. First if $\xi$ is continuously equivalent to the trivial factor then there is a continuous mapping $h: \tilde{M} \longrightarrow \mathbb{C}^*$ such that $h(Tz)h(z)^{-1} = \xi(T, z)$ for all elements $T \in \Gamma$ and points $z \in \tilde{M}$. Since $\tilde{M}$ is simply connected there exists a continuous mapping $f: \tilde{M} \longrightarrow \mathbb{C}$ such that $h(z) = \exp 2\pi i f(z)$; the functions $\sigma(T, z) = f(Tz) - f(z)$ then clearly form a continuous summand of automorphy for $\Gamma$, and since $\exp 2\pi i \sigma(T, z) = \xi(T, z)$ they are necessarily complex analytic and $c(\xi) = 0$. Conversely if $c(\xi) = 0$ there is a summand of automorphy $\sigma$ for $\Gamma$ such that $\xi(T, z) = \exp 2\pi i \sigma(T, z)$. Now it is quite easy to see that there exists a continuous mapping $f: \tilde{M} \longrightarrow \mathbb{C}$ such that $f(Tz) = f(z) + \sigma(T, z)$ for all elements $T \in \Gamma$ and all points $z \in \tilde{M}$. Merely select a finite covering of the surface $M$ by contractible open sets $U_i$ and a continuous partition of unity $\{\varepsilon_i\}$ subordinate to that covering; then if $\tilde{U}_i$ is any lifting of the set $U_i$ to $\tilde{M}$ the sets $\{\Gamma U_i\}$ where $\Gamma U_i = \bigcup_{T \in \Gamma} T U_i$ form a locally finite covering of $\tilde{M}$, and the functions $\varepsilon_i$ lift to $\Gamma$-invariant functions $\varepsilon_i$ on $\tilde{M}$ which form a continuous partition of unity subordinate to that covering of $\tilde{M}$. The functions $\sigma_i(T, z) = \varepsilon_i(z)\,\sigma(T, z)$ clearly satisfy the conditions necessary to form a continuous summand of automorphy for $\Gamma$, and the support of $\sigma_i(T, z)$ is contained in the set $\Gamma U_i$. Furthermore it is a simple exercise to verify that the mapping $f_i: \tilde{M} \longrightarrow \mathbb{C}$, defined by $f_i(z) = \sigma_i(T, T^{-1} z)$ whenever $z \in T U_i$ for some $T \in \Gamma$ and $f_i(z) = 0$ otherwise, is continuous and satisfies $f_i(Tz) = f_i(z) + \sigma_i(T, z)$ for all elements $T \in \Gamma$ and points $z \in \tilde{M}$; and the function $f = \sum_i f_i$ is then the desired function. Having found this function, the mapping $h: \tilde{M} \longrightarrow \mathbb{C}^*$ defined by $h(z) = \exp 2\pi i f(z)$ is continuous and satisfies $h(Tz)\,h(z)^{-1} = \xi(T, z)$ for all elements $T \in \Gamma$ and points $z \in \tilde{M}$; and therefore the factor of automorphy $\xi$ is continuously equivalent to the trivial factor of automorphy, and the proof of the theorem is concluded.

The same argument of course shows that two factors of automorphy are $C^\infty$ equivalent precisely when they have the same characteristic class, upon using $C^\infty$ rather than merely continuous partitions of unity. However that argument cannot be extended to analytic equivalence; indeed the situation in that case is rather different, but quite fundamental to the development of the theory. The basic result is the **weak form of Abel's Theorem,** which asserts that *a factor of automorphy of rank* 1 *for the action of the covering translation group on the universal covering space of a compact Riemann surface is analytically equivalent to a flat factor of automorphy precisely when it has zero characteristic class.* A flat factor of automorphy of rank 1 is just an element $\xi \in \mathrm{Hom}(\Gamma, \mathbb{C}^*)$; in terms of a marking of the surface such a homomorphism is completely determined by its values $\xi(A_i)$ and $\xi(B_i)$ on the canonical generators of $\Gamma$, and these values can be chosen quite arbitrarily since the relation among these canonical generators merely involves their commutators and any homomorphism into a commutative group is necessarily trivial on commutators. Choosing arbitrary values for the logarithms $\sigma(A_i) = \dfrac{1}{2\pi i} \log \xi(A_i)$ and $\sigma(B_i) = \dfrac{1}{2\pi i} \log \xi(B_i)$ it follows correspondingly that they necessarily extend to a homomorphism $\sigma \in \mathrm{Hom}(\Gamma, \mathbb{C})$ which can be viewed as a constant summand of automorphy; and therefore $c(\xi) = 0$. The argument in the converse direction is rather deeper and will not be included here, although some comments on two approaches to the proof may not be out of place. Note that in view of Theorem 2a the characteristic class of a factor of automorphy of rank 1 can be identified with the characteristic class or Chern class of the complex analytic line bundle described by that factor of automorphy; and with this observation in mind, a sheaf-theoretic proof of the theorem will be found on page 134 of [10]. Alternatively if $c(\xi) = 0$ then $\xi(T, z) = \exp 2\pi i\,\sigma(T, z)$ for some summand of automorphy $\sigma$, and as in the proof of Theorem 2b there is a $C^\infty$ function $f$ on $\tilde{M}$ such that $f(Tz) = f(z) + \sigma(T, z)$ for all elements $T \in \Gamma$ and points $z \in \tilde{M}$. Since the functions $\sigma(T, z)$ are complex analytic on $\tilde{M}$ it follows that $\bar{\partial} f(Tz) = \bar{\partial} f(z)$ for all elements $T \in \Gamma$, hence that this can be viewed as a differential form on $M$. The function $f$ can be modified by adding to it any $C^\infty$ function on $M$, viewed as a $\Gamma$-invariant $C^\infty$ function on $\tilde{M}$; and a potential-theoretic argument as in [24] shows that after such a modification it can be assumed that $\bar{\partial} f(z)$ is a harmonic differential form, that is, that $\partial\bar{\partial} f(z)=0$. The differential form $\partial f(z)$ is then a closed complex analytic differential form on the simply connected Riemann surface $\tilde{M}$, hence can be written $\partial f(z)=dg(z)$ for some complex analytic function $g$ on $\tilde{M}$; and since $dg(Tz)=\partial f(Tz)=\partial f(z)+\partial\sigma(T,z)=dg(z)+d\sigma(T,z)$ it follows that $g(Tz)=g(z)+\sigma(T,z)-\sigma'(T)$ for some constant $\sigma'(T)$, for each element $T\in\Gamma$. The function $h(z)=\exp 2\pi i\, g(z)$ is then a complex analytic nowhere vanishing function on $\tilde{M}$ and $h(Tz)\xi'(T)=\xi(T,z)h(z)$ for each element $T\in\Gamma$, where $\xi'(T)=\exp 2\pi i\,\sigma'(T)$; but that shows that the factor of automorphy $\xi$ is analytically equivalent to the flat factor of automorphy $\xi'$.

Note that as a consequence of the weak form of Abel's theorem an arbitrary factor of automorphy of characteristic class $r$ is analytically equivalent to a factor of automorphy of the form $\rho\cdot\zeta^r$ where $\zeta$ is a fixed factor of automorphy of characteristic class 1, such as the factor of automorphy associated to a point bundle $\zeta_p$, and $\rho$ is a flat factor of automorphy. The further description of analytic equivalence classes of factors of automorphy then reduces to the classification of flat factors of automorphy under equivalence; and that in turn leads to the more detailed discussion of complex analytic differential forms over Riemann surfaces.

## § 5. Abelian Differentials and the Jacobi Variety of a Riemann Surface

It will be taken as known that on an arbitrary complex manifold a complex-valued differential form $\phi$ of degree $r$ can be decomposed uniquely into a sum $\phi=\sum_{p+q=r}\phi^{(p,q)}$ of complex-valued differential forms $\phi^{(p,q)}$ of bidegree $(p,q)$ where $p\geq 0$, $q\geq 0$, and $p+q=r$; in a coordinate neighborhood $U$ with local coordinates $(z_1,\ldots,z_n)$ these component forms have expressions

$$\phi^{(p,q)}\,|\,U=\sum_{i_1,\ldots,i_p,j_1\ldots j_q}\phi_{i_1\ldots i_p;j_1\ldots j_q}(z)\,dz_{i_1}\wedge\cdots\wedge dz_{i_p}\wedge d\bar{z}_{j_1}\wedge\cdots\wedge d\bar{z}_{j_q}\,.$$

The exterior differential operator $d$ can be decomposed correspondingly into a sum $d=\partial+\bar{\partial}$ of two first order linear differential operators such that $\partial\phi^{(p,q)}$ is of bidegree $(p+1,q)$ and $\bar{\partial}\phi^{(p,q)}$ is of bidegree $(p,q+1)$. For a Riemann surface these reduce to the observations that a differential form $\phi$ of degree 1 can be decomposed into a sum $\phi=\phi^{(1,0)}+\phi^{(0,1)}$ of differential forms of bidegrees (1,0) and (0,1) and that a differential form of degree 2 must be of bidegree (1,1); and that the exterior differential operator $d$ can be decomposed correspondingly. Thus in terms of a complex analytic coordinate covering $\{U_\alpha, z_\alpha\}$ a differential form $\phi^{(1,0)}$ of bidegree (1,0) has the expressions $\phi^{(1,0)}\,|\,U_\alpha=\phi_\alpha(z)\,dz_\alpha$, where $\phi_\alpha$ is a complex-valued function in $U_\alpha$; and the exterior derivative $d\phi^{(1,0)}=\bar{\partial}\phi^{(1,0)}$

has the expressions $d\phi^{(1,0)}|U_\alpha=\bar{\partial}\phi^{(1,0)}|U_\alpha=(\partial\phi_\alpha(z)/\bar{\partial}z_\alpha)d\bar{z}_\alpha\wedge dz_\alpha$. Thus the differential form $\phi^{(1,0)}$ is closed precisely when the functions $\phi_\alpha$ are complex analytic functions in each coordinate neighborhood. The closed differential forms of bidegree (1,0) on a Riemann surface $M$ are called the *Abelian differentials* on $M$. Note that if $\phi^{(0,1)}$ is a closed differential form of bidegree (0,1) then its complex conjugate $\bar{\phi}^{(0,1)}$ is an Abelian differential; but a closed differential form $\phi$ of degree 1 is not necessarily the sum of an Abelian differential and a conjugate Abelian differential, since $d(\phi^{(1,0)}+\phi^{(0,1)})=\bar{\partial}\phi^{(1,0)}+\partial\phi^{(0,1)}=0$ does not imply that $\bar{\partial}\phi^{(1,0)}=\partial\phi^{(0,1)}=0$.

If $\phi$ is any Abelian differential on a Riemann surface $M$ and $\{U_\alpha, z_\alpha\}$ is a complex analytic coordinate covering of $M$ then $\phi|U_\alpha=\phi_\alpha(z)dz_\alpha$ and the local functions $\{\phi_\alpha(z)\}$ must be such that $\phi_\alpha(z)dz_\alpha=\phi_\beta(z)dz_\beta$ whenever $z\in U_\alpha\cap U_\beta$, hence such that $\phi_\alpha(z)=(dz_\beta/dz_\alpha)\phi_\beta(z)$ whenever $z\in U_\alpha\cap U_\beta$. The derivatives $\kappa_{\alpha\beta}(z)=dz_\beta/dz_\alpha$ are nowhere vanishing complex analytic functions in the intersections $U_\alpha\cap U_\beta$, and the chain rule for differentiation implies that $\kappa_{\alpha\beta}(z)\kappa_{\beta\gamma}(z)=\kappa_{\alpha\gamma}(z)$ whenever $z\in U_\alpha\cap U_\beta\cap U_\gamma$; these functions are thus the coordinate transformations for a complex analytic line bundle $\kappa$ over $M$, and the coefficient functions $\{\phi_\alpha(z)\}$ for an Abelian differential describe a complex analytic cross-section of that line bundle $\kappa$. The complex analytic line bundle $\kappa$ is called the *canonical bundle* of the Riemann surface $M$. When $M$ is described as the quotient of its universal covering space $\tilde{M}$ by the group of covering translations $\Gamma$ then the canonical bundle $\kappa$ is described by a factor of automorphy $\kappa$ for the action of the group $\Gamma$ on $\tilde{M}$, and that factor of automorphy is called the *canonical factor of automorphy*; the Abelian differentials on $M$ are then described by the complex analytic relatively automorphic functions for that factor of automorphy. Assuming the general theorem of uniformization, the universal covering space of any Riemann surface other than the Riemann sphere is a subset of the complex plane and hence has a single coordinate neighborhood; and the Abelian differentials can then be written in the form $\phi(z)dz$ for a complex analytic function $\phi(z)$ on $\tilde{M}$. The Abelian differentials on $M$ are just the $\Gamma$-invariant Abelian differentials on $\tilde{M}$, hence are those for which $\phi(Tz)T'(z)dz=\phi(z)dz$; and consequently the canonical factor of automorphy is just $\kappa(T,z)=(T'(z))^{-1}$.

When an Abelian differential $\phi$ on the Riemann surface $M$ is viewed as a $\Gamma$-invariant Abelian differential on the universal covering space $\tilde{M}$ it is apparent that there exists a complex analytic function $f$ on $\tilde{M}$ such that $df=\phi$, since $\tilde{M}$ is simply connected; such a function $f$ is called an *Abelian integral* for the Riemann surface $M$. Note that the Abelian integrals associated to a given Abelian differential differ merely by an additive constant; if a base point $z_0\in\tilde{M}$ is specified a normalized Abelian integral can be chosen by imposing the restriction that $f(z_0)=0$, and that Abelian integral is given explicitly by $f(z)=\int_{z_0}^{z}\phi$. Note further that since the Abelian differential $\phi$ is $\Gamma$-invariant any associated Abelian integral $f$ has the property that $f(Tz)=f(z)+\phi(T)$ for some complex constant $\phi(T)$ whenever $T\in\Gamma$, and that constant is the same for all the Abelian integrals associated to the Abelian differential. Conversely of course any complex analytic function $f$ on $\tilde{M}$ such that $f(Tz)=f(z)+\phi(T)$ for some complex constant $\phi(T)$ whenever $T\in\Gamma$ is an Abelian integral for $M$, since $df=\phi$ is necessarily $\Gamma$-invariant. The mapping $\phi:\Gamma\longrightarrow\mathbb{C}$ which associates to each element $T\in\Gamma$ the constant $\phi(T)$ is called the *period class* of the Abelian differential $\phi$; and since clearly $\phi(ST)=\phi(S)+\phi(T)$ for any two elements $S$, $T$ of $\Gamma$, the period class is an element $\phi\in\mathrm{Hom}(\Gamma,\mathbb{C})$ of the group of homomorphisms from $\Gamma$ into the additive group of the complex numbers $\mathbb{C}$. The terminology is suggested by the observation that $\phi(T)=\int_{z_0}^{Tz_0}\phi$ for any choice of path from $z_0$ to $Tz_0$ in $\tilde{M}$. Note finally that an Abelian differential on a compact Riemann surface $M$ is determined uniquely by its period class; for if $\phi'=df'$ and $\phi''=df''$ are Abelian differentials on $M$ with the same period class then $f'-f''$ is necessarily a $\Gamma$-invariant complex analytic function on $\tilde{M}$, and since $M$ is compact that function must be constant.

It is convenient to introduce at this point another of the basic results about Riemann surfaces which will be used without proof in this book, the **Riemann-Roch Theorem**; that is the assertion that if $M$ is a compact Riemann surface of genus $g$ with universal covering space $\tilde{M}$ and covering translation group $\Gamma$ and if $\xi$ is a factor of automorphy of rank 1 for the action of the group $\Gamma$ on $\tilde{M}$, then *the complex analytic relatively automorphic functions for the factor of automorphy $\xi$ form a finite-dimensional complex vector space and the dimension $\gamma(\xi)$ of this vector space satisfies*

$$\gamma(\xi)-\gamma(\kappa\xi^{-1})=c(\xi)+1-g$$

*where $\kappa$ is the canonical factor of automorphy.* This result can be considered as a more precise version of the basic existence theorem on compact Riemann surfaces, since it quite easily implies that every factor of automorphy admits a nontrivial meromorphic relatively automorphic function and conveys some more precise information about the set of complex analytic relatively automorphic functions. Proofs of this result can be found in [10] or in [24], among other places. Note that if $\xi$ is the trivial bundle then $\gamma(\xi)=1$; for the complex analytic relatively automorphic functions for $\xi$ are just the complex analytic functions on $M$, and all such are constant since $M$ is compact. Then setting $\xi$ equal to the trivial bundle in the Riemann-Roch formula it follows that $\gamma(\kappa)=g$; that is to say, *the Abelian differentials on a compact Riemann surface of genus $g$ form a $g$-dimensional complex vector space.*

Now fix a compact Riemann surface $M$ of genus $g>0$, and choose a marking $(p_0;\alpha_1,\ldots,\alpha_g,\beta_1,\ldots,\beta_g)$ for that surface and a basis $\phi_1,\ldots,\phi_g$ for the vector space of Abelian differentials on that surface. The period class $\phi_i\in\mathrm{Hom}(\Gamma,\mathbb{C})$ of the Abelian differential $\phi_i$ is described by its values on the canonical generators $A_1,\ldots,A_g,B_1,\ldots,B_g$ of the covering translation group $\Gamma$ associated to the choice of marking for $M$. All of these values can be grouped together to form the *period matrix* $\Phi$ for the marked surface $M$ associated to that choice of basis for the space of Abelian differentials. The period matrix is the $g\times 2g$ matrix $\Phi=(\Phi',\Phi'')$ where $\Phi'$ is the $g\times g$ matrix with entries $\phi_i(A_j)$ and $\Phi''$ is the $g\times g$ matrix with entries $\phi_i(B_j)$; thus row $i$ of the matrix $\Phi$ consists of the periods

$$(\phi_i(A_1),\ldots,\phi_i(A_g),\phi_i(B_1),\ldots,\phi_i(B_g))$$

of the Abelian differential $\phi_i$. This matrix has the following special properties.

**Theorem 3.** *The period matrix $\Phi=(\Phi', \Phi'')$ associated to any choices of marking and of basis for the space of Abelian differentials on a compact Riemann surface of genus $g>0$ satisfies the conditions*

(i) *Riemann's equality:* $\Phi'\,{}^t\Phi''-\Phi''\,{}^t\Phi'=0$,

(ii) *Riemann's inequality:* $i\Phi'\,{}^t\bar{\Phi}''-i\Phi''\,{}^t\bar{\Phi}'$ *is positive definite Hermitian.*

*Proof.* Choose a base point $z_0\in\tilde{M}$ and let $f_i(z)=\int_{z_0}^{z}\phi_i$ be the normalized Abelian integrals for the Riemann surface $M$. The differential forms $\phi_i\wedge\phi_j=d(f_i\phi_j)$ are of course trivial since they are of bidegree (2,0), hence have trivial integrals over the canonical fundamental polygon $\tilde{\Delta}\subseteq\tilde{M}$. Applying Stokes' theorem and recalling the form of the boundary of $\tilde{\Delta}$ it follows that

$$\begin{aligned}
0&=\int_{\tilde{\Delta}}\phi_i\wedge\phi_j=\int_{\partial\tilde{\Delta}}f_i\phi_j\\
&=\sum_{k=1}^{g}\int_{C_1\ldots C_{k-1}\tilde{\alpha}_k+C_1\ldots C_{k-1}A_k\tilde{\beta}_k-C_1\ldots C_kB_k\tilde{\alpha}_k-C_1\ldots C_k\tilde{\beta}_k}f_i(z)\,\phi_j(z)\\
&=\sum_{k=1}^{g}\int_{\tilde{\alpha}_k}[f_i(C_1\ldots C_{k-1}z)\,\phi_j(C_1\ldots C_{k-1}z)-f_i(C_1\ldots C_kB_kz)\,\phi_j(C_1\ldots C_kB_kz)]\\
&\quad+\sum_{k=1}^{g}\int_{\tilde{\beta}_k}[f_i(C_1\ldots C_{k-1}A_kz)\,\phi_j(C_1\ldots C_{k-1}A_kz)-f_i(C_1\ldots C_kz)\,\phi_j(C_1\ldots C_kz)]\\
&=\sum_{k=1}^{g}\int_{\tilde{\alpha}_k}[f_i(z)\,\phi_j(z)-(f_i(z)+\phi_i(B_k))\,\phi_j(z)]\\
&\quad+\sum_{k=1}^{g}\int_{\tilde{\beta}_k}[(f_i(z)+\phi_i(A_k))\,\phi_j(z)-f_i(z)\,\phi_j(z)]\\
&=\sum_{k=1}^{g}[-\phi_i(B_k)\int_{\tilde{\alpha}_k}\phi_j(z)+\phi_i(A_k)\int_{\tilde{\beta}_k}\phi_j(z)]\\
&=\sum_{k=1}^{g}[-\phi_i(B_k)\,\phi_j(A_k)+\phi_i(A_k)\,\phi_j(B_k)]\,,
\end{aligned}$$

hence that $-\Phi''\,{}^t\Phi'+\Phi'\,{}^t\Phi''=0$ as desired. On the other hand in a coordinate neighborhood $U\subseteq\tilde{M}$ with coordinate mapping $z$ an Abelian differential $\phi$ can be written $\phi|U=f(z)dz$ for some complex analytic function $f$; therefore

$$(\phi\wedge\bar{\phi})|U=|f(z)|^2\,dz\wedge d\bar{z}=-2i|f(z)|^2\,dx\wedge dy\,,$$

so that $i\int_U\phi\wedge\bar{\phi}\geq 0$ with equality holding only when $\phi=0$. Applying this observation to the differential form $\phi=\sum_{i=1}^{g}c_i\phi_i$, where $c_i$ are arbitrary complex constants not all of which are zero, it follows that

$$0<i\int_{\tilde{\Delta}}\phi\wedge\bar{\phi}=i\sum_{i,j=1}^{g}\int_{\tilde{\Delta}}c_i\bar{c}_j\phi_i\wedge\bar{\phi}_j\,;$$

and consequently the matrix $P$ with entries $p_{ij}=i\int_{\tilde{\Delta}}\phi_i\wedge\bar{\phi}_j$ is positive definite. That matrix is easily seen to be Hermitian as well. Since $\phi_i\wedge\bar{\phi}_j=d(f_i\bar{\phi}_j)$ it follows exactly as in the calculation used in the first part of the proof that

$$p_{ij}=i\sum_{k=1}^{g}[-\phi_i(B_k)\,\bar{\phi}_j(A_k)+\phi_i(A_k)\,\bar{\phi}_j(B_k)]\,,$$

or equivalently that $P=i(\Phi'\,{}^t\bar{\Phi}''-\Phi''\,{}^t\bar{\Phi}')$; and that suffices to conclude the proof of the theorem.

It follows immediately from Riemann's inequality that the matrices $\Phi'$ and $\Phi''$ are nonsingular. For if $\Phi'$ were singular there would exist a nonzero vector

$v \in \mathbb{C}^g$ such that ${}^t v \Phi' = 0$, and then ${}^t v(\Phi' {}^t\bar{\Phi}'' - \Phi'' {}^t\bar{\Phi}')\bar{v} = 0$ contradicting (ii) of Theorem 3; and similarly for the matrix $\Phi''$. Any other basis for the space of Abelian differentials on $M$ consists of differentials of the form $\omega_i = \sum_{j=1}^{g} c_{ij}\phi_j$, where $c_{ij} \in \mathbb{C}$ are constants such that the matrix $C = \{c_{ij}\}$ is nonsingular; and the period matrix associated to this new basis for the space of Abelian differentials is evidently the matrix $C(\Phi', \Phi'')$. It follows that there is a unique basis $\omega_1, \ldots, \omega_g$ for the space of Abelian differentials with period matrix of the form $(I, \Omega)$, where $I$ is the $g \times g$ identity matrix and $\Omega = \{\omega_{ij}\} = \{\omega_i(B_j)\}$; this is obtained from the given basis by the linear transformation with $C = (\Phi')^{-1}$. This basis is called the *canonical basis* for the space of Abelian differentials on the marked Riemann surface $M$, and its elements are called the *canonical Abelian differentials* on the marked Riemann surface $M$; and the associated matrix $(I, \Omega)$ is called the *canonical period matrix* for the marked Riemann surface $M$. Note that for the canonical period matrix the conditions of Theorem 3 take the simpler form that *the matrix $\Omega$ is symmetric and its imaginary part is positive definite.*

It is apparent from Theorem 1(a) that the canonical period matrix $(I, \Omega)$ of a marked Riemann surface $M$ of genus $g$ can be taken as the period matrix of a compact complex torus of dimension $g$; that torus is called the *Jacobi variety* of the marked Riemann surface $M$ and is denoted by $J(M)$. Thus $J(M) = \mathbb{C}^g/\mathscr{L}$ where $\mathscr{L} \subset \mathbb{C}^g$ is the lattice subgroup $\mathscr{L} = (I, \Omega) \cdot \mathbb{Z}^{2g}$. Actually of course a change in the basis of the space of Abelian differentials on $M$ replaces the period matrix $(I, \Omega)$ by the period matrix $C(I, \Omega)$ for some $C \in GL(g, \mathbb{C})$; and a change in the marking of $M$ replaces the cycles represented by the paths of the given canonical dissection of $M$ by cycles forming another basis for the homology group $H^1(M, \mathbb{Z})$, and consequently replaces the period matrix $(I, \Omega)$ by the period matrix $(I, \Omega)N$ for some $N \in GL(2g, \mathbb{Z})$. Therefore, recalling Theorem 1(b), all period matrices for the surface $M$ lead to analytically equivalent complex tori; so that the Jacobi variety $J(M)$ when considered just as a complex torus can equally well be described by an arbitrary period matrix for the Riemann surface $M$.

Now in terms of the canonical generators $A_1, \ldots, A_g$, $B_1, \ldots, B_g$ for the covering translation group $\Gamma$ of a marked Riemann surface $M$ consider the mapping

$$\rho \colon \mathbb{C}^g \longrightarrow \operatorname{Hom}(\Gamma, \mathbb{C}^*)$$

which associates to a vector $t = {}^t(t_1, \ldots, t_g) \in \mathbb{C}^g$ that group homomorphism $\rho_t \in \operatorname{Hom}(\Gamma, \mathbb{C}^*)$ for which

$$\rho_t(A_j) = 1 \quad \text{and} \quad \rho_t(B_j) = \exp 2\pi i t_j .$$

It is apparent that the mapping $\rho$ is itself a group homomorphism. The space $\mathbb{C}^g$ can be viewed as the universal covering space of the Jacobi variety $J(M)$, so that $J(M) = \mathbb{C}^g/\mathscr{L}$ where $\mathscr{L} \subset \mathbb{C}^g$ is the lattice subgroup $\mathscr{L} = (I, \Omega)\mathbb{Z}^{2g}$ in terms of the canonical period matrix $(I, \Omega)$ of the marked Riemann surface $M$; and the group $\operatorname{Hom}(\Gamma, \mathbb{C}^*)$ can be viewed as the group of flat factors of automorphy for the Riemann surface $M$. The homomorphism $\rho$ then has the following interesting properties.

**Theorem 4.** *Any element* $\rho \in \mathrm{Hom}(\Gamma, \mathbb{C}^*)$, *considered as a flat factor of automorphy on the marked compact Riemann surface* $M$, *is analytically equivalent to a flat factor of automorphy of the form* $\rho_t$ *for some* $t \in \mathbb{C}^g$; *and the flat factors of automorphy* $\rho_s$ *and* $\rho_t$ *associated to two vectors* $s, t \in \mathbb{C}^g$ *are analytically equivalent precisely when* $s - t \in \mathcal{L}$, *or what is the same thing, precisely when* $s$ *and* $t$ *represent the same point in the Jacobi variety* $J(M)$ *of* $M$.

*Proof.* As noted before any homomorphism $\rho \in \mathrm{Hom}(\Gamma, \mathbb{C}^*)$ can be written as $\rho(T) = \exp 2\pi i \sigma(T)$ for some homomorphism $\sigma \in \mathrm{Hom}(\Gamma, \mathbb{C})$. Now in terms of the canonical Abelian differentials $\omega_i$ and their associated integrals $w_i$ introduce the complex analytic function $f(z) = \sum_{i=1}^{g} \sigma(A_i) w_i(z)$ on $\tilde{M}$, and note that

$$f(A_j z) = f(z) + \sigma(A_j), \qquad f(B_j z) = f(z) + \sum_{i=1}^{g} \sigma(A_i) \omega_i(B_j).$$

The function $h(z) = \exp 2\pi i f(z)$ is then complex analytic and nowhere vanishing on $\tilde{M}$, and is easily seen to satisfy the conditions

$$h(A_j z)\rho_t(A_j) = \rho(A_j) h(z), \qquad h(B_j z)\rho_t(B_j) = \rho(B_j) h(z)$$

where

$$t = (t_1, \dots, t_g), \qquad t_j = \sigma(B_j) - \sum_{i=1}^{g} \sigma(A_i)\omega_i(B_j);$$

and consequently the factors of automorphy $\rho$ and $\rho_t$ are analytically equivalent. Next consider a flat factor of automorphy of the form $\rho_t$, and note that it can be written as $\rho_t(T) = \exp 2\pi i \sigma_t(T)$ where $\sigma_t \in \mathrm{Hom}(\Gamma, \mathbb{C})$ is that homomorphism for which $\sigma_t(A_j) = 0$ and $\sigma_t(B_j) = t_j$. This factor of automorphy is analytically equivalent to the trivial factor precisely when there exists a complex analytic nowhere vanishing function $h$ on $\tilde{M}$ such that $h(Tz)\rho_t(T) = h(z)$ for every element $T \in \Gamma$; and since any such function $h$ can be written as $h(z) = \exp 2\pi i f(z)$ for some complex analytic function $f$ on $\tilde{M}$, this condition is equivalent to the existence of a complex analytic function $f$ on $\tilde{M}$ such that $f(Tz) - f(z) + \sigma_t(T)$ is an integer constant for every element $T \in \Gamma$. Any such function $f$ must in turn be an Abelian integral for $M$, and hence must be expressible in terms of the canonical Abelian integrals as $f(z) = \sum_{i=1}^{g} c_i w_i(z)$ for some constants $c_i \in \mathbb{C}$; therefore the analytic triviality of the factor of automorphy $\rho_t$ is equivalent to the existence of some complex constants $c_i$ such that the period relation

$$\sigma_t(T) + \sum_{i=1}^{g} c_i \omega_i(T) \in \mathbb{Z}$$

holds for all $T \in \Gamma$. If this period relation holds then setting $T = A_j$ it follows that $c_j \in \mathbb{Z}$, and setting $T = B_j$ it further follows that $t_j = -\sum_{i=1}^{g} c_i \omega_i(B_j) + d_j$ for some $d_j \in \mathbb{Z}$; and since $\omega_i(B_j) = \omega_j(B_i)$ this is just the condition that $t \in \mathcal{L}$. Conversely the period relation is clearly satisfied whenever $t \in \mathcal{L}$. Altogether the factor of automorphy $\rho_t$ is analytically equivalent to the trivial factor of automorphy precisely when $t \in \mathcal{L}$; and since $\rho_s \rho_t^{-1} = \rho_{s-t}$ it is clear that the factors of automorphy $\rho_s$ and $\rho_t$ are analytically equivalent precisely when $s - t \in \mathcal{L}$. That suffices to conclude the proof of the theorem.

An immediate consequence of this theorem and of the weak form of Abel's theorem is the assertion that analytic equivalence classes of factors of automorphy for a compact Riemann surface $M$ can be put into one-to-one correspondence with the group $\mathbb{Z} \times J(M)$ where $J(M)$ is the Jacobi variety of $M$; the integer is the characteristic class of the factor of automorphy, which describes the continuous equivalence class of that factor of automorphy, and $J(M)$ is in one-to-one correspondence with the set of analytic equivalence classes within any one continuous equivalence class. This correspondence can be made canonical with the choice of a marking for the Riemann surface $M$; for the analytic equivalence classes are evidently represented uniquely by the factors of automorphy $\rho_t \zeta^r$ as $t$ varies over $\mathbb{C}^g/\mathscr{L} = J(M)$ and $r$ varies over $\mathbb{Z}$, where $\zeta$ is a factor of automorphy corresponding to the point bundle $\zeta_{p_0}$ for the base point $p_0 \in M$.

It is interesting to examine more closely the relationship between the description of factors of automorphy provided by the preceding theorem and that provided by considering the factors of automorphy associated to divisors. The basic result is the following.

**Theorem 5.** *If $p_+$ and $p_-$ are distinct points on a marked compact Riemann surface $M$ and $z_+$ and $z_-$ are any points of the universal covering space $\tilde{M}$ lying over $p_+$ and $p_-$ respectively, then associated to the divisor $\mathfrak{d} = 1 \cdot p_+ - 1 \cdot p_-$ is a complex analytic line bundle over $M$ represented by the factor of automorphy $\rho_t$, where $t \in \mathbb{C}^g$ has the coordinates $t_i = w_i(z_+) - w_i(z_-)$ in terms of the canonical Abelian integrals $w_i$ for $M$.*

*Proof.* Any factor of automorphy representing the line bundle associated to the divisor $\mathfrak{d}$ must have zero characteristic class as a consequence of Theorem 2a; and it then follows from the weak form of Abel's theorem and Theorem 4 that this factor of automorphy is analytically equivalent to a flat factor of automorphy $\rho_t$ for some $t \in \mathbb{C}^g$. The factor of automorphy $\rho_t$ admits a meromorphic relatively automorphic function $f$ having divisor $\mathfrak{d}(f) = \mathfrak{d}$ on $M$. It can be assumed that the points $p_+$ and $p_-$ do not lie on any of the paths of the canonical dissection of the marked Riemann surface $M$; hence there are points $z_+$ and $z_-$ in the canonical fundamental polygon $\tilde{\Delta} \subseteq \tilde{M}$ lying over $p_+$ and $p_-$ respectively. For any canonical Abelian differential $\omega_i = dw_i$ for $M$ note that

$$w_i(z_+) - w_i(z_-) = \frac{1}{2\pi i} \int_{\partial \tilde{\Delta}} w_i(z) f(z)^{-1} df(z)$$

$$= \frac{1}{2\pi i} \sum_{k=1}^{g} \int_{C_1 \ldots C_{k-1}\tilde{\alpha}_k + C_1 \ldots C_{k-1} A_k \tilde{\beta}_k - C_1 \ldots C_k B_k \tilde{\alpha}_k - C_1 \ldots C_k \tilde{\beta}_k} w_i(z) f(z)^{-1} df(z)$$

$$= \frac{1}{2\pi i} \sum_{k=1}^{g} \left[ -\omega_i(B_k) \int_{\tilde{\alpha}_k} f(z)^{-1} df(z) + \omega_i(A_k) \int_{\tilde{\beta}_k} f(z)^{-1} df(z) \right].$$

Choosing any branch of $\log f(z)$ along the path $\tilde{\alpha}_k$ note further that $\int_{\tilde{\alpha}_k} f(z)^{-1} df(z) = \int_{\tilde{\alpha}_k} d \log f(z) = \log f(A_k z_0) - \log f(z_0) = \log \rho_t(A_k)$ for some branch of $\log \rho_t(A_k)$, and that $\int_{\tilde{\beta}_k} f(z)^{-1} df(z) = \log \rho_t(B_k)$ for some branch of $\log \rho_t(B_k)$;

and since $\rho_t(A_k)=1$ and $\rho_t(B_k)=\exp 2\pi i t_k$ it follows that $\int_{\tilde{\alpha}_k} f(z)^{-1}df(z)=2\pi i m_k$ and $\int_{\tilde{\beta}_k} f(z)^{-1}df(z)=2\pi i(t_k+n_k)$ for some integers $m_k$ and $n_k$. Therefore

$$w_i(z_+)-w_i(z_-)=\sum_{k=1}^g[\omega_i(A_k)(t_k+n_k)-\omega_i(B_k)m_k]$$
$$=t_i+n_i-\sum_{k=1}^g\omega_{ik}m_k$$

so that $w_i(z_+)-w_i(z_-)-t_i$ are the coordinates of a point of the lattice subgroup $\mathscr{L}\subset\mathbb{C}^g$ defined by $\mathscr{L}=(I,\Omega)\cdot\mathbb{Z}^{2g}$. Since the factors of automorphy $\rho_t$ and $\rho_{t+\lambda}$ are analytically equivalent whenever $\lambda\in\mathscr{L}$ as a consequence of Theorem 4, it follows that for any elements $S,T\in\Gamma$ the factor of automorphy $\rho_t$ where $t_i=w_i(Tz_+)-w_i(Sz_-)$ represents the line bundle associated to the divisor $\mathfrak{d}$; and the proof of the theorem is thereby concluded.

The traditional form of Abel's theorem is a simple consequence of the preceding theorem, as follows. Consider the complex analytic mapping $\tilde{\phi}: \tilde{M}\longrightarrow\mathbb{C}^g$ defined by $\tilde{\phi}(z)=(w_1(z),\ldots,w_g(z))$ in terms of the canonical Abelian integrals $w_i$ for $M$. Since $\tilde{\phi}(Tz)-\tilde{\phi}(z)\in\mathscr{L}$ whenever $T\in\Gamma$, where $\mathscr{L}\subset\mathbb{C}^g$ is as usual the lattice subgroup $\mathscr{L}=(I,\Omega)\mathbb{Z}^{2g}$, it follows that the mapping $\tilde{\phi}$ induces a complex analytic mapping $\phi: M\longrightarrow J(M)$, where $M=\tilde{M}/\Gamma$ and $J(M)=\mathbb{C}^g/\mathscr{L}$. These mappings are called the *Jacobi mappings*. The mapping $\phi$ can in turn be extended to a homomorphism from the group of divisors on $M$ into $J(M)$, called the *Jacobi homomorphism*, by setting $\phi(\sum_i \nu_i\cdot p_i)=\sum_i \nu_i\phi(p_i)$. Now if $\mathfrak{d}_1=1\cdot p_1+\cdots+1\cdot p_r$ and $\mathfrak{d}_2=1\cdot q_1+\cdots+1\cdot q_r$ are two divisors of the same order on $M$, and if $\tilde{p}_i$, $\tilde{q}_i$ are any points of $\tilde{M}$ lying over $p_i$, $q_i$, then the image $\phi(\mathfrak{d}_1-\mathfrak{d}_2)\in J(M)$ is represented by the point $t\in\mathbb{C}^g$ with coordinates $t_i=\sum_{j=1}^r[w_i(\tilde{p}_j)-w_i(\tilde{q}_j)]$; and it follows from Theorem 5 that the factor of automorphy representing the complex analytic line bundle associated to the divisor $\mathfrak{d}_1-\mathfrak{d}_2$ is analytically equivalent to $\rho_t$. The divisors $\mathfrak{d}_1$ and $\mathfrak{d}_2$ are linearly equivalent, or what is the same thing the divisor $\mathfrak{d}_1-\mathfrak{d}_2$ is the divisor of a meromorphic function on $M$, precisely when this factor of automorphy is analytically trivial; and it therefore follows from Theorem 4 that *the divisors $\mathfrak{d}_1$ and $\mathfrak{d}_2$ are linearly equivalent precisely when $\phi(\mathfrak{d}_1)=\phi(\mathfrak{d}_2)$ in $J(M)$*. It is apparent that the same conclusion holds no matter what basis is chosen for the space of Abelian differentials on $M$.

The Jacobi mappings play an important role in the study of Riemann surfaces, as indicated for instance in [12]; it is useful and perhaps interesting to examine a few of their simpler properties here. Consider first the mapping $\phi_1: M\longrightarrow J(M)$ defined by $\phi_1(p)=\phi(p)-\phi(a_1)$, where $a_1$ is any fixed point of $M$ and $M$ is of genus $g>0$. This is of course a complex analytic mapping, and since $M$ is compact it follows from Remmert's proper mapping theorem [3] that the image of $M$ is a complex analytic subvariety $\phi_1(M)\subseteq J(M)$. The mapping $\phi_1$ is actually one-one. Indeed if $\phi_1(p_1)=\phi_1(p_2)$ for two distinct points $p_1$ and $p_2$ of $M$ it follows readily from Abel's theorem that $1\cdot p_1-1\cdot p_2$ is the divisor of a meromorphic function $f$ on $M$; but the function $f$ has a simple pole, hence takes every value precisely once and is consequently a complex analytic homeomorphism from $M$ onto the Riemann sphere $\mathbb{P}^1$, contradicting the assumption that $M$ is of genus $g>0$. The mapping $\phi_1$ is moreover nonsingular. Indeed if $d\phi_1(p)=0$ for some point $p\in M$ then all Abelian differentials on $M$ must vanish at $p$, since the components of $d\phi_1$ are just the canonical Abelian differentials on $M$; and that easily implies that $\gamma(\kappa\zeta_p^{-1})\geq g$, where $\kappa$ is the canonical factor of automorphy and $\zeta_p$ is a factor of automorphy representing the point bundle associated to the divisor $1\cdot p$, hence by the Riemann-Roch theorem that $\gamma(\zeta_p)\geq 2$. If $f_1$ and $f_2$ are two linearly independent relatively automorphic functions for the factor of automorphy $\zeta_p$ then $f=f_1/f_2$ is a nonconstant meromorphic function on $M$ having a single simple pole as its sole singularity; and as before that leads directly to a contradiction to the assumption that $M$ is of genus $g>0$. Altogether then the mapping $\phi_1$ is a complex analytic homeomorphism between the Riemann surface $M$ and the complex analytic submanifolds $\phi_1(M)\subseteq J(M)$, so that the Riemann surface can be viewed as being embedded in its Jacobi variety. The choice of a different base point $a_1\in M$ has the effect of translating the image $\phi_1(M)$, so that the embedding is unique up to translation.

In a similar manner for any integer $r\geq 1$ consider the mapping $\phi_r\colon M^r\longrightarrow J(M)$ defined by $\phi_r(p_1,\ldots,p_r)=\phi(p_1)+\cdots+\phi(p_r)-\phi(a_1)-\cdots-\phi(a_r)$, where $a_i$ are any fixed points of $M$. This is also a complex analytic mapping, and its image is a complex analytic subvariety $\phi_r(M^r)\subseteq J(M)$. These mappings are rather more complicated for $r>1$ than for $r=1$, no longer being one-one nor everywhere nonsingular; their detailed properties reflect some of the subtler structure of the Riemann surface. For the case $r=g$ both manifolds $M^g$ and $J(M)$ have dimension $g$, so that it might be expected at least that the mapping $\phi_g\colon M^g\longrightarrow J(M)$ is surjective; that is indeed so, and that result is known as the Jacobi inversion theorem. To see that that is so note that the surjectivity of $\phi_g$ is equivalent to the assertion that for any point $t\in\mathbb{C}^g$ the factor of automorphy $\rho_t\zeta_{a_1}\cdots\zeta_{a_g}$ admits a nontrivial complex analytic relatively automorphic function; for if $f$ is such a function then $\mathfrak{d}(f)=1\cdot p_1+\cdots+1\cdot p_g$ for some points $p_i\in M$ and the factors of automorphy $\rho_t\zeta_{a_1}\cdots\zeta_{a_g}$ and $\zeta_{p_1}\cdots\zeta_{p_g}$ are analytically equivalent, or what is the same thing the factors of automorphy $\rho_t$ and $\zeta_{p_1}\cdots\zeta_{p_g}\zeta_{a_1}^{-1}\cdots\zeta_{a_g}^{-1}$ are analytically equivalent, so that as a consequence of Theorem 5, $\phi_g(p_1,\ldots,p_g)=t+\lambda$ for some $\lambda\in\mathscr{L}$. The Riemann-Roch Theorem shows that $\gamma(\rho_t\zeta_{a_1}\cdots\zeta_{a_g})=\gamma(\kappa\rho_t^{-1}\zeta_{a_1}^{-1}\cdots\zeta_{a_g}^{-1})+1\geq 1$, and that suffices to prove the assertion. The mapping $\phi_g$ is clearly not one-one, since it is evidently unchanged by any permutation of the points $p_i$; even when factored through the quotient manifold $M^g/\mathfrak{S}_g$, where $\mathfrak{S}_g$ is the symmetric group on $g$ letters acting by permutation of the factors, the mapping $\phi_g$ fails to be one-one, although it can be demonstrated that $\phi_g$ is a birational equivalence even if not a complex analytic homeomorphism between the manifolds $M^g/\mathfrak{S}_g$ and $J(M)$. The mapping $\phi_g$ is also not everywhere nonsingular; but as a surjective complex analytic mapping between two complex manifolds of the same dimension it must be nonsingular outside a complex analytic subvariety of $M^g$. If $\phi_g$ is nonsingular at some point $(p_1,\ldots,p_g)\in M^g$ then it is a complex analytic homeomorphism between an open neighborhood $U_1\times\cdots\times U_g$ of $(p_1,\ldots,p_g)$ in $M^g$ and an open neighborhood of $\phi_g(p_1,\ldots,p_g)$ in $J(M)$; the points $p_i$ must of course be distinct, and the sets $U_i$ must be disjoint open neighborhoods of the points $p_i$ in $M$. This provides an occasionally useful coordinatization for an open neighborhood of $\phi_g(p_1,\ldots,p_g)$ in $J(M)$; since $\phi_g$ is surjective such points $\phi_g(p_1,\ldots,p_g)$ can be located arbitrarily in $J(M)$ by a suitable choice of base points $a_i$.

## § 6. Meromorphic Abelian Differentials and the Prime Function of a Riemann Surface

On a compact Riemann surface $M$ of genus $g$ the Abelian differentials can be identified with the complex analytic relatively automorphic functions for the canonical factor of automorphy $\kappa$. Applying the Riemann-Roch theorem to the factor of automorphy $\kappa$ note that $\gamma(\kappa)-\gamma(\kappa\kappa^{-1})=c(\kappa)+1-g$, and hence that $c(\kappa)=2g-2$; the divisor of an Abelian differential thus has order $2g-2$. Now in addition to the regular Abelian differentials it is of some interest to consider meromorphic Abelian differentials, or equivalently meromorphic relatively automorphic functions for the factor of automorphy $\kappa$. The divisors of such differentials are of course also of order $2g-2$ on $M$, and all are linearly equivalent. It is also clear that the total residue of any meromorphic Abelian differential is zero; for the total residue is just the integral of the differential form around the boundary of any canonical fundamental polygon, and it is evident upon recalling the form of the boundary that such an integral is necessarily zero. Thus the simplest meromorphic Abelian differentials are those with either two simple poles having opposite residues or a single double pole having zero residue; and the divisor of zeros of such a differential is of order $2g$ in either case.

The first question to be asked is whether the poles of these particularly simple meromorphic Abelian differentials can be specified quite arbitrarily. To answer this question consider any two points $p_+$ and $p_-$ on the surface $M$, not necessarily distinct; and let $\zeta_+$ and $\zeta_-$ be factors of automorphy associated to the divisors $p_+$ and $p_-$ respectively, so that there are complex analytic relatively automorphic functions $f_+$ for $\zeta_+$ and $f_-$ for $\zeta_-$ such that $\mathfrak{d}(f_+)=p_+$ and $\mathfrak{d}(f_-)=p_-$ on $M$. Applying the Riemann-Roch theorem to the factor of automorphy $\kappa\zeta_+\zeta_-$ note that $\gamma(\kappa\zeta_+\zeta_-)-\gamma(\zeta_+^{-1}\zeta_-^{-1})=c(\kappa\zeta_+\zeta_-)+1-g=g+1$; and since $c(\zeta_+^{-1}\zeta_-^{-1})=-2$ so that $\gamma(\zeta_+^{-1}\zeta_-^{-1})=0$, it follows that $\gamma(\kappa\zeta_+\zeta_-)=g+1$. Now for any Abelian differential $\phi$, viewed as a complex analytic relatively automorphic function for $\kappa$, the product $\phi f_+f_-$ is a complex analytic relatively automorphic function for $\kappa\zeta_+\zeta_-$; there are $g$ linearly independent relatively automorphic functions of this sort, and they are precisely those complex analytic relatively automorphic functions $\psi$ such that $\psi f_+^{-1}f_-^{-1}$ remains complex analytic. Thus since $\gamma(\kappa\zeta_+\zeta_-)=g+1$, there exists a complex analytic relatively automorphic function $\psi$ for $\kappa\zeta_+\zeta_-$ such that $\psi f_+^{-1}f_-^{-1}$ has singularities, and the singularities are at most at the points $p_+$ and $p_-$; and if $p_+\neq p_-$ these singularities are simple poles, while if $p_+=p_-$ it is a single double pole. Thus the poles can be specified quite arbitrarily; and the argument extends quite easily to arbitrary singularities of higher order.

Now let $\phi$ be a meromorphic Abelian differential on $M$ having nontrivial simple poles at two distinct points $p_+$ and $p_-$; and after multiplying by a suitable constant, assume that $\phi$ has residues $+1$ at the point $p_+$ and $-1$ at the point $p_-$ on $M$. The most general such differential is obtained from $\phi$ by adding a complex analytic Abelian differential to $\phi$; and a canonical such differential can be specified by imposing some period restrictions, but some care must be taken in describing the periods of $\phi$ since that differential form has singularities at which it is not locally exact. Choose a simple path $\delta$ from $p_-$ to $p_+$ on $M$, and let $\tilde{\delta}$ be any lifting of $\delta$ to the universal covering space $\tilde{M}$ of $M$. The differential form $\phi$ can be viewed as a $\Gamma$-invariant complex analytic Abelian differential on $\tilde{M}-\Gamma\tilde{\delta}$; and it clearly has zero integral around any closed path in $\tilde{M}-\Gamma\tilde{\delta}$ hence can be written as $\phi=df$ for some complex analytic function $f$ on $\tilde{M}-\Gamma\tilde{\delta}$, where the function $f$ is unique up to an additive constant. It then follows just as in the case of the Abelian differentials that $f(Tz)=f(z)+\phi_\delta(T)$ for some constant $\phi_\delta(T)$ whenever $T\in\Gamma$; and the set of all these constants can be viewed as an element $\phi_\delta\in\mathrm{Hom}(\Gamma,\mathbb{C})$, which will be called the *period class* of the meromorphic Abelian differential $\phi$ with respect to the path $\delta$. If a marking is chosen on the Riemann surface $M$ then the period class $\phi_\delta$ is uniquely determined by its values $\phi_\delta(A_i)$ and $\phi_\delta(B_i)$ on the canonical generators $A_1,\ldots,A_g, B_1,\ldots,B_g$ for $\Gamma$ associated to that marking. Now there is a unique Abelian differential having the periods $\phi_\delta(A_i)$ on the generators $A_1,\ldots,A_g$, namely the differential $\sum_{j=1}^g\phi_\delta(A_j)\omega_j$ where $\omega_j$ are the canonical Abelian differentials; and the difference

$$\omega_\delta=\phi-\textstyle\sum_{j=1}^g\phi_\delta(A_j)\omega_j$$

is the uniquely determined meromorphic Abelian differential on the marked surface $M$ having simple poles at $p_+$ and $p_-$ with residues $+1$ and $-1$ respectively and having periods

$$\omega_\delta(A_1)=\cdots=\omega_\delta(A_g)=0\,.$$

The differential form $\omega_\delta$ is called the *canonical Abelian differential of the third kind* associated to the path $\delta$ on the marked Riemann surface $M$. Note that this is only defined when $\delta$ is a simple path on $M$ having distinct end points, but the path $\delta$ does not have to be disjoint from the marking on the surface. Correspondingly a *canonical Abelian integral of the third kind* associated to the path $\delta$ on the marked Riemann surface $M$ is any complex analytic function $w_\delta$ on $\tilde{M}-\Gamma\tilde{\delta}$ such that $dw_\delta=\omega_\delta$; this function is only determined up to an additive constant, but for any determination $w_\delta(Tz)-w_\delta(z)=\omega_\delta(T)$ where $\omega_\delta\in\mathrm{Hom}(\Gamma,\mathbb{C})$ is the period class of the Abelian differential $\omega_\delta$. The function $w_\delta$ can of course be continued analytically across each arc $T\tilde{\delta}$, but is then a multiple-valued function on $\tilde{M}$ with logarithmic branch points at all points of $\tilde{M}$ lying over $p_+$ or $p_-$.

**Theorem 6.** *Let $M$ be a marked compact Riemann surface, $\omega_j$ be the canonical Abelian differentials on $M$, and $\omega_\delta$ be the canonical Abelian differentials of the third kind on $M$. If $\delta$ is disjoint from the marking then*

$$\omega_\delta(B_j)=2\pi i\textstyle\int_\delta\omega_j\,.$$

*If $\delta'$ and $\delta''$ are disjoint from one another and from the marking then*

$$\textstyle\int_{\delta'}\omega_{\delta''}=\int_{\delta''}\omega_{\delta'}\,.$$

*Proof.* If $\delta$ is disjoint from the marking of $M$ it can be lifted to a unique path $\tilde{\delta}$ in the canonical fundamental polygon $\tilde{\Delta} \subseteq \tilde{M}$; and $\partial \tilde{\delta} = z_+ - z_-$ where $z_+ \in \tilde{\Delta}$ lies over $p_+$ and $z_- \in \tilde{\Delta}$ lies over $p_-$, recalling that the paths are all directed paths. Then by the residue principle

$$\int_{\partial \tilde{\Delta}} w_j(z)\, \omega_\delta(z) = 2\pi i [w_j(z_+) - w_j(z_-)] = 2\pi i \int_{\tilde{\delta}} \omega_j(z);$$

and referring again to the detailed description of the canonical fundamental polygon,

$$\begin{aligned}\int_{\partial \tilde{\Delta}} w_j(z)\, \omega_\delta(z) &= \sum_{k=1}^{g} \int_{C_1 \dots C_{k-1} \tilde{\alpha}_k + C_1 \dots C_{k-1} A_k \tilde{\beta}_k - C_1 \dots C_k B_k \tilde{\alpha}_k - C_1 \dots C_k \tilde{\beta}_k} w_j(z)\, \omega_\delta(z) \\ &= \sum_{k=1}^{g} \left[\omega_j(A_k) \int_{\tilde{\beta}_k} \omega_\delta(z) - \omega_j(B_k) \int_{\tilde{\alpha}_k} \omega_\delta(z)\right] \\ &= \sum_{k=1}^{g} \left[\delta_j^k \omega_\delta(B_k) - \omega_{jk} \omega_\delta(A_k)\right] = \omega_\delta(B_j),\end{aligned}$$

since $\omega_\delta(A_k) = 0$. That demonstrates the first desired result. If $\delta'$ and $\delta''$ are both disjoint from the marking of $M$ they can be lifted to unique paths $\tilde{\delta}'$ and $\tilde{\delta}''$ in the canonical fundamental polygon $\tilde{\Delta} \subseteq \tilde{M}$, with $\partial \tilde{\delta}'' = z''_+ - z''_-$ and $\partial \tilde{\delta}' = z'_+ - z'_-$; and if $\delta'$ and $\delta''$ are disjoint from one another there are disjoint simple closed curves $\sigma'$ and $\sigma''$ in $\tilde{\Delta}$ encircling $\tilde{\delta}'$ and $\tilde{\delta}''$ respectively. The differential form $w_{\delta'} \omega_{\delta''}$ is regular analytic in $\tilde{\Delta} - (\tilde{\delta}' \cup \tilde{\delta}'')$ and therefore

$$\int_{\partial \tilde{\Delta}} w_{\delta'} \omega_{\delta''} = \int_{\sigma'} w_{\delta'} \omega_{\delta''} + \int_{\sigma''} w_{\delta'} \omega_{\delta''}.$$

The function $w_{\delta'}$ is complex analytic inside $\sigma''$, while the differential form $\omega_{\delta''}$ has simple poles at the points $z''_+$ and $z''_-$ with residues $+1$ and $-1$ respectively; consequently

$$\int_{\sigma''} w_{\delta'} \omega_{\delta''} = 2\pi i [w_{\delta'}(z''_+) - w_{\delta'}(z''_-)] = 2\pi i \int_{\tilde{\delta}''} \omega_{\delta'}.$$

Both functions $w_{\delta'}$ and $w_{\delta''}$ are differentiable along $\sigma'$, so by Stokes' theorem

$$\int_{\sigma'} w_{\delta'} \omega_{\delta''} = \int_{\sigma'} [d(w_{\delta'} w_{\delta''}) - w_{\delta''} \omega_{\delta'}] = -\int_{\sigma'} w_{\delta''} \omega_{\delta'};$$

and it then follows as above that

$$\int_{\sigma'} w_{\delta''} \omega_{\delta'} = 2\pi i \int_{\tilde{\delta}'} \omega_{\delta''}.$$

Combining these observations,

$$\int_{\partial \tilde{\Delta}} w_{\delta'} \omega_{\delta''} = 2\pi i \int_{\delta''} \omega_{\delta'} - 2\pi i \int_{\delta'} \omega_{\delta''}.$$

On the other hand referring yet again to the detailed description of the canonical fundamental polygon it follows as in the first part of the proof that

$$\begin{aligned}\int_{\partial \tilde{\Delta}} w_{\delta'} \omega_{\delta''} &= \sum_{k=1}^{g} [\omega_{\delta'}(A_k)\, \omega_{\delta''}(B_k) - \omega_{\delta'}(B_k)\, \omega_{\delta''}(A_k)] \\ &= 0\end{aligned}$$

since $\omega_{\delta'}(A_k)=\omega_{\delta''}(A_k)=0$. That demonstrates the second desired result and concludes the proof of the theorem.

Let $p_+$ and $p_-$ be any two distinct points on the marked Riemann surface $M$, $\delta$ be any simple path from $p_-$ to $p_+$, and $w_\delta(z)$ be any associated canonical Abelian integral of the third kind. As noted earlier the function $w_\delta(z)$ can be continued analytically across the interior of each arc $T\tilde{\delta}$, but will then have logarithmic branch points at the ends of each arc. However since the residues of $dw_\delta$ are $\pm 1$ at each branch point, the function $p_\delta(z)=\exp w_\delta(z)$ remains single valued on $\tilde{M}$; this function has simple zeros at the points $\Gamma z_+$ covering $p_+$, simple poles at the points $\Gamma z_-$ covering $p_-$, and trivial divisor otherwise, and is determined uniquely up to a nonzero constant factor by the path $\delta$. These functions form a useful auxiliary class of basic functions for the Riemann surface $M$ called the *local prime functions* for that marked surface. It follows from the periodicity properties defining the canonical Abelian differentials of the third kind that

$$p_\delta(A_j z)=p_\delta(z), \qquad p_\delta(B_j z)=p_\delta(z)\cdot\exp\omega_\delta(B_j),$$

where $A_1,\dots,A_g$, $B_1,\dots,B_g$ are the canonical generators of $\Gamma$ associated to the marking of $M$. If the path $\delta$ is disjoint from the marking of $M$ and if $\tilde{\delta}$ is the path from $z_-$ to $z_+$ covering $\delta$ and lying in the canonical fundamental polygon $\tilde{\Delta}\subseteq\tilde{M}$ then it follows from Theorem 6 that $\omega_\delta(B_j)=2\pi i\int_\delta\omega_j=2\pi i\int_{\tilde{\delta}}\omega_j=2\pi i[w_j(z_+)-w_j(z_-)]$; and the periodicity properties of the function $p_\delta$ can be expressed more simply in terms of the homomorphisms $\rho_t\in\operatorname{Hom}(\Gamma,\mathbb{C}^*)$ as

$$p_\delta(Tz)=\rho_t(T)\,p_\delta(z) \quad \text{whenever} \quad T\in\Gamma,$$

where $t\in\mathbb{C}^g$ is the point having coordinates $t_j=w_j(z_+)-w_j(z_-)$ or equivalently where $t=\tilde{\phi}(\partial\tilde{\delta})=\tilde{\phi}(z_+-z_-)=\tilde{\phi}(z_+)-\tilde{\phi}(z_-)$ in terms of the Jacobi mapping or homomorphism $\tilde{\phi}$. This simple formulation of the periodicity properties of the function $p_\delta$ can fairly easily be extended to the case of functions associated to arbitrary paths in $M$; but again some care must be taken and it is perhaps more convenient to use a somewhat more indirect extension.

Let $\delta'$ and $\delta''$ be two simple paths on $M$ disjoint from the marking of $M$ and from one another, with $\partial\delta'=p'_+-p'_-$ and $\partial\delta''=p''_+-p''_-$; and let $\tilde{\delta}'$ and $\tilde{\delta}''$ be the liftings of these paths to the canonical fundamental polygon $\tilde{\Delta}\subseteq\tilde{M}$, with $\partial\tilde{\delta}'=z'_+-z'_-$ and $\partial\tilde{\delta}''=z''_+-z''_-$. Thus $z'_+, z'_-, z''_+, z''_-$ are four distinct points of $\tilde{\Delta}$. It also follows from Theorem 6 that $\int_{\delta'}\omega_{\delta''}=\int_{\delta''}\omega_{\delta'}$, or in terms of the Abelian integrals that

$$w_{\delta''}(z'_+)-w_{\delta''}(z'_-)=w_{\delta'}(z''_+)-w_{\delta'}(z''_-);$$

and upon exponentiation therefore

$$p_{\delta''}(z'_+)/p_{\delta''}(z'_-)=p_{\delta'}(z''_+)/p_{\delta'}(z''_-).$$

Now introduce the function $p(z'_+, z'_-, z''_+, z''_-)$ defined for any four distinct points $z'_+, z'_-, z''_+, z''_-$ of $\tilde{\Delta}$ by

$$p(z'_+, z'_-, z''_+, z''_-)=p_{\delta''}(z'_+)/p_{\delta''}(z'_-)=p_{\delta'}(z''_+)/p_{\delta'}(z''_-),$$

where $\tilde{\delta}'$, $\tilde{\delta}''$ are any disjoint simple paths in the interior of $\tilde{\Delta}$ such that $\partial\tilde{\delta}' = z'_+ - z'_-$ and $\partial\tilde{\delta}'' = z''_+ - z''_-$. It is clear that the values of this function are independent of the choices of paths, since the expression $p_{\delta''}(z'_+)/p_{\delta''}(z'_-)$ does not depend on the choice of the path $\tilde{\delta}'$ and the other expression does not depend on the choice of the path $\tilde{\delta}''$; and the values are also clearly independent of the choices of the functions $p_{\delta'}$ and $p_{\delta''}$, since other choices multiply both numerator and denominator of the defining expressions by the same nonzero constant. Moreover the function $p(z'_+, z'_-, z''_+, z''_-) = p_{\delta''}(z'_+)/p_{\delta''}(z'_-)$ is meromorphic in $(z'_+, z'_-)$ for each fixed $(z''_+, z''_-)$, and is similarly meromorphic in $(z''_+, z''_-)$ for each fixed $(z'_+, z'_-)$; it is a general result from the theory of functions of several complex variables that a function which is meromorphic in each variable separately is meromorphic in all variables [3], and hence $p(z'_+, z'_-, z''_+, z''_-)$ is a meromorphic function of four complex variables on the subset of the complex manifold $\tilde{\Delta}^4 = \tilde{\Delta} \times \tilde{\Delta} \times \tilde{\Delta} \times \tilde{\Delta}$ consisting of points with distinct projections to the four factors. That this function extends to a meromorphic function on the entire manifold $\tilde{\Delta}^4$ is a straightforward application of standard techniques in function theory as well; the extended function has simple zeros along the subvarieties $z'_+ = z''_+$ or $z'_- = z''_-$, simple poles along the subvarieties $z'_+ = z''_-$ or $z'_- = z''_+$, and the value 1 along the subvarieties $z'_+ = z'_-$ or $z''_+ = z''_-$ outside the set of indeterminacy. Furthermore though the function

$$p(z'_+, z'_-, z''_+, z''_-) = p_{\delta''}(z'_+)/p_{\delta''}(z'_-)$$

clearly extends to a meromorphic function of the variables $(z'_+, z'_-)$ in the entire manifold $\tilde{M}^2$ for any fixed points $(z''_+, z''_-) \in \tilde{\Delta}^2$; and for any elements $T'_+$ and $T'_-$ of $\Gamma$ this extension has the property that

$$p(T'_+ z'_+, T'_- z'_-, z''_+, z''_-) = \Theta(z''_+, z''_-)\, p(z'_+, z'_-, z''_+, z''_-)$$

where

$$\Theta(z''_+, z''_-) = \rho_{\tilde{\phi}(z''_+ - z''_-)}(T'_+)/\rho_{\tilde{\phi}(z''_+ - z''_-)}(T'_-)$$

is a complex analytic function on $\tilde{M}^2$, indeed where

$$\Theta(z''_+, z''_-) = \exp 2\pi i \textstyle\sum_{j=1}^{g} v_j [w_j(z''_+) - w_j(z''_-)]$$

for some integers $v_j$ depending on the transformations $T'_+$ and $T'_-$. Thus the function $p(z'_+, z'_-, z''_+, z''_-)$ extends to a meromorphic function on the open subset $\tilde{M}^2 \times \tilde{\Delta}^2 \subseteq \tilde{M}^4$. For this extension and for any elements $T''_+$ and $T''_-$ of $\Gamma$ it follows still further that

$$\begin{aligned} &p(T'_+ z'_+, T'_- z'_-, T''_+ z''_+, T''_- z''_-) \\ &\quad = \Theta(T''_+ z''_+, T''_- z''_-)\, p(z'_+, z'_-, T''_+ z''_+, T''_- z''_-) \\ &\quad = \Theta(T''_+ z''_+, T''_- z''_-)\, \Theta'(z'_+, z'_-)\, p(z'_+, z'_-, z''_+, z''_-), \end{aligned}$$

where

$$\Theta'(z'_+, z'_-) = \rho_{\tilde{\phi}(z'_+ - z'_-)}(T''_+)/\rho_{\tilde{\phi}(z'_+ - z'_-)}(T''_-)$$

is a complex analytic function on $\tilde{M}^2$ of a form similar to that of $\Theta(z''_+, z''_-)$. This provides an extension of the function $p(z'_+, z'_-, z''_+, z''_-)$ to a meromorphic function on the entire complex manifold $\tilde{M}^4$; that extension is called the *prime function* of the marked Riemann surface. It is left as an interesting exercise for the reader to verify, by considering in more detail the terms $\Theta(z''_+, z''_-)$ and $\Theta'(z'_+, z'_-)$, that the apparent dissymetry in the above formula for the transform $p(T'_+z'_+, T'_-z'_-, T''_+z''_+, T''_-z''_-)$ of the prime function is purely notational.

It is a simple matter now to compile a catalogue of the properties of the prime function $p(z'_+, z'_-, z''_+, z''_-)$ of the marked Riemann surface $M$. This function is meromorphic on the four dimensional complex manifold $\tilde{M}^4$, with simple zeros along the subvarieties $z'_+ = Tz''_+$ or $z'_- = Tz''_-$ for all $T \in \Gamma$, simple poles along the subvarieties $z'_+ = Tz''_-$ or $z'_- = Tz''_+$ for all $T \in \Gamma$, and trivial divisor otherwise. It satisfies the obvious symmetry properties

$$\begin{aligned} p(z'_+, z'_-, z''_+, z''_-) &= p(z''_+, z''_-, z'_+, z'_-) \\ &= p(z'_-, z'_+, z''_+, z''_-)^{-1} = p(z'_+, z'_-, z''_-, z''_+)^{-1}, \end{aligned}$$

and the transformation properties

$$p(Tz'_+, z'_-, z''_+, z''_-) = \rho_{\tilde{\phi}(z''_+ - z''_-)}(T)\, p(z'_+, z'_-, z''_+, z''_-) \quad \text{whenever} \quad T \in \Gamma,$$

and the others that can be deduced from this and the symmetry properties. Note that the symmetry properties imply that the prime function has the value 1 at all points of determinacy along the subvarieties $z'_+ = z'_-$ or $z''_+ = z''_-$. Furthermore if $z'_-, z''_+, z''_-$ are fixed points on $\tilde{M}$ which represent distinct points on $M$ then $p(z) = p(z, z'_-, z''_+, z''_-)$ is a nontrivial meromorphic function of the single complex variable $z$ on the Riemann surface $\tilde{M}$. This function has simple zeros at the points $\Gamma z''_+$, simple poles at the points $\Gamma z''_-$, the value 1 at the point $z'_-$, and satisfies

$$p(Tz) = \rho_{\tilde{\phi}(z''_+ - z''_-)}(T)\, p(z) \quad \text{whenever} \quad T \in \Gamma.$$

Letting $\tilde{\delta}$ be any simple path from $z''_-$ to $z''_+$ on $\tilde{M}$ which projects to a simple path $\delta$ on $M$, there is a single valued branch of

$$w_\delta(z) = \log p(z)$$

defined on $\tilde{M} - \Gamma\tilde{\delta}$ for which $w_\delta(z'_-) = 0$; that is the unique canonical Abelian integral of the third kind associated to the path $\delta$ normalized to vanish at the point $z'_-$. The exterior derivative

$$\omega_\delta(z) = dw_\delta(z) = p(z)^{-1}\, dp(z)$$

is the canonical Abelian differential of the third kind associated to the path $\delta$.

# Chapter III. Generalized Theta Functions

## § 7. Theta Factors of Automorphy and Generalized Theta Functions

If $M$ is a marked compact Riemann surface of genus $g>0$ and $\xi$ is a factor of automorphy with characteristic class $c(\xi)=r$ then by the Riemann-Roch theorem $\gamma(\xi)=\gamma(\kappa\xi^{-1})+r+1-g$, where $\kappa$ is the canonical factor of automorphy; and if $r\geq 2g-1$ then $c(\kappa\xi^{-1})=2g-2-r<0$ so that $\gamma(\kappa\xi^{-1})=0$ and $\gamma(\xi)=r+1-g$. Thus all factors of automorphy with characteristic class $r$ admit equally many complex analytic relatively automorphic functions whenever $r\geq 2g-1$. Now the set of factors of automorphy with characteristic class $r$ can be parametrized by the complex manifold $\mathbb{C}^g$, by associating to any point $t\in\mathbb{C}^g$ the factor of automorphy $\rho_t\zeta^r$ where $\rho\colon \mathbb{C}^g \longrightarrow \operatorname{Hom}(\Gamma, \mathbb{C}^*)$ is the canonical homomorphism considered previously and $\zeta$ is a factor of automorphy representing the point bundle associated to the divisor $1\cdot p_0$ for the base point $p_0$ of the marked surface $M$; and it can be asked whether the set of all complex analytic relatively automorphic functions for these factors of automorphy can be parametrized accordingly by the complex manifold $\mathbb{C}^g\times\mathbb{C}^{r+1-g}$ if $r\geq 2g-1$. The answer is provided by the following result.

**Theorem 7.** *Let $M$ be a marked compact Riemann surface of genus $g>0$ with universal covering space $\tilde{M}$, and $\zeta$ be a factor of automorphy representing the point bundle associated to the divisor $1\cdot p_0$ where $p_0$ is the base point of $M$. Then if $r\geq 2g-1$ there are $r+1-g$ complex analytic functions $f_i$ on $\mathbb{C}^g\times\tilde{M}$ such that*

$$(1)\qquad f_i(t, Tz)=\rho_t(T)\zeta(T, z)^r f_i(t, z) \quad \textit{whenever} \quad T\in\Gamma$$

*and such that for each fixed $t\in\mathbb{C}^g$ these are linearly independent functions on $\tilde{M}$ hence form a basis for the space of complex analytic relatively automorphic functions for the factor of automorphy $\rho_t\zeta^r$.*

*Proof.* The first step in the proof is the demonstration of the local form of the theorem, the assertion that for any point $t_0\in\mathbb{C}^g$ there exist an open neighborhood $U$ of $t_0$ in $\mathbb{C}^g$ and complex analytic functions $f_i(t, z)$ in $U\times\tilde{M}$ such that for each fixed point $t\in U$ these functions form a basis for the space of complex analytic relatively automorphic functions for the factor of automorphy $\rho_t\zeta^r$. For this purpose it is convenient to use local coordinates in $U$ provided by the Jacobi mapping. Choose fixed base points $a_i\in\tilde{M}$ and open sets $\tilde{U}_i\subseteq\tilde{M}$ such that

$$-\tilde{\phi}_g\colon \tilde{U}_1\times\cdots\times\tilde{U}_g\longrightarrow U$$

is a complex analytic homeomorphism, where as earlier $\tilde{\phi}_g(z_1, \ldots, z_g) = \tilde{\phi}(z_1) + \cdots + \tilde{\phi}(z_g) - \tilde{\phi}(a_1) - \cdots - \tilde{\phi}(a_g)$ for any points $z_i \in \tilde{U}_i$ and $\tilde{\phi}: \tilde{M} \longrightarrow \mathbb{C}^g$ is the Jacobi mapping; the sets $U_i$ can be taken to be disjoint. In these terms the desired local assertion is that there are $r+1-g$ complex analytic functions $f_i(z_1, \ldots, z_g, z)$ in $\tilde{U}_1 \times \cdots \times \tilde{U}_g \times \tilde{M}$ such that

$$f_i(z_1, \ldots, z_g, Tz) = \rho_{\tilde{\phi}(z_1, \ldots, z_g)}(T)^{-1} \zeta(T, z)^r f_i(z_1, \ldots, z_g; z) \quad \text{whenever} \quad T \in \Gamma,$$

and that for each fixed $(z_1, \ldots, z_g)$ these functions are linearly independent.

For a useful auxiliary construction let $\zeta_a$ be a factor of automorphy of characteristic class $g$ representing the line bundle associated to the divisor $1 \cdot a_1 + \cdots + 1 \cdot a_g$; thus $\zeta_a$ admits a complex analytic relatively automorphy function $h$ for which $\mathfrak{d}(h) = 1 \cdot a_1 + \cdots + 1 \cdot a_g$. Consider then the factor of automorphy $\zeta_a \zeta^r$ of characteristic class $r + g \geq 3g - 1$, and note that as a consequence of the Riemann-Roch theorem $\gamma(\zeta_a \zeta^r) = r + 1$. There are therefore $r+1$ linearly independent complex analytic functions $h_1, \ldots, h_r$ on $\tilde{M}$ such that

$$h_j(Tz) = \zeta_a(T, z) \zeta(T, z)^r h_j(z) \quad \text{whenever} \quad T \in \Gamma.$$

It further follows from the Riemann-Roch theorem that for any fixed points $z_i \in \tilde{U}_i$ the $(r+1) \times g$ matrix $(h_j(z_i))$ must be of rank $g$. Indeed if that matrix had rank $\rho < g$ there would exist $r+1-\rho > r+1-g$ linearly independent vectors $c^k \in \mathbb{C}^{r+1}$, $1 \leq k \leq r+1-\rho$, such that $\sum_{j=1}^{r+1} c_j^k h_j(z_i) = 0$; hence there would exist $r+1-\rho > r+1-g$ linearly independent complex analytic relatively automorphic functions $\sum_{j=1}^{r+1} c_j^k h_j$ for the factor of automorphy $\zeta_a \zeta^r$, all of which vanish at the points $z_1, \ldots, z_g$. The points $z_i$ represent distinct points on $M$ since the neighborhoods $U_i$ are disjoint, so it follows readily that there are at least $r+1-\rho > r+1-g$ linearly independent complex analytic relatively automorphic functions for the factor of automorphy $\zeta_a \zeta^r \zeta_z^{-1}$ where $\zeta_z$ is a factor of automorphy of characteristic class $g$ representing the line bundle associated to the divisor $1 \cdot z_1 + \cdots + 1 \cdot z_g$; but from the Riemann-Roch theorem $\gamma(\zeta_a \zeta^r \zeta_z^{-1}) = r+1-g$, which is a contradiction. The matrix $(h_j(z_i))$ is thus of rank $g$ for any points $z_i \in \tilde{U}_i$; hence after shrinking the neighborhoods further if necessary this matrix can be taken to consist of the last $g$ columns of a matrix $C(z_1, \ldots, z_g)^{-1}$ where $C: \tilde{U}_1 \times \cdots \times \tilde{U}_g \longrightarrow GL(r+1, \mathbb{C})$ is a complex analytic mapping. Thus there are complex analytic functions $c_{ij}(z_1, \ldots, z_g)$ in $\tilde{U}_1 \times \cdots \times \tilde{U}_g$ such that $\sum_{j=1}^{r+1} c_{ij}(z_1, \ldots, z_g) h_j(z_k) = 0$ for $1 \leq i \leq r+1-g$, $1 \leq k \leq g$, and all points $z_i \in \tilde{U}_i$. The functions

$$g_i(z_1, \ldots, z_g, z) = \sum_{j=1}^{r+1} c_{ij}(z_1, \ldots, z_g) h_j(z) \quad \text{for} \quad 1 \leq i \leq r+1-g$$

are therefore complex analytic functions on $\tilde{U}_1 \times \cdots \times \tilde{U}_g \times \tilde{M}$, and for each fixed $(z_1, \ldots, z_g) \in \tilde{U}_1 \times \cdots \times \tilde{U}_g$ are $r+1-g$ linearly independent relatively automorphic functions for the factor of automorphy $\zeta_a \zeta^r$ and vanish at the points $\Gamma z_i$.

Returning then to the proof of the local form of the theorem, choose a point $z_0 \in \tilde{M}$ which is not contained in any of the sets $\Gamma \tilde{U}_i$ or $\Gamma a_i$ and consider the prime function $p(z, z_0, z_j, a_j)$ as a meromorphic function of $z$ and $z_j$. For each fixed $z_j \in \tilde{U}_j$ this is a nontrivial meromorphic function of $z \in \tilde{M}$ which has simple zeros at the points $\Gamma z_j$, simple poles at the points $\Gamma a_j$, and satisfies

$$p(Tz, z_0, z_j, a_j) = \rho_{\tilde{\phi}(z_j - a_j)}(T)\, p(z, z_0, z_j, a_j) \quad \text{whenever} \quad T \in \Gamma.$$

The functions

$$f_i(z_1, \ldots, z_g, z) = g_i(z_1, \ldots, z_g, z)\, h(z)^{-1} \textstyle\prod_{j=1}^{g} p(z, z_0, z_j, a_j)^{-1} \quad \text{for} \quad 1 \le i \le r+1-g$$

are then just the desired functions. They are meromorphic in $\tilde{U}_1 \times \cdots \times \tilde{U}_g \times \tilde{M}$, and for each fixed $(z_1, \ldots, z_g) \in \tilde{U}_1 \times \cdots \times \tilde{U}_g$ are actually nontrivial complex analytic functions on $\tilde{M}$ since by construction the zeros of $g_i(z_1, \ldots, z_g, z)$ and $p(z, z_0, z_j, a_j)$ cancel the poles of $h(z)^{-1}$ and $p(z, z_0, z_j, a_j)^{-1}$; and they are linearly independent as functions of $z$, and transform properly under the action of the covering translation group since $\prod_{j=1}^{g+1} \rho_{\tilde{\phi}(z_j - a_j)} = \rho_{\tilde{\phi}(z_1, \ldots, z_g)}$.

Having proved the local form of the theorem, choose a covering of $\mathbb{C}^g$ by open subsets $U_\alpha$ such that there are $r+1-g$ complex analytic functions $\{f_i^\alpha(t, z)\}$ on $U_\alpha \times \tilde{M}$ satisfying the desired conditions. Then for each $t \in U_\alpha \cap U_\beta$ the functions $\{f_i^\alpha(t, z)\}$ and $\{f_i^\beta(t, z)\}$ are two bases for the space of complex analytic relatively automorphic functions for the factor of automorphy $\rho_t \zeta^r$, and consequently $f_i^\alpha(t, z) = \sum_{j=1}^{r+1-g} \phi_{ij}^{\alpha\beta}(t) f_j^\beta(t, z)$ for a uniquely determined nonsingular matrix $\phi^{\alpha\beta}(t) = (\phi_{ij}^{\alpha\beta}(t)) \in GL(r+1-g, \mathbb{C})$; it is clear that these matrices are complex analytic functions of $t$ and that $\phi^{\alpha\beta}(t)\phi^{\beta\gamma}(t) = \phi^{\alpha\gamma}(t)$ whenever $t \in U_\alpha \cap U_\beta \cap U_\gamma$, so that these matrices describe a complex analytic vector bundle of rank $r+1-g$ over $\mathbb{C}^g$. That bundle must be topologically trivial since $\mathbb{C}^g$ is contractible, so by Grauert's theorem [8] it must also be analytically trivial; hence there are complex analytic mappings $\phi^\alpha: U_\alpha \longrightarrow GL(r+1-g, \mathbb{C})$ such that $\phi^\alpha(t)\phi^{\alpha\beta}(t) = \phi^\beta(t)$ whenever $t \in U_\alpha \cap U_\beta$. There are then well defined complex analytic functions determined on all of $\mathbb{C}^g \times \tilde{M}$ by setting $f_i(t, z) = \sum_{j=1}^{r+1-g} \phi_{ij}^\alpha(t) f_j^\alpha(t, z)$ whenever $t \in U_\alpha$; and these clearly satisfy the desired conditions, so concluding the proof of the theorem.

A set of functions $f_i$ satisfying the conditions of Theorem 7 can be viewed as a single vector-valued function $\theta: \mathbb{C}^g \times \tilde{M} \longrightarrow \mathbb{C}^{r+1-g}$; and any such function will be called a *generalized theta function* of rank $r+1-g$ associated to the marked Riemann surface $M$. Such a function is of course not uniquely determined; but it is clear that if $f_i$ and $f_i'$ are any two sets of functions satisfying the conditions of Theorem 7 then

$$f_i'(t, z) = \textstyle\sum_{j=1}^{g} c_{ij}(t) f_j(t, z)$$

for some uniquely determined complex analytic functions $c_{ij}$ on $\mathbb{C}^g$ forming a nonsingular matrix at each point $t$, since these are two bases for the same vector space of functions on $\tilde{M}$ for each fixed $t \in \mathbb{C}^g$. Hence if $\theta$ and $\theta'$ are any two generalized theta functions of rank $r+1-g$ there is a uniquely determined complex analytic mapping $C: \mathbb{C}^g \longrightarrow GL(r+1-g, \mathbb{C})$ such that $\theta'(t, z) = C(t)\theta(t, z)$ for all $(t, z) \in \mathbb{C}^g \times \tilde{M}$. These generalized theta functions have an interesting transformational property reflecting the observation made in Theorem 4 that the flat factors of automorphy $\rho_t$ and $\rho_{t+\lambda}$ are analytically equivalent whenever $\lambda \in \mathbb{C}^g$ belongs to the lattice subgroup $\mathscr{L} = (I, \Omega)\mathbb{Z}^{2g}$ defining the Jacobi variety $J(M) = \mathbb{C}^g/\mathscr{L}$. To make this analytic equivalence conveniently explicit note that any $\lambda \in \mathscr{L}$ has components $\lambda_i = m_i + \sum_{j=1}^g \omega_{ij} n_j$ for some uniquely determined integers $m_i, n_i$ and then associate to $\lambda$ the complex analytic function

(2) $$\rho(\lambda, z) = \exp 2\pi i \textstyle\sum_{j=1}^g n_j [w_j(z) - w_j(z_0)]$$

on $\tilde{M}$, where $w_j(z)$ are the canonical Abelian integrals and $z_0$ is the base point of $\tilde{M}$. It is clear that $\rho(\lambda, A_i z) = \rho(\lambda, z) = \rho(\lambda, z)\rho_\lambda(A_i)$ and that $\rho(\lambda, B_i z) = \rho(\lambda, z) \exp 2\pi i \sum_{j=1}^g n_j \omega_{ji} = \rho(\lambda, z)\rho_\lambda(B_i)$ for the canonical generators of $\Gamma$, hence that

(3) $$\rho(\lambda, Tz) = \rho(\lambda, z)\rho_\lambda(T) \quad \text{for all} \quad T \in \Gamma;$$

and since $\rho(\lambda, z)$ is nowhere vanishing it follows that the flat factor of automorphy $\rho_\lambda$ is analytically trivial, hence that the flat factors of automorphy $\rho_t \rho_\lambda = \rho_{t+\lambda}$ and $\rho_t$ are analytically equivalent for any $t \in \mathbb{C}^g$. Now for each fixed point $z \in \tilde{M}$ it is evident that $\rho(\lambda' + \lambda'', z) = \rho(\lambda', z)\rho(\lambda'', z)$ for any lattice vectors $\lambda'$ and $\lambda''$ in $\mathscr{L}$; thus the mapping $\rho_z$ which associates to each lattice vector $\lambda \in \mathscr{L}$ the constant $\rho_z(\lambda) = \rho(\lambda, z) \in \mathbb{C}^*$ is a group homomorphism $\rho_z \in \operatorname{Hom}(\mathscr{L}, \mathbb{C}^*)$, and can also be viewed as a flat factor of automorphy for the action of the lattice subgroup $\mathscr{L}$ on the complex manifold $\mathbb{C}^g$ by translation.

**Corollary 1 to Theorem 7.** *To each generalized theta function $\theta$ of rank $r+1-g \geq g$ there is associated a unique complex analytic factor of automorphy $\chi$ of rank $r+1-g$ for the action of the lattice subgroup $\mathscr{L}$ on the complex manifold $\mathbb{C}^g$ such that*

(4) $$\theta(t+\lambda, z) = \rho_z(\lambda)\chi(\lambda, t)\theta(t, z) \quad \text{for all} \quad \lambda \in \mathscr{L} \quad \text{and} \quad (t, z) \in \mathbb{C}^g \times \tilde{M};$$

*and the factors of automorphy thus associated to any two generalized theta functions of the same rank are analytically equivalent.*

*Proof.* The component functions $f_i$ of a generalized theta function $\theta$ of rank $r+1-g$ form for each fixed $t \in \mathbb{C}^g$ a basis for the vector space of complex analytic functions on $\tilde{M}$ satisfying (1). If $\lambda \in \mathscr{L}$ then the products $\rho(\lambda, z) f_i(t, z)$ remain linearly independent as functions of $z$ and, recalling (3), satisfy

$$\rho(\lambda, Tz) f_i(t, Tz) = \rho_\lambda(T)\rho_t(T)\zeta(T, z)^r \cdot \rho(\lambda, z) f_i(t, z) \quad \text{for all} \quad T \in \Gamma,$$

hence form a basis for the vector space of all such relatively automorphic functions; but the functions $f_i(t+\lambda, z)$ form another basis for the same vector space, and therefore

$$f_i(t+\lambda, z) = \textstyle\sum_{j=1}^{r+1-g} \chi_{ij}(\lambda, t)\rho(\lambda, z) f_j(t, z)$$

for a uniquely determined nonsingular matrix $\chi(\lambda, t) = (\chi_{ij}(\lambda, t))$ depending on $t$ and $\lambda$. It is evident that this matrix is complex analytic in $t$; and since $\rho(\lambda, z) = \rho_z(\lambda)$ is a factor of automorphy of rank one for the action of the lattice subgroup $\mathscr{L}$ on the complex manifold $\mathbb{C}^g$ it is also evident from uniqueness that $\chi(\lambda, t)$ is a factor of automorphy of rank $r+1-g$ for the same group. If $\theta$ and $\theta'$ are two generalized theta functions of rank $r+1-g$ with associated factors of automorphy $\chi$ and $\chi'$ then $\theta'(t, z) = C(t)\,\theta(t, z)$ for a uniquely determined complex analytic mapping $C: \mathbb{C}^g \longrightarrow GL(r+1-g, \mathbb{C})$; replacing $t$ by $t+\lambda$ and recalling (4) it follows that

$$\rho_z(\lambda)\,\chi'(\lambda, t)\,\theta'(t, z) = C(\lambda+t)\,\rho_z(\lambda)\,\chi(\lambda, t)\,\theta(t, z)$$

and hence that

$$\chi'(\lambda, t)^{-1}\, C(\lambda+t)\,\chi(\lambda, t) = C(t)\,, \tag{5}$$

which implies that the factors of automorphy $\chi$ and $\chi'$ are analytically equivalent. That suffices to conclude the proof of the corollary.

Merely by change of emphasis the preceding corollary can be viewed as giving for each integer $s = r+1-g \geq g$ a unique complex analytic equivalence class of complex analytic factors of automorphy of rank $s$ for the action of the lattice subgroup $\mathscr{L}$ on $\mathbb{C}^g$, or equivalently a unique complex analytic equivalence class of complex analytic vector bundles of rank $s$ over the Jacobi variety $J(M) = \mathbb{C}^g/\mathscr{L}$. This class or any element of it will be called the *theta factor of rank* $s$ for the Jacobi variety, and will be denoted by $\chi$ or by $\chi_s$ as necessary. Furthermore for each point $z \in \tilde{M}$ there are also given nontrivial complex analytic relatively automorphic functions for the factor of automorphy $\rho_z \otimes \chi$, namely those generalized theta functions $\theta(t, z)$ to which that factor of automorphy is associated as in the corollary. As noted before there are many different generalized theta functions of rank $s$, to all of which are associated analytically equivalent factors of automorphy and to some of which may even be associated the same factor of automorphy. If $\theta$ and $\theta'$ are two generalized theta functions of rank $s$ to which are associated the same theta factor $\chi$ then $\theta'(t, z) = C(t)\,\theta(t, z)$ for some complex analytic mapping $C: \mathbb{C}^g \longrightarrow GL(s, \mathbb{C})$, and as in Eq. (5) in the proof of the corollary $\chi(\lambda, t)\,C(t) = C(t+\lambda)\,\chi(\lambda, t)$ for all $\lambda \in \mathscr{L}$. The mapping $C$ thus describes a complex analytic automorphism of the complex analytic vector bundle described by the factor of automorphy $\chi$, or for short $C$ is a complex analytic automorphism of the factor of automorphy $\chi$; and that automorphism transforms one theta factor into the other. Thus up to complex analytic automorphism there is actually a unique generalized theta function to which is associated any particular theta factor within the equivalence class. All of these observations can be summarized as follows.

**Corollary 2 to Theorem 7.** *For each integer $s \geq g$ there is a unique complex analytic equivalence class of complex analytic factors of automorphy for the action of the lattice subgroup $\mathscr{L}$ on the complex manifold $\mathbb{C}^g$, the theta factor of rank $s$;*

*and for each particular theta factor* $\chi$ *and each point* $z \in \tilde{M}$ *there is a complex analytic relatively automorphic function* $\theta(t, z)$ *for the factor of automorphy* $\rho_z \otimes \chi$, *the generalized theta function of rank s, which is uniquely determined up to a complex analytic automorphism of the factor of automorphy* $\chi$. *These functions are characterized by the property that for each fixed* $t \in \mathbb{C}^g$ *their components are a basis for the space of complex analytic relatively automorphic functions for the factor of automorphy* $\rho_t \otimes \zeta^{s+g-1}$ *for the action of the covering translation group* $\Gamma$ *on the complex manifold* $\tilde{M}$.

Thus the generalized theta function $\theta(t, z)$ is a relatively automorphic function in each variable separately, satisfying the transformational properties

(6) $$\theta(t+\lambda, z) = \rho_z(\lambda)\chi(\lambda, t)\theta(t, z) \quad \text{for all} \quad \lambda \in \mathscr{L},$$

(7) $$\theta(t, Tz) = \rho_t(T)\zeta(T, z)^{s+g-1}\theta(t, z) \quad \text{for all} \quad T \in \Gamma;$$

or equivalently $\theta(t, z)$ is a relatively automorphic function for the action of the transformation group $\mathscr{L} \times \Gamma$ on the product manifold $\mathbb{C}^g \times \tilde{M}$. Having fixed the factor of automorphy $\chi$ in its equivalence class and chosen a relatively automorphic function $\theta$, any other generalized theta function for that theta factor $\chi$ must be of the form $\theta'(t, z) = C(t)\theta(t, z)$ where $C: \mathbb{C}^g \longrightarrow GL(s, \mathbb{C})$ is a complex analytic mapping such that

(8) $$C(t+\lambda) = \chi(\lambda, t)C(t)\chi(\lambda, t)^{-1} \quad \text{for all} \quad \lambda \in \mathscr{L}.$$

In these formulas $\zeta$ is a fixed factor of automorphy for the action of the group $\Gamma$ on the complex manifold $\tilde{M}$, representing the line bundle associated to the divisor $1 \cdot p_0$ where $p_0$ is the base point of the marked surface $M$; $\chi$ is the theta factor, defined by these formulas; and $\rho_t \in \operatorname{Hom}(\Gamma, \mathbb{C}^*)$ and $\rho_z \in \operatorname{Hom}(\mathscr{L}, \mathbb{C}^*)$ are flat factors of automorphy for the action of the group $\Gamma$ on the complex manifold $\tilde{M}$ and the action of the group $\mathscr{L}$ on the complex manifold $\mathbb{C}^g$ respectively, as defined quite explicitly in the preceding discussion. Indeed $\rho_t$ was defined by

(9) $$\rho_t(A_j) = 1, \quad \rho_t(B_j) = \exp 2\pi i t_j,$$

where $t = (t_1, \ldots, t_g)$ is any point of $\mathbb{C}^g$ and $A_1, \ldots, A_g$, $B_1, \ldots, B_g$ are the canonical generators of $\Gamma$; and letting $\phi: \Gamma \longrightarrow \mathscr{L}$ be the homomorphism which maps the element $T \in \Gamma$ to the vector $\phi(T) = (\omega_1(T), \ldots, \omega_g(T)) \in \mathscr{L}$ consisting of the periods of the canonical differentials and letting $\tilde{\phi}: \tilde{M} \longrightarrow \mathbb{C}^g$ be the Jacobi mapping $\tilde{\phi}(z) = (w_1(z), \ldots, w_g(z))$ it follows that $\rho_z$ can be defined by

(10) $$\rho_z(\phi(T)) = \rho_{\tilde{\phi}(z-z_0)}(T) \quad \text{whenever} \quad T \in \Gamma,$$

where $z_0$ is the base point of $\tilde{M}$.

The vector bundles described by theta factors are quite closely related to the Jacobi mappings; but to see that it is necessary first to examine the Jacobi mappings more closely. Note that the set of positive divisors of order $r$ on the Riemann surface $M$ can be identified with the quotient space of the complex analytic manifold $M^r = M \times \cdots \times M$ by the natural action of the symmetric group on $r$ letters $\mathfrak{S}_r$ as permutations of the factors; for the mapping which associates to each point $(p_1, \ldots, p_r) \in M^r$ the divisor $1 \cdot p_1 + \cdots + 1 \cdot p_r$ identifies precisely those points of $M^r$ which differ by a permutation of the factors. This quotient space $M^r/\mathfrak{S}_r$ can be given the structure of a complex analytic manifold in a manner making the natural mapping $M^r \longrightarrow M^r/\mathfrak{S}_r$ a complex analytic mapping. Indeed if $(p_1, \ldots, p_r) \in M^r$ and if $U_i$ are open coordinate neighborhoods of $p_i$ in $M$ with local coordinates $z_i$ then the mapping $\sigma: U_1 \times \cdots \times U_r \longrightarrow \mathbb{C}^r$ defined by $\sigma(z_1, \ldots, z_r) = (\sigma_1(z), \ldots, \sigma_r(z))$ where $\sigma_k(z) = z_1^k + \cdots + z_r^k$ is easily seen to determine a coordinate mapping on the open subset $\mathfrak{S}_r \cdot (U_1 \times \cdots \times U_r)/\mathfrak{S}_r \subseteq M^r/\mathfrak{S}_r$, and these coordinate neighborhoods evidently describe the desired complex analytic structure. The Jacobi mapping $\phi_r: M^r \longrightarrow J(M)$ defined by $\phi_r(p_1, \ldots, p_r) = \phi(p_1) + \cdots + \phi(p_r) - \phi(a_1) - \cdots - \phi(a_r)$, where $a_i$ are some fixed points of $M$, clearly commutes with the action of the permutation group $\mathfrak{S}_r$ on $M^r$ and hence induces a complex analytic mapping $\phi_r^*: M^r/\mathfrak{S}_r \longrightarrow J(M)$; this is in many ways a more interesting and natural mapping to consider than $\phi_r$ itself.

**Theorem 8.** *For any integer $r \geq 2g-1$ the complex manifold $M^r/\mathfrak{S}_r$ of positive divisors of order $r$ on the marked Riemann surface $M$ of genus $g > 0$, under the Jacobi mapping $\phi_r^*: M^r/\mathfrak{S}_r \longrightarrow J(M)$ defined by*

$$\phi_r^*(p_1, \ldots, p_r) = \phi(p_1) + \cdots + \phi(p_r) - r\phi(p_0),$$

*can be given the structure of the complex analytic projective space bundle over $J(M)$ described by the theta factor of automorphy ${}^t\chi_{r+1-g}^{-1}$.*

*Proof.* Although complex analytic projective space bundles have not previously been defined in this book, the appropriate definition should be quite apparent from what has been said. The assertion of the theorem is in part that for any sufficiently small open neighborhood $U$ of any point of $J(M)$ there is a complex analytic homeomorphism $\tau_U: U \times \mathbb{P}_{r-g} \longrightarrow (\phi_r^*)^{-1}(U)$, where $\mathbb{P}_{r-g}$ is the complex projective space of dimension $r-g$, such that $\phi_r^* \circ \tau_U$ coincides with the natural projection mapping $U \times \mathbb{P}_{r-g} \longrightarrow U$ and that whenever $U \cap V \neq \emptyset$ the restricted homeomorphism $\tau_V^{-1} \circ \tau_U: (U \cap V) \times \mathbb{P}_{r-g} \longrightarrow (U \cap V) \times \mathbb{P}_{r-g}$ is a projective transformation over each fixed point of $U \cap V$. The assertion of the theorem is further that the projective space bundle is that described by the theta factor ${}^t\chi_{r+1-g}^{-1}$ in the following manner. If $\mathbb{C}_*^s$ denotes the complement of the origin in the complex vector space $\mathbb{C}^s$ with coordinates $(u_1, \ldots, u_s)$, where $s = r+1-g$, then the multiplicative group $\mathbb{C}^*$ acts on $\mathbb{C}_*^s$ by $t \cdot (u_1, \ldots, u_s) = (tu_1, \ldots, tu_s)$ and the quotient manifold $\mathbb{C}_*^s/\mathbb{C}^*$ is the complex projective space $\mathbb{P}_{r-g} = \mathbb{P}_{s-1}$; the coordinates $(u_1, \ldots, u_s)$ are homogeneous coordinates for that projective space. Now consider the product manifold $\mathbb{C}^g \times \mathbb{P}_{r-g}$, with coordinates $(t_1, \ldots, t_g)$ in $\mathbb{C}^g$ and homogeneous coordinates $(u_1, \ldots, u_s)$ in $\mathbb{P}_{r-g}$, and introduce the action of the lattice subgroup $\mathcal{L} = (I, \Omega)\mathbb{Z}^{2g} \subseteq \mathbb{C}^g$ as a group of complex analytic automorphisms of $\mathbb{C}^g \times \mathbb{P}_{r-g}$ defined by

$$\lambda \cdot (t, u) = (t + \lambda, {}^t\chi_s(\lambda, t)^{-1} u)$$

for any $\lambda \in \mathscr{L}$; these are of course automorphisms of $\mathbb{C}^g \times \mathbb{P}_{r-g}$ as a complex analytic projective space bundle over $\mathbb{C}^g$. The quotient space $(\mathbb{C}^g \times \mathbb{P}_{r-g})/\mathscr{L}$ is then a complex analytic projective space bundle over the quotient space $\mathbb{C}^g/\mathscr{L} = P(M)$, and this is the bundle of interest. In order to prove the theorem it suffices merely to show that there is a surjective complex analytic mapping $\tau: \mathbb{C}^g \times \mathbb{P}_{r-g} \longrightarrow M^r/\mathfrak{S}_r$, commuting with the above action of the lattice subgroup $\mathscr{L}$, such that $\phi_r \circ \tau(t,u) \equiv t \bmod \mathscr{L}$ and $\tau(t,u) = \tau(t,u')$ precisely when $u = u'$; for then $\tau$ obviously induces a bundle isomorphism $\tau: (\mathbb{C}^g \times \mathbb{P}_{r-g})/\mathscr{L} \longrightarrow M^r/\mathfrak{S}_r$. This uses the known result that a one-one complex analytic mapping between complex manifolds is necessarily an analytic homeomorphism.

To carry out the remainder of the proof let $\theta(t, z) = (f_i(t, z))$ be a generalized theta function of rank $s$ associated to the theta factor $\chi_s$, and for any point $(t, u) \in \mathbb{C}^g \times \mathbb{C}^s_*$ consider the complex analytic function $f(u, t, z) = \sum_{i=1}^s u_i f_i(t, z)$ on $\tilde{M}$. This is a nontrivial complex analytic relatively automorphic function for the factor of automorphy $\rho_t \zeta^r$, and its divisor can be viewed as a point of $M^r/\mathfrak{S}_r$. The mapping which associates to the point $(t, u)$ the divisor $\tau(t, u) = \mathfrak{d}(f(u, t, z))$ is thus a well defined mapping $\tau: \mathbb{C}^g \times \mathbb{C}^s_* \longrightarrow M^r/\mathfrak{S}_r$. Since the functions $f(u,t,z)$ and $c f(u, t, z) = f(cu, t, z)$ evidently have the same divisor whenever $c \in \mathbb{C}^*$ the mapping $\tau$ really induces a mapping $\tau: \mathbb{C}^g \times \mathbb{P}_{r-g} \longrightarrow M^r/\mathfrak{S}_r$. This mapping is surjective, since the line bundle associated to any positive divisor of order $r$ can be described by the factor of automorphy $\rho_t \zeta^r$ for some $t \in \mathbb{C}^g$ and is therefore the divisor of the function $f(u, t, z)$ for some $u \in \mathbb{C}^s_*$. If $\lambda \in \mathscr{L}$ and $(t', u') = \lambda \cdot (t, u) \in \mathbb{C}^g \times \mathbb{C}^s_*$, so that $t' = t + \lambda$ and $u' = {}^t\chi(\lambda, t)^{-1} u$ or equivalently $u_j = \sum_{i=1}^s u'_i \chi_{ij}(\lambda, t)$, then

$$\begin{aligned} f(u', t', z) &= \textstyle\sum_{i=1}^s u'_i f_i(t + \lambda, z) \\ &= \textstyle\sum_{i=1}^s u'_i \rho_z(\lambda) \chi_{ij}(\lambda, t) f_j(t, z) \\ &= \rho_z(\lambda) \textstyle\sum_{j=1}^s u_j f_j(t, z) \\ &= \rho(\lambda, z) f(u, t, z), \end{aligned}$$

where $\rho(\lambda, z)$ is a complex analytic nowhere vanishing function of $z \in \tilde{M}$; and consequently $\tau(t', u') = \mathfrak{d}(f(u', t', z)) = \mathfrak{d}(f(u, t, z)) = \tau(t, u)$, so that the mapping $\tau$ does commute with the action of the lattice subgroup $\mathscr{L}$ on $\mathbb{C}^g \times \mathbb{P}_{r-g}$. Since the line bundle associated to the divisor $\mathfrak{d}(f(u, t, z)) = 1 \cdot p_1 + \cdots + 1 \cdot p_r$ is represented by the factor of automorphy $\rho_t \zeta^r$ then the line bundle associated to the divisor $1 \cdot p_1 + \cdots + 1 \cdot p_r - r \cdot p_0$ is represented by the factor of automorphy $\rho_t$; it then follows from Theorem 5 that $t \equiv \phi(p_1) + \cdots + \phi(p_r) - r\phi(p_0) \bmod \mathscr{L}$, or equivalently $\phi_r \circ \tau(t, u) \equiv t \bmod \mathscr{L}$. Moreover if $\mathfrak{d}(f(u, t, z)) = \mathfrak{d}(f(u', t, z))$ for some points $(t, u)$ and $(t, u')$ in $\mathbb{C}^g \times \mathbb{C}^s_*$ then necessarily $f(u', t, z) = c \cdot f(u, t, z)$ for some constant $c \in \mathbb{C}^*$ and all points $z \in \tilde{M}$; and since $f_i(t, z)$ are linearly independent functions on $\tilde{M}$ it follows that $u' = cu$, hence that $u'$ and $u$ are homogeneous coordinates of the same point of $\mathbb{P}_{s-1}$. All that remains to be verified in order to conclude the proof of the theorem is just that the mapping $\tau: \mathbb{C}^g \times \mathbb{C}^s_* \longrightarrow M^r/\mathfrak{S}_r$ is a complex analytic mapping.

In order to demonstrate analyticity consider a fixed point $(t_0, u_0) \in \mathbb{C}^g \times \mathbb{C}^s_*$ with image $\tau(t_0, u_0) = 1 \cdot p_1 + \cdots + 1 \cdot p_r \in M^r/\mathfrak{S}_r$, and let $U_i$ be a relatively compact coordinate neighborhood of $p_i$ with local coordinate $z_i$; to be more precise, if $p_i = p_j$ assume that $U_i = U_j$ and that the same local coordinate $z_i = z_j$ is chosen in both, and if $p_i \neq p_j$ assume that $U_i \cap U_j = \emptyset$. Thus none of the points $p_i$ lie on any boundary arc $\partial U_j$, so that $f(u_0, t_0, z_j) \neq 0$ for $z_j \in \partial U_j$; it follows from continuity that $f(u, t, z_j) \neq 0$ for $(t, u)$ sufficiently near $(t_0, u_0)$ and all $z_j \in \partial U_j$, and hence from the argument principle that $f(u, t, z_j)$ has as many zeros in $U_j$ as has $f(u_0, t_0, z_j)$ whenever $(t, u)$ is sufficiently near $(t_0, u_0)$. Therefore whenever $(t, u)$ is sufficiently near $(t_0, u_0)$ necessarily $\mathfrak{d}(f(u, t, z)) = 1 \cdot z_1 + \cdots + 1 \cdot z_r$ where $z_j \in U_j$; and from the residue principle it follows that

$$z_1^k + \cdots + z_r^k = \frac{1}{2\pi i} \sum_{U_j} \int_{\partial U_j} (\partial f(u, t, z)/\partial z) f(u, t, z)^{-1} z^k dz ,$$

where the sum is extended over all the distinct coordinate neighborhoods. This expression is obviously a complex analytic function of $(t, u)$; but for $k = 1, \ldots, r$ these are exactly the local coordinates on $M^r/\mathfrak{S}_r$, and consequently the mapping $\tau: \mathbb{C}^g \times \mathbb{C}^s_* \longrightarrow M^r/\mathfrak{S}_r$ is complex analytic. That suffices to conclude the proof of the theorem.

A note about the terminology should perhaps be included to conclude this section. The classical theta functions describe in normal form the complex analytic cross-sections of complex analytic line bundles over complex tori; the functions considered here have been called generalized theta functions merely because they describe some complex analytic cross-sections of some complex analytic vector bundles over complex tori. Whether these functions exhibit any analogues of the subtle and interesting finer properties of the classical theta functions, and indeed whether there are any corresponding normal forms which reflect these properties, remain to be investigated.

## § 8. Generalized Theta Functions and Canonical Subvarieties of the Jacobi Variety

The generalized theta functions can be used to describe in an interesting alternative way the images of the normalized Jacobi mappings $\phi_\mu: M^\mu \longrightarrow J(M)$ defined by $\phi_\mu(p_1, \ldots, p_\mu) = \phi(p_1) + \cdots + \phi(p_\mu) - \mu\phi(p_0)$ where $p_0$ is the base point of the marked Riemann surface $M$. The notation $\phi_\mu$ will be used through most of this and the next sections to denote this particular normalization of the Jacobi mapping. As noted before, in the discussion of the Jacobi inversion theorem, the mapping $\phi_\mu$ is surjective if $\mu = g$, and hence of course also for $\mu \geq g$. If $1 \leq \mu \leq g-1$ then the image $\phi_\mu(M^\mu) = W_\mu \subseteq J(M)$ is a complex analytic subvariety of $J(M)$ of dimension at most $\mu$, as a consequence of the Remmert proper mapping theorem [3]. These subvarieties are irreducible, in the sense that any meromorphic function on $J(M)$ which vanishes on a nonempty relatively open subset of a subvariety $W_\mu$ necessarily vanishes on all of $W_\mu$; for the composition of the restriction of any such meromorphic function to $W_\mu$ and the complex analytic mapping $\phi_\mu$ is a meromorphic function on the complex manifold $M^\mu$, and if that function vanishes on an open subset it vanishes identically. In terms of local coordinates $z_j$ on $M$ the Jacobian matrix of the mapping $\phi_\mu: M^\mu \longrightarrow J(M)$ at a point $(z_1, \dots, z_m)$ is $(w_i'(z_j))$ where $w_i(z)$ are the canonical Abelian integrals; and since the Abelian differentials $\omega_i(z) = w_i'(z)\,dz$ are linearly independent it is evident that the rank of that Jacobian matrix is its maximal value $\mu$ on a dense open subset of $M^\mu$. The image $\phi_\mu(M^\mu) = W_\mu \subseteq J(M)$ thus has dimension equal to $\mu$ for $1 \le \mu \le g-1$. These subvarieties $W_\mu$ for $1 \le \mu \le g-1$ are called the *subvarieties of positive divisors* of order $\mu$ on $J(M)$. As also noted before the subvariety $W_1 \subseteq J(M)$ is a complex analytic submanifold analytically homeomorphic to the Riemann surface $M$ itself.

As a preliminary observation recall from Theorem 5 that for any point $t \in \mathbb{C}^g$

$$t \equiv \phi_\mu(p_1, \dots, p_\mu) \equiv (\phi(p_1) - \phi(p_0)) + \cdots + (\phi(p_\mu) - \phi(p_0)) \bmod \mathcal{L}$$

if and only if the flat factor of automorphy $\rho_t$ is analytically equivalent to $\zeta_{p_1} \cdots \zeta_{p_\mu} \zeta^{-\mu}$, or equivalently if and only if $\rho_t \zeta^\mu$ is analytically equivalent to $\zeta_{p_1} \cdots \zeta_{p_\mu}$, where as usual $\zeta_{p_i}$ is a complex analytic factor of automorphy describing the line bundle associated to the divisor $1 \cdot p_i$ and $\zeta = \zeta_{p_0}$. Now the factor of automorphy $\rho_t \zeta^\mu$ has characteristic class $\mu$ and is analytically equivalent to a factor of automorphy of the form $\zeta_{p_1} \cdots \zeta_{p_\mu}$ for some points $p_i \in M$ precisely when $\gamma(\rho_t \zeta^\mu) \ge 1$; for $\gamma(\zeta_{p_1} \cdots \zeta_{p_\mu}) \ge 1$, and if $\gamma(\rho_t \zeta^\mu) \ge 1$ and $f$ is any complex analytic relatively automorphic function for the factor of automorphy $\rho_t \zeta^\mu$ then $\mathfrak{d}(f) = 1 \cdot p_1 + \cdots + 1 \cdot p_\mu$ for some points $p_i \in M$ and $\rho_t \zeta^\mu$ is analytically equivalent to $\zeta_{p_1} \cdots \zeta_{p_\mu}$. Thus $t \in \mathbb{C}^g$ represents a point of $W_\mu \subseteq J(M) = \mathbb{C}^g/\mathcal{L}$ precisely when $\gamma(\rho_t \zeta^\mu) \ge 1$, or more succintly

$$W_\mu = \{t \in \mathbb{C}^g \mid \gamma(\rho_t \zeta^\mu) \ge 1\}/\mathcal{L}\,. \tag{11}$$

This formula can then be used to extend the definition of the sets $W_\mu$ to any integer $\mu \in \mathbb{Z}$, whereas they were originally defined only for $1 \le \mu \le g-1$. Note though that if $\mu \ge g$ then by the Riemann-Roch theorem $\gamma(\rho_t \zeta^\mu) = \gamma(\kappa \rho_t^{-1} \zeta^{-\mu}) + \mu + 1 - g \ge 1$ for all points $t \in \mathbb{C}^g$, and consequently

$$W_\mu = J(M) \quad \text{whenever} \quad \mu \ge g\,;$$

that is of course quite compatible with the preceding definition. If $\mu = 0$ then since $\rho_t$ has characteristic class 0 and consequently $\gamma(\rho_t) \ge 1$ precisely when the factor of automorphy $\rho_t$ is analytically trivial it follows from Theorem 4 that

$$W_0 = \mathcal{L}/\mathcal{L} \subseteq \mathbb{C}^g/\mathcal{L} = J(M)\,;$$

thus $W_0$ consists of a single point of $J(M)$. Finally if $\mu<0$ then $c(\rho_t\zeta^\mu)=\mu<0$ so that $\gamma(\rho_t\zeta^\mu)=0$ for all points $t\in\mathbb{C}^g$; thus

$$W_\mu=\phi \quad \text{whenever} \quad \mu<0.$$

The subsets $W_\mu\subseteq J(M)$ are therefore well defined complex analytic subvarieties of $J(M)$ for any $\mu\in\mathbb{Z}$. Any translate $W_\mu+c$ of one of these sets by an element $c$ of the group $J(M)$ is of course an analytically homeomorphic complex analytic subvariety of $J(M)$, as is the negative subvariety $-W_\mu$; here $W_\mu+c=\{t+c\,|\,t\in W_\mu\}$ and $-W_\mu=\{-t\,|\,t\in W_\mu\}$. As a final bit of additional notation the canonical factor of automorphy $\kappa$ has characteristic class $2g-2$ and is therefore analytically equivalent to a factor of automorphy of the form $\rho_k\zeta^{2g-2}$ for some point $k\in\mathbb{C}^g$; the point of $J(M)$ represented by $k$, which is uniquely determined by the marking of the Riemann surface $M$, is called the *canonical point* of $J(M)$.

Now let $\theta(t,z)=(f_i(t,z))$ be a generalized theta function of rank $s=r+1-g\geq g$ for the marked Riemann surface $M$, and let $z_1,\dots,z_n$ be points of $\tilde{M}$ representing points $p_1,\dots,p_n$ of $M$. The column vectors $\theta(t,z_j)$ can be taken as the columns of an $s\times n$ matrix

$$\theta(t;z_1,\dots,z_n)=(\theta(t,z_1),\dots,\theta(t,z_n)),$$

so that

$$\theta(t;z_1,\dots,z_n)=(f_i(t,z_j)),\qquad 1\leq i\leq s,\qquad 1\leq j\leq n.$$

**Theorem 9.** *If* $z_1,\dots,z_n$ *on* $\tilde{M}$ *represent distinct points* $p_1,\dots,p_n$ *on* $M$ *and* $s=r+1-g\geq g$ *then*

$$W_{r-n}+\phi_n(p_1,\dots,p_n)=\{t\in\mathbb{C}^g\,|\,\operatorname{rank}\theta(t;z_1,\dots,z_n)<s\}/\mathscr{L} \tag{12}$$

*and*

$$-W_{2g-2+n-r}+k+\phi_n(p_1,\dots,p_n)=\{t\in\mathbb{C}^g\,|\,\operatorname{rank}\theta(t;z_1,\dots,z_n)<n\}/\mathscr{L}. \tag{13}$$

*Proof*. Setting $\rho=\operatorname{rank}\theta(t;z_1,\dots,z_n)$ note first that

$$\begin{aligned} s-\rho&=\dim\{c\in\mathbb{C}^s\,|\,{}^tc\,\theta(t;z_1,\dots,z_n)=0\}\\ &=\dim\{c\in\mathbb{C}^s\,|\,\textstyle\sum_{i=1}^s c_if_i(t,z_j)=0 \text{ for } 1\leq j\leq n\}.\end{aligned}$$

Since the functions $f_i(t,z)$ form a basis for the vector space of complex analytic relatively automorphic functions for the factor of automorphy $\rho_t\zeta^r$ it follows that $s-\rho$ is just the dimension of the subspace of those relatively automorphic functions which vanish at all of the points $z_1,\dots,z_n$. If $f(z)$ is such a relatively automorphic function and if $h_i(z)$ are nontrivial analytic relatively automorphic functions for the factors of automorphy $\zeta_{p_i}$, so that $\mathfrak{d}(h_i)=1\cdot p_i$, then $f(z)h_1(z)^{-1}\dots h_n(z)^{-1}$ is a complex analytic relatively automorphic function for the factor of automorphy $\rho_t\zeta^r\zeta_{p_1}^{-1}\dots\zeta_{p_n}^{-1}$ if the points $p_1,\dots,p_n$ are distinct; and conversely if $g(z)$ is a complex analytic relatively automorphic function for the factor of automorphy $\rho_t\zeta^r\zeta_{p_1}^{-1}\ldots\zeta_{p_n}^{-1}$ then $g(z)h_1(z)\ldots h_n(z)$ is a complex analytic relatively automorphic function for the factor of automorphy $\rho_t\zeta^r$ and it vanishes at all of the points $z_1,\ldots,z_n$. It therefore follows that

$$s-\rho=\gamma(\rho_t\zeta^r\zeta_{p_1}^{-1}\ldots\zeta_{p_n}^{-1}); \tag{14}$$

and consequently $\rho<s$ precisely when $\gamma(\rho_t\zeta^r\zeta_{p_1}^{-1}\ldots\zeta_{p_n}^{-1})\geq 1$. Now recalling Theorem 5 note that

$$\gamma(\rho_t\zeta^r\zeta_{p_1}^{-1}\ldots\zeta_{p_n}^{-1})=\gamma(\rho_t\zeta\zeta_{p_1}^{-1}\ldots\zeta\zeta_{p_n}^{-1}\zeta^{r-n})=\gamma(\rho_{t-\phi_n(p_1,\ldots,p_n)}\zeta^{r-n});$$

and it then follows from (11) that $\gamma(\rho_t\zeta^r\zeta_{p_1}^{-1}\ldots\zeta_{p_n}^{-1})\geq 1$ precisely when $t-\phi_n(p_1,\ldots,p_n)\in W_{r-n}$, which suffices to prove (12). On the other hand it follows from the Riemann-Roch theorem that

$$\gamma(\rho_t\zeta^r\zeta_{p_1}^{-1}\ldots\zeta_{p_n}^{-1})=\gamma(\kappa\rho_t^{-1}\zeta^{-r}\zeta_{p_1}\ldots\zeta_{p_n})+s-n,$$

and hence by (14)

$$n-\rho=\gamma(\kappa\rho_t^{-1}\zeta^{-r}\zeta_{p_1}\ldots\zeta_{p_n});$$

consequently $\rho<n$ precisely when $\gamma(\kappa\rho_t^{-1}\zeta^{-r}\zeta_{p_1}\ldots\zeta_{p_n})\geq 1$. Again recalling Theorem 5 note that

$$\begin{aligned}\gamma(\kappa\rho_t^{-1}\zeta^{-r}\zeta_{p_1}\ldots\zeta_{p_n})&=\gamma(\kappa\zeta^{2-2g}\rho_t^{-1}\zeta^{-1}\zeta_{p_1}\ldots\zeta^{-1}\zeta_{p_n}\zeta^{2g-2+n-r})\\&=\gamma(\rho_{k-t+\phi_n(p_1,\ldots,p_n)}\zeta^{2g-2+n-r});\end{aligned}$$

and it then follows from (11) that $\gamma(\kappa\rho_t^{-1}\zeta^{-r}\zeta_{p_1}\ldots\zeta_{p_n})\geq 1$ precisely when $k-t+\phi_n(p_1,\ldots,p_n)\in W_{2g-2+n-r}$, which suffices to prove (13) and hence to conclude the proof of the theorem.

If $z_1,\ldots,z_n$ are fixed points of $\tilde{M}$, representing distinct points $p_1,\ldots,p_n$ of $M$, then $\theta(t;z_1,\ldots,z_n)$ is a complex analytic mapping $\theta:\mathbb{C}^g\longrightarrow\mathbb{C}^{s\times n}$ when viewed as a function of $t\in\mathbb{C}^g$ and the preceding theorem can be applied. If $s\leq n$ then the set of points $t\in\mathbb{C}^g$ at which the matrix $\theta(t;z_1,\ldots,z_n)$ has rank strictly less than $s$ is the complex analytic subvariety of $\mathbb{C}^g$ defined as the set of zeros of all $s\times s$ subdeterminants formed from the matrix $\theta(t;z_1,\ldots,z_n)$; and if $n\leq s$ then the set of points $t\in\mathbb{C}^g$ at which the matrix $\theta(t;z_1,\ldots,z_n)$ has rank strictly less than $n$ is the complex analytic subvariety of $\mathbb{C}^g$ defined as the set of zeros of all $n\times n$ subdeterminants formed from the matrix $\theta(t;z_1,\ldots,z_n)$. Thus fixed translates of the subvarieties $W_{r-n}\subseteq J(M)$ and $-W_{2g-2+n-r}\subseteq J(M)$ are complex analytic subvarieties defined quite explicitly in terms of the generalized theta functions. The simplest special cases of the preceding theorem, in the sense of involving the fewest auxiliary points $z_1,\ldots,z_n$ to describe the subvarieties $W_\mu$, are those given by Eq. (13) when $s$ taken its minimal value $s=g$ and when $1\leq n\leq g$; and the result can be restated as follows.

**Corollary 1 to Theorem 9.** *Let $\theta(t, z)$ be a generalized theta function of rank $g$ for the marked Riemann surface $M$, and let $z_1, \ldots, z_g$ be any fixed points of $\tilde{M}$ which represent distinct points $p_1, \ldots, p_g$ of $M$. Then for $1 \leq n \leq g$*

$$-W_{n-1}+k+\phi_n(p_1, \ldots, p_n)=\{t \in \mathbb{C}^g \mid \operatorname{rank}(\theta(t, z_1), \ldots, \theta(t, z_n))<n\}/\mathscr{L} . \tag{15}$$

To examine this result in somewhat more detail note first that for $n=1$ the assertion is that

$$k+\phi_1(p_1)=\{t \in \mathbb{C}^g \mid \operatorname{rank} \theta(t, z_1)<1\}/\mathscr{L}$$

for any point $z_1 \in \tilde{M}$ representing a point $p_1 \in M$. For fixed $z_1$ the function $\theta(t, z_1)$ can be viewed as a complex analytic vector field on $\mathbb{C}^g$, that is as a complex analytic mapping $\theta(z_1)\colon \mathbb{C}^g \longrightarrow \mathbb{C}^g$; and this vector field is singular only at the points $k+\phi_1(p_1) \bmod \mathscr{L}$, that is, $\theta(t, z_1)=0$ if and only if $t \equiv k+\phi_1(p_1) \bmod \mathscr{L}$. The vanishing of this vector field is of course just the simultaneous vanishing of its $g$ component functions; and as is familiar a single point of $\mathbb{C}^g$ can only be described analytically by the vanishing of at least $g$ complex analytic functions in an open neighborhood of that point. By translation this provides a rather simple and explicit analytic description of any single point $c \in J(M)$; for when $\tilde{c} \in \mathbb{C}^g$ represents that point and $\tilde{k} \in \mathbb{C}^g$ represents the canonical point then the complex analytic vector field $\theta(t+\tilde{k}+\tilde{\phi}_1(z_1)-\tilde{c}, z_1)$ vanishes precisely at the points $\tilde{c}+\mathscr{L} \subseteq \mathbb{C}^g$ for any fixed point $z_1 \in \mathbb{C}^g$. Next note that for $n=2$ the assertion is that

$$-W_1+k+\phi_2(p_1, p_2)=\{t \in \mathbb{C}^g \mid \operatorname{rank}(\theta(t, z_1), \theta(t, z_2))<2\}/\mathscr{L}$$

for any points $z_1, z_2 \in \tilde{M}$ representing distinct points $p_1, p_2 \in M$. For fixed $z_1$ and $z_2$ the functions $\theta(t, z_1)$ and $\theta(t, z_2)$ are two complex analytic vector fields on $\mathbb{C}^g$, which are singular at the disjoint discrete point sets representing $k+\phi_1(p_1)$ and $k+\phi_2(p_2)$ respectively; and these vector fields are linearly dependent precisely along the one-dimensional complex analytic submanifold $-W_1+k+\phi_2(p_1, p_2)$ passing through these singular points, when that submanifold is viewed as lying in $\mathbb{C}^g$. This submanifold is thus described analytically as the set of common zeros of the $\binom{g}{2}$ determinants of all the $2 \times 2$ submatrices of the $g \times 2$ matrix $(\theta(t, z_1), \theta(t, z_2))$. Actually of course at least one of the vector fields $\theta(t, z_1)$ or $\theta(t, z_2)$ is nonsingular at each point $t \in \mathbb{C}^g$, so that

$$-W_1+k+\phi_2(p_1, p_2)=\{t \in \mathbb{C}^g \mid \operatorname{rank}(\theta(t, z_1), \theta(t, z_2))=1\}/\mathscr{L} .$$

Therefore in some open neighborhood of any given point $t_0 \in -W_1+k+\phi_2(p_1, p_2)$ at least one row of the matrix $(\theta(t, z_1), \theta(t, z_2))$ is nonsingular; and in that neighborhood the condition that rank $(\theta(t, z_1) \theta(t, z_2))=1$ is equivalent to the vanishing of the $g-1$ determinants of all the $2 \times 2$ submatrices containing that fixed row. Thus in an open neighborhood of any point the submanifold $-W_1+k+\phi_2(p_1, p_2)$ can be described analytically as the set of common zeros of $g-1$ of these $\binom{g}{2}$ determinants, but different sets of these determinants may well be required near different points of the submanifold. It is again familiar that in an open neighborhood of each point a one-dimensional submanifold of $\mathbb{C}^g$ can be described by the vanishing of $g-1$ complex analytic functions in that neighborhood. By translation of course this provides a rather simple and explicit analytic description of the submanifolds $W_1$ or $-W_1$ or any translate of either; thus for example the submanifold $W_1 \subseteq J(M)$, representing the canonical imbedding of the marked Riemann surface $M$ in its Jacobi variety, can be described as the set of points of $\mathbb{C}^g$ at which the complex analytic vector fields $\theta(\tilde{k}+\phi_2(z_1, z_2)-t, z_1)$ and $\theta(\tilde{k}+\phi_2(z_1, z_2)-t, z_2)$ are linearly dependent, whenever $z_1, z_2 \in \tilde{M}$ represent distinct points of $M$. In general the assertion is that

$$-W_{n-1}+k+\phi_n(p_1, \dots, p_n)=\{t \in \mathbb{C}^g \mid \operatorname{rank}(\theta(t, z_1), \dots, \theta(t, z_n))<n\}/\mathcal{L}$$

for any points $z_1, \dots, z_n \in \tilde{M}$ representing distinct points $p_1, \dots, p_n \in M$; thus the subvariety $-W_{n-1}+k+\phi_n(p_1, \dots, p_n) \subset J(M)$ can be described analytically as the set of common zeros of the $\binom{g}{n}$ determinants of all $n \times n$ submatrices of the $g \times n$ matrix $(\theta(t, z_1), \dots, \theta(t, z_n))$ for $1 \leq n \leq g$. The vector fields $\theta(t, z_1), \dots, \theta(t, z_{n-1})$ are linearly dependent on the proper analytic subvariety

$$-W_{n-2}+k+\phi_{n-1}(p_1, \dots, p_{n-1}) \subset -W_{n-1}+k+\phi_n(p_1, \dots, p_n)$$

for $n \geq 2$, hence they are linearly independent on a dense open subset of $-W_{n-1}+k+\phi_n(p_1, \dots, p_n)$ since this latter subvariety is irreducible; therefore at a dense open subset of the subvariety $-W_{n-1}+k+\phi_n(p_1, \dots, p_n)$ the matrix $(\theta(t, z_1), \dots, \theta(t, z_{n-1}))$ has rank exactly $n-1$, and arguing as above $g-n+1$ of the determinants suffice to describe that subvariety locally at those points. Of course in the extreme case $n=g$

$$-W_{g-1}+k+\phi_g(p_1, \dots, p_g)=\{t \in \mathbb{C}^g \mid \det(\theta(t, z_1), \dots, \theta(t, z_g))=0\}/\mathcal{L}.$$

One rather straightforward extension of the preceding results should perhaps be mentioned here. As a natural modification of the alternative definition (11) of the analytic subvarieties $W_\mu \subseteq J(M)$ introduce the further subsets

$$W_\mu^\nu=\{t \in \mathbb{C}^g \mid \gamma(\rho_t \zeta^\mu) \geq \nu\}/\mathcal{L} \tag{16}$$

for any integer $\mu$ and any positive integer $\nu$; thus $W_\mu^1=W_\mu$ and $W_\mu^{\nu+1} \subseteq W_\mu^\nu$, and it is easily verified that $W_\mu^\nu=\phi$ for sufficiently large $\nu$.

**Corollary 2 to Theorem 9.** *Let $\theta(t, z)$ be a generalized theta function of rank $g$ for the marked Riemann surface $M$, and let $z_1, \dots, z_g$ be any fixed points of $\tilde{M}$ which represent distinct points $p_1, \dots, p_g$ of $M$. Then for $1 \leq n \leq g$, $1 \leq \nu$,*

$$-W_{n-1}^\nu+k+\phi_n(p_1, \dots, p_n)=\{t \in \mathbb{C}^g \mid \operatorname{rank}(\theta(t, z_1), \dots, \theta(t, z_n)) \leq n-\nu\}/\mathcal{L}. \tag{17}$$

*Proof*. As in the proof of Theorem 9, for the special case $r=2g-1$, it follows that

$$n-\operatorname{rank}(\theta(t,z_1),\dots,\theta(t,z_n))=\gamma(\rho_{k-t+\phi_n(p_1,\dots,p_n)}\zeta^{n-1});$$

and consequently (17) follows immediately from the definition (16) of the subsets $W_\mu^\nu$, to conclude the proof of the corollary.

It is a direct consequence of this corollary that the subsets $W_\mu^\nu$ are indeed complex analytic subvarieties of the Jacobi variety $J(M)$. Note further that in the description (15) of the set $-W_{n-1}+k+\phi_n(p_1,\dots,p_n)$ the dense open subset consisting of those points at which $\operatorname{rank}(\theta(t,z_1),\dots,\theta(t,z_n))=n-1$ is precisely the complement of the proper analytic subvariety

$$-W_{n-1}^2+k+\phi_n(p_1,\dots,p_n)\subset -W_{n-1}+k+\phi_n(p_1,\dots,p_n).$$

There are a number of relations known to hold among these various subvarieties $W_\mu^\nu$, and many have quite interesting interpretations and consequences in the study of properties of Riemann surfaces, as indicated partially in [12].

It is interesting at this point to return to the construction of the generalized theta functions given in the proof of Theorem 7 in order to obtain a more primitive local description of the analytic subvarieties $W_\mu\subseteq J(M)$ which is equivalent to the global description provided by Corollary 1 to Theorem 9 but which is in some instances rather easier to use. Considering the special case $r=2g-1$ and slightly changing the notation used in the proof of Theorem 7, consider a point $t_0\in J(M)$ and an open neighborhood $U$ of $t_0$ in $J(M)$ parametrized by the complex analytic homeomorphism

(18) $$-\tilde\phi_g\colon \tilde U_1\times\cdots\times\tilde U_g\longrightarrow U,$$

where $\tilde U_i\subseteq\tilde M$ represent disjoint coordinate neighborhoods $U_i\subseteq M$ and

(19) $$\tilde\phi_g(s_1,\dots,s_g)=\tilde\phi(s_1)+\cdots+\tilde\phi(s_g)-\tilde\phi(a_1)-\cdots-\tilde\phi(a_g)$$

for variable points $s_i\in\tilde U_i$ and fixed base points $a_i\in\tilde M$. Let $\zeta$ be a factor of automorphy representing the complex analytic line bundle associated to the divisor $1\cdot p_0$, where $p_0$ is the base point of the marked Riemann surface $M$, and let $\zeta_a$ be a factor of automorphy representing the complex analytic line bundle associated to the divisor $1\cdot a_1+\cdots+1\cdot a_g$, viewed as a divisor on $M$; and let $h$ be a complex analytic relatively automorphic function for the factor of automorphy $\zeta_a$ with $\mathfrak{d}(h)=1\cdot a_1+\cdots+1\cdot a_g$, and let $h_j$ be a basis for the space of complex analytic relatively automorphic functions for the factor of automorphy $\zeta_a\zeta^{2g-1}$ for $1\le j\le g$. As a temporary notational convenience the column vector of length $2g$ having the functions $h_j$ as entries will be denoted by $\mathbf{h}$. Then, as observed in the proof of Theorem 7, if the neighborhood $U$ is sufficiently small there is a complex analytic mapping $C\colon\tilde U_1\times\cdots\times\tilde U_g\longrightarrow GL(2g,\mathbb{C})$ such that the vectors $\mathbf{h}(s_i)$ form the last $g$ columns of the matrix $C(s_1,\dots,s_g)^{-1}$, hence such that

(20) $$C(s_1,\dots,s_g)^{-1}=(*,\dots,*,\mathbf{h}(s_1),\dots,\mathbf{h}(s_g)).$$

Indeed the mapping $C$ can be so chosen that the generalized theta function $\theta(t, z) = (f_i(t, z))$ of rank $g$ is given for $t \in U$ by

$$f_i(-\tilde{\phi}_g(s_1, \ldots, s_g), z) = \textstyle\sum_{j=1}^{2g} c_{ij}(s_1, \ldots, s_g) h_j(z) / p(z, s_1, \ldots, s_g) h(z) \tag{21}$$

for $1 \leq i \leq g$, where $p(z, s_1, \ldots, s_g) = \prod_{i=1}^{g} p(z, z_0, s_i, a_i)$ in terms of the prime function of the marked Riemann surface $M$. The function $p(z, s_1, \ldots, s_g)$ is a complex analytic nowhere vanishing function on the subset

$$(\tilde{M} - \textstyle\bigcup_i \Gamma \tilde{U}_i - \bigcup_i \Gamma a_i) \times \tilde{U}_1 \times \cdots \times \tilde{U}_g \subset \tilde{M} \times \tilde{U}_1 \times \cdots \times \tilde{U}_g\,,$$

provided that $z_0 \notin \Gamma \tilde{U}_i$ and $z_0 \notin \Gamma a_i$ for any $i$. Now (20) can be rewritten in matrix form as

$$C(s_1, \ldots, s_g) \mathbf{h}(s_k) = \begin{pmatrix} 0 \\ \delta_k \end{pmatrix},$$

where $0, \delta_k \in \mathbb{C}^g$ are the zero vector and column $k$ of the $g \times g$ identity matrix respectively; and (21) can be rewritten correspondingly as

$$C(s_1, \ldots, s_g) \mathbf{h}(z) = \begin{pmatrix} A \\ B \end{pmatrix},$$

where $A, B \in \mathbb{C}^g$ are some vectors depending analytically on the variables $s_1, \ldots, s_g, z$, in particular where

$$A = p(z, s_1, \ldots, s_g) h(z) \theta(-\tilde{\phi}_g(s_1, \ldots, s_g), z)\,.$$

Combining these observations it follows that for any points $z_1, \ldots, z_n \in \tilde{M}$

$$\begin{aligned} &C(s_1, \ldots, s_g)(\mathbf{h}(z_1), \ldots, \mathbf{h}(z_n), \mathbf{h}(s_1), \ldots, \mathbf{h}(s_g)) \\ &\quad = \begin{pmatrix} \theta(-\tilde{\phi}_g(s_1, \ldots, s_g), z_1, \ldots, z_n) & 0 \\ * & I \end{pmatrix} \begin{pmatrix} D(s, z) & 0 \\ 0 & I \end{pmatrix}, \end{aligned} \tag{22}$$

where $\theta(t, z_1, \ldots, z_n) = (\theta(t, z_1), \ldots, \theta(t, z_n))$ is a $g \times n$ matrix, $D(s, z)$ is the $n \times n$ diagonal matrix with entries $p(z_k, s_1, \ldots, s_g) h(z_k)$ along the diagonal for $1 \leq k \leq n$, $I$ is the $g \times g$ identity matrix, and 0 is a zero matrix of the appropriate size. The $g \times n$ matrix block indicated by the asterisk is of little interest here.

If it is assumed that $z_0 \notin \bigcup_i \Gamma \tilde{U}_i \cup \bigcup_i \Gamma a_i$ and that $z_k \notin \bigcup_i \Gamma \tilde{U}_i \cup \bigcup_i \Gamma a_i$ for $1 \leq k \leq n$ then the matrix $D(s, z)$ is a complex analytic nonsingular matrix-valued function of $(s_1, \ldots, s_g)$ in the set $\tilde{U}_1 \times \cdots \times \tilde{U}_g$; and in that case it is apparent from (22) that

$$\begin{aligned} &\operatorname{rank}(\mathbf{h}(z_1), \ldots, \mathbf{h}(z_n), \mathbf{h}(s_1), \ldots, \mathbf{h}(s_g)) \\ &\quad = \operatorname{rank} \theta(-\tilde{\phi}_g(s_1, \ldots, s_g), z_1, \ldots, z_n) + g\,. \end{aligned} \tag{23}$$

Upon recalling Corollary 1 to Theorem 9 it follows that if $z_1, \ldots, z_a$ are fixed points of $\tilde{M}$ which satisfy the restriction described in the preceding sentence and which represent distinct points $p_1, \ldots, p_n$ of $M$ for $1 \leq n \leq g$ then under the complex analytic homeomorphism (18)

$$(24) \qquad (-W_{n-1}+k+\phi(p_1)+\cdots+\phi(p_n)-n\phi(p_0)) \cap U$$
$$\cong \{(s_1, \ldots, s_g) \in \tilde{U}_1 \times \cdots \times \tilde{U}_g \,|\, \operatorname{rank}(\mathbf{h}(z_1), \ldots, \mathbf{h}(z_n)\, \mathbf{h}(s_1), \ldots, \mathbf{h}(s_g)) < n+g\}.$$

Actually this consequence of (22), which provides an occasionally useful local description of the subvariety $W_{n-1}$, can be derived rather more simply directly from the Riemann-Roch theorem as follows; but there is a finer point to this consequence of (22) which will then be taken up. If $\operatorname{rank}(\mathbf{h}(z_1), \ldots, \mathbf{h}(z_n), \mathbf{h}(s_1), \ldots, \mathbf{h}(s_g)) = r < n+g$ then there are $2g-r \geq g-n+1$ linearly independent vectors $c_i \in \mathbb{C}^{2g}$ such that ${}^t c_i \cdot (\mathbf{h}(z_1), \ldots, \mathbf{h}(z_n), \mathbf{h}(s_1), \ldots, \mathbf{h}(s_g)) = 0$; and consequently the $2g-r$ functions ${}^t c_i \cdot \mathbf{h}$ for $1 \leq i \leq 2g-r$ are linearly independent complex analytic relatively automorphic functions for the factor of automorphy $\zeta_a \zeta^{2g-1}$, and they vanish at the points $z_1, \ldots, z_n, s_1, \ldots, s_g$. These points of $M$ represent distinct points $p_1, \ldots, p_n, q_1, \ldots, q_g$ of $M$ by assumption, and therefore by the customary argument

$$\gamma(\zeta_a \zeta^{2g-1} \zeta_{p_1}^{-1} \ldots \zeta_{p_n}^{-1} \zeta_{q_1}^{-1} \ldots \zeta_{q_g}^{-1}) \geq 2g-r \geq g-n+1\,.$$

It then follows from the Riemann-Roch theorem that

$$\gamma(\kappa^{-1} \zeta_a^{-1} \zeta^{1-2g} \zeta_{p_1} \ldots \zeta_{p_n} \zeta_{q_1} \ldots \zeta_{q_g}) \geq 1\,;$$

and since the characteristic class of this factor of automorphy is $n-1$ it further follows, again by the customary argument, that

$$-k+\phi(p_1+\cdots+p_n+q_1+\cdots+q_g-a_1-\cdots-a_g) \in W_{n-1}\,,$$

and consequently that

$$-\tilde{\phi}_g(s_1, \ldots, s_g) = \phi(a_1+\cdots+a_g-q_1-\cdots-q_g) \in -W_{n-1}+k+\phi(p_1+\cdots+p_n)$$

as desired.

There is, as already mentioned, a finer point to the equality (22) than is expressed by (23). Viewing the matrices $H(z, s) = (\mathbf{h}(z_1), \ldots, \mathbf{h}(z_n), \mathbf{h}(s_1), \ldots, \mathbf{h}(s_g))$ and $\theta(s, z) = \theta(-\tilde{\phi}_g(s_1, \ldots, s_g), z_1, \ldots, z_n)$ as complex analytic functions of $s = (s_1, \ldots, s_g) \in \tilde{U}_1 \times \cdots \times \tilde{U}_g$ for fixed $z = (z_1, \ldots, z_n)$, it of course follows from (22) or (23) that

$$\{s \in \tilde{U}_1 \times \cdots \times \tilde{U}_g \,|\, \operatorname{rank} H(z, s) < n+g\} = \{s \in \tilde{U}_1 \times \cdots \times \tilde{U}_g \,|\, \operatorname{rank} \theta(s, z) < n\}.$$

These two expressions thus yield analytic descriptions of the same complex analytic subvariety $V \subseteq \tilde{U}_1 \times \cdots \times \tilde{U}_g$; for $V$ can be described either as the set of common zeros of the determinants of all $(n+g) \times (n+g)$ submatrices of $H(z, s)$

or as the set of common zeros of the determinants of all $n \times n$ submatrices of $\theta(s, z)$. More than this, though, it further follows from (22) that the determinants of all the $(n+g) \times (n+g)$ submatrices of $H(z, s)$ generate the same ideal in the ring of all complex analytic functions in $\tilde{U}_1 \times \cdots \times \tilde{U}_g$ as do the determinants of all the $n \times n$ submatrices of $\theta(s, z)$; that is to say, the determinant of any particular $(n+g) \times (n+g)$ submatrix of $H(z, s)$ can be expressed as a linear combination of the determinants of all the $n \times n$ submatrices of $\theta(s, z)$ with coefficients which are complex analytic functions in $\tilde{U}_1 \times \cdots \times \tilde{U}_{g_1}$ and the determinant of any particular $n \times n$ submatrix of $\theta(s, z)$ can be expressed similarly in terms of the determinants of all the $(n+g) \times (n+g)$ submatrices of $H(s, z)$. To see that that is the case consider a particular $n \times n$ submatrix $\theta_0(s, z)$ of $\theta(s, z)$ and the corresponding submatrix

$$\tilde{\theta}_0(s, z) = \begin{pmatrix} \theta_0(s, z) D(s, z) & 0 \\ * & I \end{pmatrix} \quad \text{of} \quad \begin{pmatrix} \theta(s, z) D(s, z) & 0 \\ * & I \end{pmatrix}.$$

It follows from (22) that each row of $\tilde{\theta}_0(s, z)$ is a linear combination of the rows of $H(z, s)$ with coefficients from the matrix $C(s_1, \ldots, s_g)$, hence with coefficients which are complex analytic functions in $\tilde{U}_1 \times \cdots \times \tilde{U}_g$; and hence $\det\tilde{\theta}_0(s, z) = \det(\theta_0(s, z)) \cdot \det(D(s, z))$, which is really just the exterior product of the rows of $\tilde{\theta}_0(s, z)$, is a linear combination of the various $(n+g)$-fold exterior products of the rows of $H(z, s)$ with coefficients which are complex analytic functions in $\tilde{U}_1 \times \cdots \times \tilde{U}_g$, or equivalently is a linear combination of the determinants of all the $(n+g) \times (n+g)$ submatrices of $H(z, s)$ with coefficients which are complex analytic functions in $\tilde{U}_1 \times \cdots \times \tilde{U}_g$. Since the matrix $C(s_1, \ldots, s_g)$ is nonsingular in $\tilde{U}_1 \times \cdots \times \tilde{U}_g$ then after multiplying (22) on the left by the inverse matrix the same argument shows that the determinant of any particular $(n+g) \times (n+g)$ submatrix of $H(z, s)$ is a linear combination of the determinants of all the $(n+g) \times (n+g)$ submatrices of

$$\begin{pmatrix} \theta(s, z) D(s, z) & 0 \\ * & I \end{pmatrix}$$

with coefficients which are complex analytic functions in $\tilde{U}_1 \times \cdots \times \tilde{U}$; and of course the only nontrivial such determinants are equal to $\det(\theta_0(s, z)) \cdot \det(D(s, z))$ where $\theta_0(s, z)$ is an $n \times n$ submatrix of $\theta(s, z)$. Since $\det D(s, z)$ is complex analytic and nowhere vanishing in $\tilde{U}_1 \times \cdots \times \tilde{U}_g$, in view of the restrictions imposed on the points $z_k$, that suffices to prove the desired assertion. This assertion can then be used to demonstrate the following extension of Corollary 1 to Theorem 9.

**Theorem 10.** *Let $\theta(t, z)$ be a generalized theta function of rank $g$ for the marked Riemann surface $M$, and let $z_1, \ldots, z_g$ be any fixed points of $\tilde{M}$ which represent distinct points $p_1, \ldots, p_g$ of $M$. If*

$$\operatorname{rank}(\theta(t_0, z_1), \ldots, \theta(t_0, z_n)) = n - 1$$

*for some point* $t_0 \in \mathbb{C}^g$ *and some index* $1 \leq n \leq g$ *then* $t_0$ *represents a regular point of the complex analytic subvariety*

$$-W_{n-1}+k+\phi_n(p_1,\dots,p_n)\subseteq J(M)$$

*and there are* $g-n+1$ *determinants of* $n\times n$ *submatrices of the matrix function* $(\theta(t,z_1),\dots,\theta(t,z_n))$ *of* $t$ *which generate the proper ideal of that subvariety at the point* $t_0$.

*Proof.* In view of the preceding observation it suffices to prove the corresponding assertion for the local matrix function $H(z,s)$ in terms of any parametrization (18) of an open neighborhood $U$ of the point $t_0$ in $J(M)$ for which the base points $z_0, a_1, \dots, a_g$ and coordinate neighborhoods $\tilde{U}_1, \dots, \tilde{U}_g$ are so chosen that $z_0 \notin \bigcup_i \Gamma \tilde{U}_i \cup \bigcup_i \Gamma a_i$ and that $z_k \notin \bigcup_i \Gamma \tilde{U}_i \cup \bigcup_i \Gamma a_i$ for $1 \leq k \leq n$. Thus consider such a parametrization, where $-\tilde{\phi}_g(s_0)=t_0$ for the point $s_0=(s_1^0,\dots,s_g^0)\in \tilde{U}_1\times\cdots\times\tilde{U}_g$, and suppose that $\operatorname{rank} H(z,s_0)=n+g-1$. Of course it then follows from (24) that $t_0\in -W_{n-1}+k+\phi_n(p_1,\dots,p_n)$. To simplify the notation relabel the functions $h_j$ so that the first $n+g-1$ rows of the matrix $H(z,s_0)$ are linearly independent; then let $G_i(z,s)$ be the $(n+g)\times(n+g)$ matrix formed of rows $1,2,\dots,n+g-1,i$ of the matrix $H(z,s)$, for $n+g\leq i\leq 2g$, and let $g_i(z,s)=\det G_i(z,s)$. The functions $g_i(z,s)$ are complex analytic functions of $s\in\tilde{U}_1\times\cdots\times\tilde{U}_g$, and as in (24) under the complex analytic homeomorphism (18)

$$(-W_{n-1}+k+\phi(p_1+\cdots+p_n-np_0))\cap\tilde{U}\cong\{s\in\tilde{U}_1\times\cdots\times\tilde{U}_g\,|\,g_i(z,s)=0,\ n+g\leq i\leq 2g\}.$$

In order to show simultaneously that $s_0$ is a regular point of this subvariety and that the $g-n+1$ functions $g_i(z,s)$ as functions of $s$ generate the proper ideal of this subvariety at the point $s_0$, and hence to prove the desired result, it suffices merely to show that the $(g-n+1)\times g$ matrix $\partial g_i/\partial s_j$ attains its maximal rank $g-n+1$ at the point $s_0$; and that can be accomplished quite easily if the parametrization (18) is chosen sufficiently carefully.

Note that if (18) is one parametrization of an open neighborhood of $t_0$ in $J(M)$, satisfying the conditions listed in the preceding paragraph, then any sufficiently small change of the base points $a_1,\dots,a_g$ will yield another parametrization satisfying the same conditions; this observation will be used to impose two additional restrictions on the parametrization to be used in the remainder of the proof. The first restriction is that

$$\phi_g(a_1,\dots,a_g)\notin -W_{g-n-1}+2t_0-\phi_n(p_1,\dots,p_n), \tag{25}$$

which can always be achieved since $g-n-1<g$ and the subvariety on the right above is consequently a proper analytic subvariety of $J(M)$. This restriction is also preserved by any sufficiently small perturbation of the base points $a_1,\dots,a_g$. Next note that since $t_0\in -W_{n-1}+k+\phi_n(p_1,\dots,p_n)$ then necessarily

$$\begin{aligned} t_0&=k+\phi_n(p_1,\dots,p_n)-\phi_{n-1}(q_1,\dots,q_{n-1})\\ &=k+\phi(p_1+\cdots+p_n-q_1-\cdots-q_{n-1}-p_0)\end{aligned} \tag{26}$$

for some points $q_i \in M$. Indeed these points $q_i$ are determined uniquely up to order; for if there were a different representation of $t_0$ in this form then $\gamma(\zeta_{q_1} \cdots \zeta_{q_{n-1}}) \geq 2$ so that $t_0 \in -W^2_{n-1} + k + \phi_n(p_1, \ldots, p_n)$, and by Corollary 2 to Theorem 9 that would imply that

$$\operatorname{rank}(\theta(t_0, z_1), \ldots, \theta(t_0, z_n)) \leq n-2$$

in contradiction to the hypothesis. The second restriction is that the base points $a_1, \ldots, a_g$ be so chosen that

$$(27) \qquad s_i^0 \neq q_j \quad \text{for} \quad 1 \leq i \leq g, \quad 1 \leq j \leq n-1,$$

and this can evidently also be attained by a small change of the base points. The matrix $H(z, s_0)$ is by hypothesis of rank $n+g-1$, indeed the first $n+g-1$ rows of that matrix are linearly independent and the remaining rows are linearly dependent on these; and the second restriction (27) implies that the matrix obtained from $H(z, s_0)$ by deleting any one of the last $g$ columns is also of rank $n+g-1$, or what is evidently equivalent, that the matrix obtained from the first $n+g-1$ rows of $H(z, s_0)$ by deleting any one of the last $g$ columns is nonsingular. To see this, suppose that rank $H^*(z, s_0) \leq n+g-2$ where $H^*(z, s_0)$ is the matrix obtained from $H(z, s_0)$ by deleting column $n+i$ for some index $1 \leq i \leq g$. That implies that there are at least $g-n+2$ linearly independent vectors $c \in \mathbb{C}^{2g}$ such that ${}^t c \cdot H^*(z, s_0) = 0$, hence that there are at least $g-n+2$ linearly independent complex analytic relatively automorphic functions for the factor of automorphy $\zeta_a \zeta^{2g-1}$, vanishing at the points $z_1, \ldots, z_n, s_1^0, \ldots, s_{i-1}^0, s_{i+1}^0, \ldots, s_g^0$; and that in turn implies as is customary that

$$\gamma(\zeta_a \zeta^{2g-1} \zeta_{s_i^0} \zeta_{z_1}^{-1} \cdots \zeta_{z_n}^{-1} \zeta_{s_1^0}^{-1} \cdots \zeta_{s_g^0}^{-1}) \geq g-n+2 .$$

It then follows from the Riemann-Roch theorem that

$$\gamma(\kappa \zeta_a^{-1} \zeta^{1-2g} \zeta_{s_i^0}^{-1} \zeta_{z_1} \cdots \zeta_{z_n} \zeta_{s_1^0} \cdots \zeta_{s_g^0}) \geq 1 ,$$

hence as is also customary that

$$k - \phi_g(a_1, \ldots, a_g) - \phi_1(s_i^0) + \phi_n(p_1, \ldots, p_n) + \phi_g(s_1^0, \ldots, s_g^0) \in W_{n-2} ;$$

and since $-\tilde{\phi}_g(s_1^0, \ldots, s_g^0) + \tilde{\phi}_g(a_1, \ldots, a_g) = t_0$ it further follows that

$$t_0 \in -W_{n-2} - \phi_1(s_i^0) + k + \phi_n(p_1, \ldots, p_n) ,$$

or equivalently that

$$t_0 = k + \phi(p_1 + \cdots + p_n - q_1^* - \cdots - q_{n-2}^* - s_i^0 - p_0)$$

for some points $q_j^* \in M$. However this is another representation of the point $t_0$ in the form (26); and since as noted above the points $q_j$ are determined uniquely up to order by (26) it follows that $s_i^0 = q_j$ for some index $j$, contradicting (27).

Having chosen a parametrization with these properties turn back again to the original course of the proof and suppose in contradiction to the desired result that the matrix $\partial g_i/\partial s_j$ has rank strictly less than $g-n+1$ at the point $s_0$. There must then exist constants $c_{n+g}, \ldots, c_{2g}$, not all of which are zero, such that

$$\sum_{i=n+g}^{2g} c_i \frac{\partial g_i}{\partial s_j}(z, s_0)=0 \quad \text{for} \quad 1 \leq j \leq g .$$

Now to simplify the notation further choose another basis $h_j^*$ for the space of complex analytic relatively automorphic functions for the factor of automorphy $\zeta_a \zeta^{2g-1}$ such that $h_j^* = h_j$ for $1 \leq j \leq n+g-1$ and $h_{n+g}^* = \sum_{i=n+g}^{2g} c_i h_i$; and note that upon carrying out the construction in the first paragraph of the proof in terms of this new basis it follows that $g_{n+g}^* = \sum_{i=n+g}^{2g} c_i g_i$. Thus, dropping the asterisks, it can be assumed without loss of generality that

$$g_{n+g}(z, s_0) = \frac{\partial g_{n+g}}{\partial s_j}(z, s_0)=0 \quad \text{for} \quad 1 \leq j \leq g ;$$

and since clearly $\partial g_{n+g}/\partial s_j = \partial(\det G_{n+g})/\partial s_j = \det(\partial G_{n+g}/\partial s_j)$ it follows that the matrices $G_{n+g}$ and $\partial G_{n+g}/\partial s_j$ for $1 \leq j \leq g$ are all singular at the point $s_0$. Actually the matrix $G_{n+g}(z, s_0)$ is of rank $n+g-1$; so there is a nonzero vector $c \in \mathbb{C}^{n+g}$ such that ${}^t c \cdot G_{n+g}(z, s_0)=0$, and that vector is unique up to a nonzero scalar factor. Now the matrix $\partial G_{n+g}/\partial s_j$ differs from the matrix $G_{n+g}$ only in column $n-j$; and since the matrix obtained from $G_{n+g}$ by deleting column $n+j$ is of rank $n+g-1$ as noted above, it follows that the matrix $\partial G_{n+g}/\partial s_j$ is also of rank $n+g-1$ at the point $s_0$, and indeed that ${}^t c \cdot (\partial G_{n+g}/\partial s_j)(z, s_0)=0$, for $1 \leq j \leq g$. The nontrivial complex analytic relatively automorphic function $\sum_{i=1}^{n+g} c_i h_i$ thus vanishes at the points $z_1, \ldots, z_n$ and moreover vanishes to at least second order at the points $s_1^0, \ldots, s_g^0$, so that

$$\gamma(\zeta_a \zeta^{2g-1} \zeta_{p_1}^{-1} \cdots \zeta_{p_n}^{-1} \zeta_{s_1^0}^{-2} \cdots \zeta_{s_g^0}^{-2}) \geq 1 .$$

It therefore follows as is customary that

$$\phi_g(a_1, \ldots, a_g) - \phi_n(p_1, \ldots, p_n) - 2\phi_g(s_1^0, \ldots, s_g^0) \in W_{g-n-1},$$

or since $-\tilde{\phi}_g(s_1^0, \ldots, s_g^0) + \tilde{\phi}_g(a_1, \ldots, a_g) = t_0$ that

$$2t_0 - \phi_g(a_1, \ldots, a_g) - \phi_n(p_1, \ldots, p_n) \in W_{g-n-1},$$

in contradiction to (25). That contradiction then serves to complete the proof of the theorem.

Actually of course it follows from Corollary 2 to Theorem 9 that if $\operatorname{rank}(\theta(t_0, z_1), \ldots, \theta(t_0, z_n)) \leq n-2$ then

$$t_0 \in -W_{n-1}^2 + k + \phi_n(p_1, \ldots, p_n) \subseteq -W_{n-1} + k + \phi_n(p_1, \ldots, p_n) ;$$

and since it is known [12] that $W_{n-1}^2$ is precisely the singular locus of $W_{n-1}$ for $1 \leq n \leq g$ it follows that $t_0$ is a regular point of $-W_{n-1}+k+\phi_n(p_1, \ldots, p_n)$ precisely when $\operatorname{rank}(\theta(t_0, z_1), \ldots, \theta(t_0, z_n)) = n-1$, to complete the first part of the conclusion of Theorem 10. It is the second part of the conclusion of Theorem 10 that is really of greater interest.

No attempt has been made here to discuss the general theory of complex vector bundles, either analytically or topologically; indeed the emphasis has been almost entirely on one particular complex analytic vector bundle arising naturally in the study of compact Riemann surfaces, and those analytic properties of that bundle which originate from its connection with Riemann surfaces. However it really must be pointed out here that Theorems 9 and 10 together imply that *the homology classes in $J(M)$ represented by the complex analytic subvarieties $-W_n \subset J(M)$ for $0 \leq n \leq g-1$ are dual to the characteristic classes (the Chern classes) of the complex analytic vector bundle over $J(M)$ described by the theta factor of automorphy of rank $g$.* No proof of that assertion will be given here, since to do so would require either an extensive digression or the blatant assumption of considerable prerequisite knowledge needed nowhere else in this book. For those readers having the appropriate interests and prerequisites, though, it will be noted that this assertion follows from the natural modification of the argument given in Appendix 2 of [6]; for $\theta(t; z_1, \ldots, z_n)$ is a complex analytic mapping from $\rho_{z_1}^{-1} \oplus \cdots \oplus \rho_{z_n}^{-1}$ to the bundle $\chi_t$, and since $\rho_{z_1}^{-1} \oplus \cdots \oplus \rho_{z_n}^{-1}$ is topologically trivial the mapping $\theta(t; z_1, \ldots, z_n)$ can be considered as given essentially by a family of sections of the bundle $\chi_t$. Alternatively Theorem 10 can be used to deduce sufficient transversality that the results of [19] apply directly.

## § 9. Relations between Theta Factors of Automorphy

The generalized theta functions of any rank $s = r+1-g \geq g$ can be used to describe the subvarieties $W_\mu$ of positive divisors of the Jacobi variety $J(M)$, as in Theorem 9. The preceding section was principally devoted to an analysis of that description for the particular case that $s = g$, while the present section will be concerned more with the cases that $s > g$. If $s = r+1-g \geq g+1$ then there are indices $n$ in the range $1 \leq n \leq s-g$; and for any such index $2g-2+n-r \leq -1$ and the left hand side of (13) in Theorem 9 is therefore the empty set. This leads to the following rather interesting consequence of Theorem 9.

**Corollary 2 to Theorem 9.** *Let $\theta(t, z)$ be a generalized theta function of rank $s \geq g+1$ for the marked Riemann surface $M$, and let $z_1, \ldots, z_{s-g}$ be any fixed points of $\tilde{M}$ which represent distinct points of $M$. Then for $1 \leq n \leq s-g$*

$$\operatorname{rank}(\theta(t, z_1), \ldots, \theta(t, z_n)) = n \quad \textit{for all} \quad t \in \mathbb{C}^g .$$

To examine the significance of this result note first that for $n=1$ the assertion is that rank $\theta(t, z_1) = 1$ for all $t \in \mathbb{C}^g$. For fixed $z_1$ the function $\theta(t, z_1)$ can be viewed as a complex analytic vector field of dimension $s$ on the manifold $\mathbb{C}^g$, that is, as a complex analytic mapping $\theta(z_1): \mathbb{C}^g \longrightarrow \mathbb{C}^s$; and this vector field is nonsingular over all of $\mathbb{C}^g$, that is, $\theta(t, z_1) \neq 0$ for all $t \in \mathbb{C}^g$. Now recall from Corollary 1 to Theorem 7 that

$$(28) \qquad \theta(t+\lambda, z_1) \cdot \rho_{z_1}(\lambda)^{-1} = \chi_s(\lambda, t) \cdot \theta(t, z_1) \quad \text{for all} \quad \lambda \in \mathscr{L},$$

where $\rho_{z_1}$ is a flat factor of automorphy of rank 1 and $\chi_s$ is a theta factor of automorphy of rank $s$, both for the action of the lattice group $\mathscr{L}$ on $\mathbb{C}^g$. Recall further from § 3 that an equation of the form (28) can be interpreted as the assertion that the complex analytic mapping $\theta(z_1): \mathbb{C}^g \longrightarrow \mathbb{C}^{s \times 1}$ determines a bundle homomorphism

$$\theta(z_1): \rho_{z_1}^{-1} \longrightarrow \chi_s$$

from the complex analytic line bundle represented by the factor of automorphy $\rho_{z_1}^{-1}$ into the complex analytic vector bundle of rank $s$ represented by the factor of automorphy $\chi_s$. Indeed since $\theta(t, z_1)$ is of maximal rank everywhere the homomorphism $\theta(z_1)$ is injective, hence exhibits the complex analytic line bundle $\rho_{z_1}^{-1}$ as a subbundle of the complex analytic vector bundle $\chi_s$. More generally for any index in the range $1 \leq n \leq s-g$ consider the complex analytic mapping $\theta(z_1, \ldots, z_n): \mathbb{C}^g \longrightarrow \mathbb{C}^{s \times n}$ defined by $\theta(t; z_1, \ldots, z_n) = (\theta(t, z_1), \ldots, \theta(t, z_n))$ where $z_1, \ldots, z_n$ represent distinct points of $M$. The assertion of the corollary is that $\operatorname{rank} \theta(t; z_1, \ldots, z_n) = n$ for all points $t \in \mathbb{C}^g$. Again recalling Corollary 1 to Theorem 7 note that in analogy with (28)

$$(29) \qquad \theta(t+\lambda; z_1, \ldots, z_n) \cdot \operatorname{diag}(\rho_{z_1}(\lambda)^{-1}, \ldots, \rho_{z_n}(\lambda)^{-1}) = \chi_s(\lambda, t) \cdot \theta(t; z_1, \ldots, z_n)$$

for all $\lambda \in \mathscr{L}$, where $\operatorname{diag}(\rho_{z_1}(\lambda)^{-1}, \ldots, \rho_{z_n}(\lambda)^{-1})$ denotes the $n \times n$ diagonal matrix with the listed entries along the diagonal. As before an equation of the form (29) can be interpreted as the assertion that the complex analytic mapping $\theta(z_1, \ldots, z_n)$ determines a bundle homomorphism

$$\theta(z_1, \ldots, z_n): \rho_{z_1}^{-1} \oplus \cdots \oplus \rho_{z_n}^{-1} \longrightarrow \chi_s$$

from the direct sum of the complex analytic line bundles represented by the factors of automorphy $\rho_{z_1}^{-1}, \ldots, \rho_{z_n}^{-1}$ into the complex analytic vector bundle of rank $s$ represented by the factor of automorphy $\chi_s$. Indeed since $\theta(t; z_1, \ldots, z_n)$ is everywhere of maximal rank the homomorphism $\theta(z_1, \ldots, z_n)$ is injective hence exhibits the complex analytic vector bundle $\rho_{z_1}^{-1} \oplus \cdots \oplus \rho_{z_n}^{-1}$ as a subbundle of the complex analytic vector bundle $\chi_s$.

Introducing the quotient bundle $Q$, the preceding observation can be rephrased as the assertion that there is an exact sequence of complex analytic vector bundles of the form

$$0 \longrightarrow \rho_{z_1}^{-1} \oplus \cdots \oplus \rho_{z_n}^{-1} \xrightarrow{\theta(z_1, \ldots, z_n)} \chi_s \longrightarrow Q \longrightarrow 0$$

for any index in the range $1 \leq n \leq s-g$. It is further possible to identify the quotient bundle $Q$ quite explicitly as follows. As a preliminary definition, note that if $\xi$ is any factor of automorphy for the action of the lattice subgroup $\mathscr{L}$ on $\mathbb{C}^g$ and if $t_0$ is any fixed point of $\mathbb{C}^g$ then the function $T_{t_0}\xi(\lambda, t) = \xi(\lambda, t-t_0)$ is also a factor of automorphy $T_{t_0}\xi$ for this action of $\mathscr{L}$; this will be called the *translate* of the factor of automorphy $\xi$ by $t_0$, and the complex analytic vector bundle $T_{t_0}\xi$ represented by this factor of automorphy will be called the translate of the complex analytic vector bundle $\xi$.

**Theorem 11.** *If $\chi_s$ is the complex analytic vector bundle over $J(M)$ associated to a theta factor of automorphy of rank $s \geq g+1$ and if $z_1, \ldots, z_{s-g}$ are fixed points of $\tilde{M}$ which represent distinct points of $M$ then for any index in the range $1 \leq n \leq s-g$ there is an exact sequence of complex analytic vector bundles over $J(M)$ of the form*

$$(30) \qquad 0 \longrightarrow \rho_{z_1}^{-1} \oplus \cdots \oplus \rho_{z_n}^{-1} \xrightarrow{\theta(z_1, \ldots, z_n)} \chi_s \longrightarrow T_{t_0}\chi_{s-n} \longrightarrow 0,$$

*where $t_0 = \tilde{\phi}(z_1 + \cdots + z_n - nz_0) \in J(M)$.*

*Proof.* Let $\theta_{s-n}(t,z) = (f_i^{s-n}(t,z))$ be a generalized theta function of rank $s-n \geq g$ and $\chi_{s-n}(\lambda, t) = (\chi_{ij}^{s-n}(\lambda, t))$ be the corresponding theta factor of automorphy, so that

$$(31) \qquad f_i^{s-n}(t+\lambda, z) = \rho_z(\lambda) \sum_{j=1}^{s-n} \chi_{ij}^{s-n}(\lambda, t) f_j^{s-n}(t,z)$$

for all $\lambda \in \mathscr{L}$. The functions $f_i^{s-n}(t,z)$ are complex analytic on $\mathbb{C}^g \times \tilde{M}$ and satisfy

$$(32) \qquad f_i^{s-n}(t, Tz) = \rho_t(T)\zeta(T,z)^{r-n} f_i^{s-n}(t,z)$$

for all $T \in \Gamma$, where $r = s+g-1$; moreover for each fixed $t \in \mathbb{C}^g$ these functions $f_i^{s-n}(t,z)$ form a basis for the space of complex analytic relatively automorphic functions for the factor of automorphy $\rho_t \zeta^{r-n}$. Further let $\zeta_z$ be a factor of automorphy representing the complex analytic line bundle associated to the divisor $1 \cdot z_1 + \cdots + 1 \cdot z_n$ on $M$; by Theorem 5 the factor of automorphy $\zeta_z \zeta^{-n}$ is analytically equivalent to the flat factor of automorphy $\rho_{t_0}$ where $t_0 = \tilde{\phi}(z_1 + \cdots + z_n - nz_0)$, hence the factor of automorphy $\zeta_z$ can be so chosen that $\zeta_z = \rho_{t_0}\zeta^n$. There is a complex analytic relatively automorphic function $f$ for the factor of automorphy $\zeta_z$ such that $\mathfrak{d}(f) = 1 \cdot z_1 + \cdots + 1 \cdot z_n$ on $M$; and this function then has the property that

$$(33) \qquad f(Tz) = \rho_{t_0}(T)\zeta(T,z)^n f(z)$$

for all $T \in \Gamma$. Upon combining Eqs. (32) and (33) note that

$$f(Tz) f_i^{s-n}(t, Tz) = \rho_{t+t_0}(T)\zeta(T,z)^r f(z) f_i^{s-n}(t,z)$$

for all $T \in \Gamma$, which of course means that for any fixed $t \in \mathbb{C}^g$ the functions $f(z) f_i^{s-n}(t-t_0, z)$ are $s-n$ linearly independent complex analytic relatively automorphic functions for the factor of automorphy $\rho_t \zeta^r$. These functions can consequently be expressed in terms of a generalized theta function $\theta_s(t, z) = (f_i^s(t, z))$ of rank $s$ in the form

$$f(z) f_i^{s-n}(t-t_0, z) = \sum_{j=1}^{s} c_{ij}(t) f_j^s(t, z), \tag{34}$$

where the $(s-n) \times s$ matrix $C(t) = (c_{ij}(t))$ is a complex analytic function of $t$ and is of rank $s-n$ at all points $t \in \mathbb{C}^g$. Combining (31) and (34) note that for any $\lambda \in \mathcal{L}$

$$\begin{aligned} f(z) f_i^{s-n}(t-t_0+\lambda, z) &= f(z) \rho_z(\lambda) \sum_j \chi_{ij}^{s-n}(\lambda, t-t_0) f_j^{s-n}(t-t_0, z) \\ &= \rho_z(\lambda) \sum_{jk} \chi_{ij}^{s-n}(\lambda, t-t_0) c_{jk}(t) f_k^s(t, z); \end{aligned}$$

and if $\chi_s(\lambda, t) = (\chi_{ij}^s(\lambda, t))$ is the theta factor of automorphy associated to the generalized theta function $\theta_s(t, z)$ then in analogy with (31) for any $\lambda \in \mathcal{L}$

$$\sum_j c_{ij}(t+\lambda) f_j^s(t+\lambda, z) = \rho_z(\lambda) \sum_{jk} c_{ij}(t+\lambda) \chi_{jk}^s(\lambda, t) f_k^s(t, z).$$

The last two formulas are equal, in view of (34); and recalling that the functions $f_k^s(t, z)$ are linearly independent it then follows that

$$\sum_j \chi_{ij}^{s-n}(\lambda, t-t_0) c_{jk}(t) = \sum_j c_{ij}(t+\lambda) \chi_{jk}^s(\lambda, t)$$

for $1 \leq i \leq s-n$, $1 \leq k \leq s$, or in matrix terms that

$$T_{t_0} \chi_{s-n}(\lambda, t) \cdot C(t) = C(t+\lambda) \cdot \chi_s(\lambda, t)$$

for all $\lambda \in \mathcal{L}$. This can in turn be interpreted as the assertion that the complex analytic mapping $C: \mathbb{C}^g \longrightarrow \mathbb{C}^{(s-n) \times s}$ determines a bundle homomorphism

$$C: \chi_s \longrightarrow T_{t_0} \chi_{s-n};$$

and this bundle homomorphism must be surjective since rank $C(t) = s-n$ for all points $t \in \mathbb{C}^g$. Considering as before the matrix $\theta(t; z_1, \ldots, z_n) = (f_i^s(t, z_k))$ determining the bundle homomorphism

$$\theta(z_1, \ldots, z_n): \rho_{z_1}^{-1} \oplus \cdots \oplus \rho_{z_n}^{-1} \longrightarrow \chi_s$$

note that as a consequence of (34) the entry in row $i$, column $k$ of the product matrix $C(t)\theta(t; z_1, \ldots, z_n)$ is

$$\sum_j c_{ij}(t) f_j^s(t, z_k) = f(z_k) f_i^{s-n}(t-t_0, z_k),$$

and this is zero since by assumption $f(z_k) = 0$. Thus $C(t)\theta(t; z_1, \ldots, z_n) = 0$ for all $t \in \mathbb{C}^g$, so that the image of the bundle homomorphism $\theta(z_1, \ldots, z_n)$ is contained in the kernel of the bundle homomorphism $C$; indeed these subbundles must coincide, since both are of rank $n$. That suffices to conclude the proof of the theorem.

To examine the significance of this result note first that any flat complex line bundle $\rho_z \in \operatorname{Hom}(\mathscr{L}, \mathbb{C}^*)$ over a complex torus $\mathbb{C}^g/\mathscr{L}$ is topologically trivial, just as are flat complex line bundles over Riemann surfaces; and note also that the complex analytic vector bundle $\chi_{s-n}$ is continuously equivalent to any translate $T_t\chi_{x-n}$ of that bundle. Recalling further that in the category of continuous complex vector bundles any exact sequence of vector bundles of the form (30) splits [11], it follows that for any indices $s \geq g+1$ and $1 \leq n \leq s-g$ the complex analytic vector bundle $\chi_s$ is continuously equivalent to the direct sum $I_n \oplus \chi_{s-n}$ of the trivial complex vector bundle of rank $n$ and the complex analytic vector bundle $\chi_{s-n}$. In particular *any complex analytic vector bundle $\chi_s$ for $s \geq g$ is continuously equivalent to the direct sum $I_{s-g} \oplus \chi_g$ of the trivial complex vector bundle of rank $s-g$ and the complex analytic vector bundle $\chi_g$*. Thus the purely topological properties of the vector bundles represented by generalized theta factors of automorphy are all merely manifestations of the topological properties of the vector bundle represented by the basic theta factor of automorphy of rank $g$.

In the complex analytic category the situation is rather more complicated. The existence of the exact sequence (30) can be rephrased in terms of factors of automorphy as the assertion that for any indices $s \geq g+1$ and $1 \leq n \leq s-g$ the factor of automorphy $\chi_s(\lambda, t)$ is analytically equivalent to a factor of automorphy of the form

$$(35) \qquad \begin{pmatrix} R_z(\lambda)^{-1} & S_z(\lambda, t) \\ 0 & \chi_{s-n}(\lambda, t-t_0) \end{pmatrix}$$

for some complex analytic mappings $S_z(\lambda): \mathbb{C}^g \longrightarrow \mathbb{C}^{n \times (s-n)}$, where $R_z(\lambda) = \operatorname{diag}(\rho_{z_1}(\lambda), \ldots, \rho_{z_n}(\lambda))$ and $t_0 = \tilde{\phi}(z_1 + \cdots + z_n - nz_0)$. The mapping $S_z(\lambda)$ describes the particular extension of the subbundle $R_z^{-1}$ by the bundle $T_{t_0}\chi_{s-n}$ which yields the bundle $\chi_s$; that extension is the trivial extension precisely when it is possible to choose $S_z(\lambda, t) = 0$ for all $t \in \mathbb{C}^g$, in the representation (35). For this to be the trivial extension is of course equivalently that the bundle $\chi_s$ be analytically equivalent to the direct sum $R_z^{-1} \oplus T_{t_0}\chi_{s-n}$; but that can never happen.

**Corollary 1 to Theorem 11.** *With the hypotheses as in the statement of Theorem* 11, *the exact sequence* (30) *exhibits the complex analytic vector bundle $\chi_s$ as a nontrivial extension of the bundle $\rho_{z_1}^{-1} \oplus \cdots \oplus \rho_{z_n}^{-1}$ by the bundle $T_{t_0}\chi_{s-n}$.*

*Proof.* Suppose conversely that for some points $z_1, \ldots, z_n$ of $\tilde{M}$ representing distinct points of $M$ the factor of automorphy $\chi_s(\lambda, t)$ is analytically equivalent to a factor of automorphy of the form (35) with $S_z(\lambda, t) = 0$ for all $t \in \mathbb{C}^g$; without loss of generality it can be assumed that $\chi_s(\lambda, t)$ is actually equal to the factor of automorphy (35) with $S_z(\lambda, t) = 0$ for all $t \in \mathbb{C}^g$. Select some fixed points $w_1, \ldots, w_s$ of $\tilde{M}$ which represent distinct points of $M$ and for which the flat line bundles

$\rho_{z_i}$ and $\rho_{w_j}$ over $J(M)$ are not analytically equivalent for any indices $1 \leq i \leq n$, $1 \leq j \leq s$; for a discussion of analytic equivalences among flat line bundles over complex tori see § 1 of the appendix. Introducing the generalized theta function $\theta(t; z) = (f_i(t, z))$ of rank $s$ associated to this theta factor $\chi_s(\lambda, t)$, the matrix-valued function

$$\vartheta(t; w_1, \ldots, w_s) = (\theta(t; w_1), \ldots, \theta(t; w_s))$$

viewed as a complex analytic mapping $\theta(w_1, \ldots, w_s)$: $\mathbb{C}^g \longrightarrow \mathbb{C}^{s \times s}$ clearly satisfies

$$\begin{aligned} \vartheta(t+\lambda; w_1, \ldots, w_s) R_w(\lambda)^{-1} &= \chi_s(\lambda, t)\theta(t; w_1, \ldots, w_s) \\ &= (R_z(\lambda)^{-1} \oplus \chi_{s-n}(\lambda, t-t_0))\theta(t; w_1, \ldots, w_s) \end{aligned}$$

for all $\lambda \in \mathscr{L}$, where $R_w(\lambda) = \operatorname{diag}(\rho_{w_1}(\lambda), \ldots, \rho_{w_s}(\lambda))$. Considering only the first $n$ rows of this matrix equality, note that

$$f_i(t+\lambda, w_j)\rho_{w_j}(\lambda)^{-1} = \rho_{z_i}(\lambda)^{-1} f_i(t, w_j)$$

for all $\lambda \in \mathscr{L}$ and for any indices $1 \leq i \leq n$, $1 \leq j \leq s$. This is just the assertion that the complex analytic function $f_i(t, w_j)$ is relatively automorphic for the flat factor of automorphy $\rho_{w_j}\rho_{z_i}^{-1}$. Since a flat factor of automorphy on a complex torus has zero characteristic class, any complex analytic relatively automorphic function for a flat factor of automorphy is either identically zero or nowhere zero; the function $f_i(t, w_j)$ must therefore be identically zero, since the flat factor of automorphy $\rho_{w_j}\rho_{z_i}^{-1}$ is by construction not analytically trivial. It then follows that $\operatorname{rank} \theta(t; w_1, \ldots, w_s) \leq s-n < s$ for all points $t \in \mathbb{C}^g$. On the other hand it follows from Theorem 9 that

$$-W_{g-1} + k + \tilde{\phi}(w_1 + \cdots + w_s - sz_0) = \{t \in \mathbb{C}^g \mid \operatorname{rank} \theta(t; w_1, \ldots, w_s) < s\}/\mathscr{L},$$

which implies that $\operatorname{rank} \theta(t; w_1, \ldots, w_s) = s$ on the complement of a proper analytic subvariety of $\mathbb{C}^g$; that contradiction serves to conclude the proof of the corollary.

Rather than attempting to analyze all possible nontrivial such extensions, which is a serious problem in itself and is further complicated by incomplete knowledge of the analytic properties of the vector bundles $\chi_{s-n}$, it will merely be noted here that Theorem 11 can be interpreted as a form of homogeneity property of the vector bundles $\chi_s$. Changing notation slightly, for any index $s \geq g$ and any points $z_1, \ldots, z_{2g} \in M$ representing distinct points of $M$ the factor of automorphy $\chi_{s+2g}(\lambda, t)$ is analytically equivalent to a factor of automorphy of the form

$$\begin{pmatrix} R_z(\lambda)^{-1} & S_z(\lambda, t) \\ 0 & \chi_s(\lambda, t-t_0) \end{pmatrix} \tag{36}$$

for some complex analytic mappings $S_z(\lambda)\colon \mathbb{C}^g \longrightarrow \mathbb{C}^{2g\times s}$, where $R_z(\lambda)=\operatorname{diag}(\rho_{z_1}(\lambda),\ldots,\rho_{z_{2g}}(\lambda))$ and $t_0=\phi(z_1+\cdots+z_{2g}-2gz_0)$. Now the point $t_0$ can be taken to represent any arbitrary point of $J(M)$ by a suitable choice of the points $z_1,\ldots,z_{2g}$. Indeed holding $z'_{g+1},\ldots,z'_{2g}$ fixed the mapping $\tilde{\phi}\colon \tilde{M}^g \longrightarrow J(M)$ which sends $(z'_1,\ldots,z'_g)\in\tilde{M}^g$ to $\tilde{\phi}(z'_1+\cdots+z'_{2g}-2gz_0)\in J(M)$ is surjective by the Jacobi inversion theorem, so that $\tilde{\phi}(z'_1+\cdots+z'_{2g}-2gz_0)\equiv t_0 \bmod \mathscr{L}$ for some points $(z'_1,\ldots,z'_g)$. Now for any arbitrary points $z_1,\ldots,z_g\in\tilde{M}$ it is clear from the Riemann-Roch theorem that $\gamma(\zeta_{z'_1}\ldots\zeta_{z'_{2g}}\zeta_{z_1}^{-1}\ldots\zeta_{z_g}^{-1})\geq 1$, hence that for some points $z_{g+1},\ldots,z_{2g}\in\tilde{M}$ the divisors $z'_1+\cdots+z'_{2g}$ and $z_1+\cdots+z_{2g}$ are linearly equivalent and $\tilde{\phi}(z_1+\cdots+z_{2g}-2gz_0)\equiv\tilde{\phi}(z'_1+\cdots+z'_{2g}-2gz_0)\bmod\mathscr{L}$; and it can of course be assumed that $z_1,\ldots,z_g$ represent distinct points of $M$. Choosing points $z^0=(z_1^0,\ldots,z_{2g}^0)$ such that $\tilde{\phi}(z_1^0+\cdots+z_{2g}^0-2gz_0)\equiv 0 \bmod \mathscr{L}$ it follows that all the factors of automorphy (36) are analytically equivalent to a factor of automorphy of the form

$$\begin{pmatrix} R(\lambda)^{-1} & S(\lambda,t)\\ 0 & \chi_s(\lambda,t)\end{pmatrix},$$

where $R(\lambda)=R_{z^0}(\lambda)$ and $S(\lambda,t)=S_{z^0}(\lambda,t)$; and in (36) the point $t_0$ can be taken to represent an arbitrary point of $J(M)$. This relation between the factors of automorphy $\chi_s(\lambda,t)$ and $T_{t_0}\chi_s(\lambda,t)=\chi_s(\lambda,t-t_0)$ is the homogeneity property represented by Theorem 11.

This homogeneity has a particularly simple form for the scalar factor of automorphy $\det\chi_s(\lambda,t)$ or for the complex analytic line bundle it represents. Since analytically equivalent factors of automorphy of any rank clearly have analytically equivalent determinants it follows that the scalar factors of automorphy $\det R(\lambda)^{-1}\cdot\det\chi_s(\lambda,t)$ and $\det R_z(\lambda)^{-1}\cdot\det\chi_s(\lambda,t-t_0)$ are analytically equivalent for any $t_0\in J(M)$. Now recalling (10)

$$\det R_z(\lambda)=\textstyle\prod_{i=1}^{2g}\rho_{z_i}(\lambda)=\rho_{t_0}(\lambda),$$

where $t_0=\tilde{\phi}(z_1+\cdots+z_{2g}-2gz_0)$ and $\rho_{t_0}(\lambda)$ stands for $\rho_{t_0}(T)$ for any element $T\in\Gamma$ such that $\phi(T)=\lambda$ under the natural period mapping $\phi\colon\Gamma\longrightarrow\mathscr{L}$. Consequently *the scalar factors of automorphy* $\det\chi_s(\lambda,t)$ *and* $\rho_{t_0}(\lambda)^{-1}\det\chi_s(\lambda,t-t_0)$ *are analytically equivalent for any point* $t_0\in J(M)$.

Finally it should be observed that Theorem 11 suggests the framework within which it is perhaps most natural to discuss the analogues of the generalized theta functions for ranks less than $g$. Of course in a sense the proof of Theorem 7 already gives the local existence of what must be the generalized theta functions of lower ranks; and the local existence is all that can really be demanded of such functions. To be more precise, if $t_0\in\mathbb{C}^g$ represents a point in the complement $X_{n-1}\subset J(M)$ of the analytic subvariety $-W_{n-1}+k\subset J(M)$ for some fixed integer in the range $1\leq n\leq g-1$ then there is an open neighborhood $U$ of $t_0$ in $\mathbb{C}^g$ such that every $t\in U$ also represents a point in $X_{n-1}$. It then follows easily from the Riemann-Roch theorem that $\gamma(\rho_t\zeta^{2g-n-1})=g-n$ for all $t\in U$; and since that dimension is constant as $t$ varies over $U$ the argument used in the proof of Theorem 7 shows that if $U$ is sufficiently small there are $g-n$ complex analytic functions $f_i^U$ on $U\times\tilde{M}$ such that

$$f_i^U(t, Tz)=\rho_t(T)\zeta(T,z)^{2g-n-1}f_i^U(t,z) \quad \text{whenever} \quad T\in\Gamma;$$

and such that for each fixed $t\in U$ these functions form a basis for the space of complex analytic relatively automorphic functions for the factor of automorphy $\rho_t\zeta^{2g-n-1}$. Choose a collection $\{U_\alpha\}$ of such sets which represent a coordinate covering of the open subset $X_{n-1}\subset J(M)$, and let $\theta^\alpha_{g-n}(t,z)=(f_i^\alpha(t,z))$ be the associated complex analytic functions on $U_\alpha\times\tilde{M}$. Note that if $t_\alpha\in U_\alpha$ and $t_\beta\in U_\beta$ represent the same point of $J(M)$ then $t_\alpha-t_\beta=\lambda_{\alpha\beta}\in\mathscr{L}$ and the flat factors of automorphy $\rho_{t_\alpha}$ and $\rho_{t_\beta}$ are analytically equivalent; indeed the analytic equivalence can be exhibited quite explicitly by using the local form of the functions (2) in § 7. If the lattice element $\lambda_{\alpha\beta}$ has coordinates

$$\lambda_{\alpha\beta i}=m_{\alpha\beta i}+\sum_{j=1}^g \omega_{ij}n_{\alpha\beta j}$$

for some integers $m_{\alpha\beta i}$, $n_{\alpha\beta i}$ then introduce the nowhere vanishing complex analytic function

$$\rho(\lambda_{\alpha\beta},z)=\exp 2\pi i\sum_{j=1}^g n_{\alpha\beta i}[w_j(z)-w_j(z_0)] \tag{37}$$

of $z\in\tilde{M}$ and note as in (3) in § 7 that

$$\rho(\lambda_{\alpha\beta},Tz)\cdot\rho_{t_\beta}(T)=\rho_{t_\alpha}(T)\cdot\rho(\lambda_{\alpha\beta},z)$$

for all $T\in\Gamma$, thus exhibiting the desired analytic equivalence. For any fixed point $z\in\tilde{M}$ the collection of functions $\rho_z^{\alpha\beta}=\rho(\lambda_{\alpha\beta},z)$ are the coordinate transformations for a flat complex line bundle $\rho_z$ over $X_{n-1}$; of course this line bundle is just the restriction to the open subset $X_{n-1}\subset J(M)$ of the flat complex line bundle over $J(M)$ defined by the element $\rho_z\in\operatorname{Hom}(\mathscr{L},\mathbb{C}^*)$ in § 7. Now whenever $t_\alpha\in U_\alpha$ and $t_\beta\in U_\beta$ do represent the same point of $J(M)$ it is easy to see that the functions $f_i^\alpha(t_\alpha,z)$ and $\rho(\lambda_{\alpha\beta},z)f_i(t_\beta,z)$ are two bases for the same space of complex analytic relatively automorphic functions for the action of the group $\Gamma$ on $\tilde{M}$. Therefore

$$f_i^\alpha(t_\alpha,z)=\sum_{j=1}^{g-n}\chi_{ij}^{\alpha\beta}(t_\beta)\rho(\lambda_{\alpha\beta},z)f_j^\beta(t_\beta,z) \tag{38}$$

for a uniquely determined complex analytic mapping $\chi_{g-n}^{\alpha\beta}\colon U_\alpha\cap U_\beta\longrightarrow GL(g-n,\mathbb{C})$, when the sets $U_\alpha$ and $U_\beta$ are viewed as subsets of $J(M)$; and the collection of these mappings $\{\chi_{g-n}^{\alpha\beta}\}$ are the coordinate transformations for a complex analytic vector bundle $\chi_{g-n}$ of rank $g-n$ over the complex manifold $X_{n-1}$. Eqs. (37) can be rewritten in the form

$$\vartheta_{g-n}^\alpha(t,z)=\rho_z^{\alpha\beta}\chi_{g-n}^{\alpha\beta}(t)\theta_{g-n}^\beta(t,z) \quad \text{whenever} \quad t\in U_\alpha\cap U_\beta\subset J(M), \tag{39}$$

which exhibits the functions $\theta^{\alpha}_{g-n}(t,z)$ quite explicitly as describing a complex analytic cross-section of the complex analytic vector bundle $\rho_z \otimes \chi_{g-n}$ over $X_{n-1}$. These functions $\theta^{\alpha}_{g-n}(t,z)$ can be considered as generalized theta functions of rank $g-n$. The analogue of Theorem 8 then holds for the vector bundle $\chi_{g-n}$ over $X_{n-1}$, with the same proof; but that matter will not be pursued further here.

For the alternative approach to these functions suggested by Theorem 11 consider first the generalized theta function $\theta_g(t,z)$ of rank $g$, which can be viewed as an analytic family of complex analytic cross-sections of the vector bundles $\rho_z \otimes \chi$ over $J(M)$. It is more convenient to consider this theta function in local terms; so choose a coordinate covering $\{U_\alpha\}$ of the complex manifold $J(M)$, and let $\rho^{\alpha\beta}_z: U_\alpha \cap U_\beta \longrightarrow \mathbb{C}^*$ and $\chi^{\alpha\beta}_g: U_\alpha \cap U_\beta \longrightarrow GL(g,\mathbb{C})$ be coordinate transformations for the line bundle $\rho_z$ and the vector bundle $\chi_g$ respectively, and let $\theta^{\alpha}_g(t,z) = (f^{\alpha}_i(t,z))$ be the complex analytic mappings $\theta^{\alpha}_g: U_\alpha \times \tilde{M} \longrightarrow \mathbb{C}^g$ representing the generalized theta function $\theta_g(t,z)$, in these local terms. More generally for any point $t_0 \in \mathbb{C}^g$ the translated theta function $T_{t_0}\theta_g(t,z) = \theta_g(t-t_0,z)$ represents an analytic family of complex analytic cross-sections of the translated vector bundle $T_{t_0}(\rho_z \otimes \chi_g) = (T_{t_0}\rho_z) \otimes (T_{t_0}\chi_g)$; and in local terms, for the same coordinate covering $\{U_\alpha\}$ as above, let $T_{t_0}\rho^{\alpha\beta}_z$ and $T_{t_0}\chi^{\alpha\beta}_g$ be coordinate transformations for these translated bundles and let $T_{t_0}\theta^{\alpha}_g(t,z)$ be the complex analytic mappings representing the translated theta function.

Now for any fixed points $z_1, \dots, z_n \in \tilde{M}$ representing $n$ distinct points $p_1, \dots, p_n \in M$ for $1 \le n \le g-1$ let

$$R^{\alpha\beta} = \operatorname{diag}(\rho^{\alpha\beta}_{z_1}, \dots, \rho^{\alpha\beta}_{z_n})$$

be coordinate transformations for the complex analytic vector bundle $R = \rho_{z_1} \oplus \cdots \oplus \rho_{z_n}$ and note that the complex analytic mappings $\theta^{\alpha}_g(z_1, \dots, z_n)$: $U_\alpha \longrightarrow \mathbb{C}^{g \times n}$ defined by

$$\theta^{\alpha}_g(t; z_1, \dots, z_n) = (\theta^{\alpha}_g(t,z_1), \dots, \theta^{\alpha}_g(t,z_n))$$

satisfy

$$\theta^{\alpha}_g(t; z_1, \dots, z_n) \cdot R^{\alpha\beta}(t)^{-1} = \chi^{\alpha\beta}_g(t) \cdot \theta^{\beta}_g(t; z_1, \dots, z_n)$$

whenever $t \in U_\alpha \cap U_\beta$; and similarly of course for any point $t_0 \in \mathbb{C}^g$

$$T_{t_0}\theta^{\alpha}_g(t; z_1, \dots, z_n) \cdot T_{t_0}R^{\alpha\beta}(t)^{-1} = T_{t_0}\chi^{\alpha\beta}_g(t) \cdot T_{t_0}\theta^{\beta}_g(t; z_1, \dots, z_n)$$

whenever $t \in U_\alpha \cap U_\beta$. This can be interpreted as the assertion that the complex analytic mappings $T_{t_0}\theta^{\alpha}_g(z_1, \dots, z_n): U_\alpha \longrightarrow \mathbb{C}^{g \times n}$ describe a bundle homomorphism

$$(40) \qquad T_{t_0}\theta_g(z_1, \dots, z_n): T_{t_0}R^{-1} \longrightarrow T_{t_0}\chi_g .$$

As in Corollary 1 to Theorem 9, rank $T_{t_0}\theta^{\alpha}_g(t; z_1, \dots, z_n) = \operatorname{rank} \theta_g(t-t_0; z_1, \dots, z_n) < n$ precisely when $t - t_0 \in -W_{n-1} + k + \phi(p_1 + \cdots + p_n - np_0)$; so upon choosing

$t_0 = -\tilde{\phi}(z_1 + \cdots + z_n - nz_0) \in \mathbb{C}^g$ the bundle homomorphism (40) is injective over the complement $X_{n-1} \subset J(M)$ of the complex analytic subvariety $-W_{n-1} + k \subset J(M)$. Thus for this choice of the point $t_0$ the bundle homomorphism (40) is part of the short exact sequence of complex analytic vector bundles over the open subset $X_{n-1} \subset J(M)$

(41) $$0 \longrightarrow R^{-1} | X_{n-1} \xrightarrow{T_{t_0}\theta_g(z_1, \ldots, z_n)} T_{t_0}\chi_g | X_{n-1} \longrightarrow \chi_{g-n} \longrightarrow 0$$

for some complex analytic vector bundle $\chi_{g-n}$ of rank $g-n$ over $X_{n-1}$.

This construction, and possibly even the complex analytic equivalence class of the vector bundle $\chi_{g-n}$, depend on the choice of the points $z_1, \ldots, z_n$; and to analyze this dependence it is convenient to derive more explicitly the coordinate transformations for the bundle $\chi_{g-n}$. Since $\operatorname{rank} T_{t_0}\theta_g^\alpha(t; z_1, \ldots, z_n) = n$ whenever $t \in U'_\alpha = U_\alpha \cap X_{n-1}$, after refining the coordinate covering if necessary there will exist complex analytic mappings $C^\alpha: U'_\alpha \longrightarrow GL(g, \mathbb{C})$ such that

(42) $$C^\alpha(t) \cdot T_{t_0}\theta_g^\alpha(t; z_1, \ldots, z_n) = \begin{pmatrix} I \\ 0 \end{pmatrix} \quad \text{whenever} \quad t \in U'_\alpha,$$

where $I$ is the $n \times n$ identity matrix. Then if $t \in U'_\alpha \cap U'_\beta$

$$\begin{aligned} \begin{pmatrix} I \\ 0 \end{pmatrix} &= C^\alpha(t) \cdot T_{t_0}\chi_g^{\alpha\beta}(t) \cdot T_{t_0}\theta_g^\beta(t; z_1, \ldots, z_n) \cdot T_{t_0}R^{\alpha\beta}(t) \\ &= C^\alpha(t) \cdot T_{t_0}\chi_g^{\alpha\beta}(t) \cdot C^\beta(t)^{-1} \cdot \begin{pmatrix} I \\ 0 \end{pmatrix} \cdot T_{t_0}R^{\alpha\beta}(t) \end{aligned}$$

so that

(43) $$C^\alpha(t) \cdot T_{t_0}\chi_g^{\alpha\beta}(t) \cdot C^\beta(t)^{-1} = \begin{pmatrix} T_{t_0}R^{\alpha\beta}(t)^{-1} & * \\ 0 & \chi_{g-n}^{\alpha\beta}(t) \end{pmatrix},$$

where the complex analytic mappings $\chi_{g-n}^{\alpha\beta}: U'_\alpha \cap U'_\beta \longrightarrow GL(g-n, \mathbb{C})$ are coordinate transformations for the vector bundle $\chi_{g-n}$. Furthermore the functions $\tilde{\theta}^\alpha(t, z) = C^\alpha(t) \cdot T_{t_0}\theta_g^\alpha(t, z)$ clearly satisfy

$$\begin{aligned} \tilde{\theta}^\alpha(t, z) &= C^\alpha(t) \cdot T_{t_0}\chi_g^{\alpha\beta}(t) \cdot T_{t_0}\theta_g^\beta(t, z) \cdot T_{t_0}\rho_z^{\alpha\beta}(t) \\ &= C^\alpha(t) \cdot T_{t_0}\chi_g^{\alpha\beta}(t) \cdot C^\beta(t)^{-1} \cdot \tilde{\theta}^\beta(t, z) \cdot T_{t_0}\rho_z^{\alpha\beta}(t) \end{aligned}$$

whenever $t \in U'_\alpha \cap U'_\beta$; so that, recalling (43), the last $g-n$ components of the vector-valued functions $\tilde{\theta}^\alpha(t, z)$ are complex analytic vector-valued functions $\tilde{\theta}_{g-n}^\alpha(t, z)$ of rank $g-n$ on $U'_\alpha \times \tilde{M}$ and satisfy

(44) $$\tilde{\theta}_{g-n}^\alpha(t, z) = \chi_{g-n}^{\alpha\beta}(t) \cdot \tilde{\theta}_{g-n}^\beta(t, z) \cdot T_{t_0}\rho_z^{\alpha\beta}(t)$$

whenever $t \in U'_\alpha \cap U'_\beta$. Now the components $\tilde{f}_i^\alpha(t, z)$ of the functions $\tilde{\theta}_{g-n}^\alpha(t, z)$ are $g-n$ linearly independent complex analytic relatively automorphic functions for the factor of automorphy $\rho_{t-t_0}\zeta^{2g-1}$ for any fixed $t\in U'_\alpha$; and it follows from (42) of course that these functions vanish at the points $z_1, \ldots, z_n$. The factor of automorphy associated to the divisor $1\cdot p_1+\cdots+1\cdot p_n$ is analytically equivalent to $\rho_{-t_0}\zeta^n$, in view of the choice of $t_0$; so there is a complex analytic relatively automorphic function $f$ for the factor of automorphy $\rho_{-t_0}\zeta^n$ such that $\mathfrak{d}(f)=1\cdot p_1+\cdots+1\cdot p_n$. The quotients $\tilde{f}_i^\alpha(t,z)/f(z)$ are therefore complex analytic relatively automorphic functions for the factor of automorphy $\rho_t\zeta^{2g-n-1}$; and they are the components of vector-valued functions $\theta^\alpha_{g-n}(t,z)$ on $U'_\alpha\times\tilde{M}$ which also satisfy (44) over $U'_\alpha\cap U'_\beta$. This identifies the functions $\theta^\alpha_{g-n}(t,z)$ with the generalized theta functions of rank $g-n$ as previously constructed by the alternative approach; and since the line bundles $\rho_z$ and $T_{t_0}\rho_z$ are analytically equivalent, the factor of automorphy $\rho_z$ being flat, a comparison of Eqs. (39) and (44) identifies the two constructions of the vector bundles over $X_{n-1}$ associated to these generalized theta functions. An immediate consequence of this is that the quotient bundle $\chi_{g-n}$ in the exact sequence (41) is up to analytic equivalence independent of the choices of the points $z_1, \ldots, z_n$.

One concluding remark should be inserted here for those readers familiar with analytic sheaf theory. The bundle homomorphism (40) is defined over the entire manifold $J(M)$, so that there is an exact sequence of coherent analytic sheaves over $J(M)$ of the form

$$0\longrightarrow\mathcal{O}(R^{-1})\xrightarrow{T_{t_0}\theta_g(z_1,\ldots,z_n)^*}\mathcal{O}(T_{t_0}\chi_g)\longrightarrow\tilde{\chi}_{g-n}\longrightarrow 0, \tag{45}$$

where $\mathcal{O}(\xi)$ denotes the sheaf of germs of complex analytic cross-sections of the vector bundle $\xi$ and $\tilde{\chi}_{g-n}$ is the sheaf defined by this sequence, the cokernel of the homomorphism $T_{t_0}\theta_g(z_1,\ldots,z_n)^*$. Upon comparing this sequence with (41) it is evident that the restriction $\tilde{\chi}_{g-n}|X_{n-1}$ is a locally free sheaf of rank $g-n$, indeed that $\tilde{\chi}_{g-n}|X_{n-1}=\mathcal{O}(\chi_{g-n})$ where $\chi_{g-n}$ is the vector bundle over $X_{n-1}$ introduced above. This vector bundle can thus always be extended over all of $J(M)$ as a coherent analytic sheaf.

## § 10. Dimensions of Spaces of Generalized Theta Functions

The classification of all scalar factors of automorphy for a Jacobi variety $J(M)$ and the detailed analysis of those factors of automorphy most closely related to the subvarieties of positive divisors in $J(M)$ comprise a well known chapter from the classical theory of theta functions. A survey of these results from the point of view adopted in the present discussion of generalized theta functions is included in an appendix for the convenience of the reader. A simple comparison of properties of the generalized theta functions with properties of the classical theta functions leads almost immediately to the identification of the determinants of the generalized theta factors of automorphy with certain classical factors of automorphy. Of course upon recalling the results obtained in § 9 it is enough merely to consider the scalar factor of automorphy $\det\chi_g(\lambda,t)$. To describe the classical theta factors most easily assume that the period matrix for the lattice subgroup $\mathscr{L} \subset \mathbb{C}^g$ defining the Jacobi variety $J(M)$ is reduced to the normal form $\Omega = (I, \Omega_2)$; the generators of $\mathscr{L}$ are thus the $2g$ column vectors $\lambda_1, \ldots, \lambda_{2g}$ of the matrix $\Omega$, and $\lambda_1, \ldots, \lambda_g$ are the standard generators for the lattice subgroup $\mathbb{Z}^g \subset \mathscr{L}$. The *Abelian theta factor of automorphy* for the action of $\mathscr{L}$ on $\mathbb{C}^g$ is the factor of automorphy defined by

$$(46) \qquad \xi(\lambda_i, t) = 1, \qquad \xi(\lambda_{i+g}, t) = \exp 2\pi i(k_i - r_i - t_i)$$

for $1 \leq i \leq g$, where $(k_1, \ldots, k_g) \in \mathbb{C}^g$ represents the canonical point $k \in J(M)$ and $(r_1, \ldots, r_g) \in \mathbb{C}^g$ represents the Riemann point $r \in J(M)$. There is up to a constant factor a unique complex analytic relatively automorphic function for the Abelian theta factor of automorphy, the Abelian theta function $\theta_0(t)$ defined by

$$\theta_0(t) = \sum_{n \in \mathbb{Z}^g} \exp 2\pi i \left[\tfrac{1}{2} {}^t n \Omega_2 n + {}^t n(t - r + \varepsilon)\right]$$

where $\varepsilon_i = \frac{1}{2}\omega_{i,i+g}$; and this function generates the proper ideal of the subvariety $W_{g-1} \subset J(M)$ at each point of $J(M)$.

**Theorem 12.** *On a Jacobi variety $J(M)$ the scalar factor of automorphy $\det \chi_g(\lambda, t)$ is analytically equivalent to the Abelian theta factor of automorphy $\xi(\lambda, t)$.*

*Proof.* Let $\theta(t, z)$ be a generalized theta function of rank $g$ and let $z_1, \ldots, z_g$ be any fixed points of $\tilde{M}$ which represent distinct points of $M$. Then as noted in Corollary 1 to Theorem 9 the function

$$f(t) = \det(\theta(t, z_1), \ldots, \theta(t, z_g))$$

vanishes precisely on the subvariety $-W_{g-1} + k + s \subset J(M)$, where $s = \tilde{\phi}(z_1 + \cdots + z_g - g z_0)$; indeed by Theorem 10 the function $f(t)$ vanishes to first order at any regular point of the subvariety $-W_{g-1} + k + s$, hence generates the proper ideal of that subvariety at all points of $J(M)$. Moreover since

$$\theta(t + \lambda, z) = \rho_z(\lambda)\chi_g(\lambda, t)\theta(t, z) \quad \text{for all} \quad \lambda \in \mathscr{L}$$

and since upon recalling (10) it is easily seen that

$$\rho_{z_1}(\lambda) \cdot \cdots \cdot \rho_{z_g}(\lambda) = \rho_s(\lambda) \quad \text{for all} \quad \lambda \in \mathscr{L}$$

it follows that

$$f(t + \lambda) = \rho_s(\lambda) \det \chi_g(\lambda, t) f(t) \quad \text{for all} \quad \lambda \in \mathscr{L}.$$

Now it is well known, as proved in the appendix, that $W_{g-1} = -W_{g-1} + k$. Thus the functions $f(t)$ and $\theta_0(t - s)$ are relatively automorphic for the factors of automorphy $\rho_s \det \chi_g$ and $T_s\xi$ respectively, where $T_s\xi(\lambda, t) = \xi(\lambda, t - s)$; and both generate the proper ideal of the subvariety $W_{g-1} + s = -W_{g-1} + k + s$ at all points of $J(M)$. The factors of automorphy $\rho_s \cdot \det \chi_g$ and $T_s\xi$ are therefore analytically equivalent; indeed the quotient $f(t)/\theta_0(t-s)$ is a nowhere vanishing complex analytic relatively automorphic function for the quotient factor of automorphy $\rho_s \cdot \det \chi_g / T_s\xi$. Since it follows immediately from the definition (46) that $T_s\xi = \rho_s \cdot \xi$, it follows that the factors of automorphy $\det \chi_g$ and $\xi$ are also equivalent as desired, and the proof of the theorem is thereby concluded.

It should be noted as a passing consequence of Theorem 12 that the homogeneity property of the Abelian theta factor of automorphy (46), that $T_s\xi = \rho_s \cdot \xi$ for any point $s \in \mathbb{C}^g$, implies the previously noted homogeneity property of the factor of automorphy $\det \chi_g$, that $T_s \det \chi_g$ is analytically equivalent to $\rho_s \cdot \det \chi_g$ for any point $s \in \mathbb{C}^g$. A more interesting consequence of Theorem 12 and of the classical observation that the space of complex analytic relatively automorphic functions for the Abelian theta factor of automorphy (46) is one dimensional is the following.

**Theorem 13.** *For any integer $s \geq g$ and any point $z \in \tilde{M}$ the space of complex analytic relatively automorphic functions for the factor of automorphy $\rho_z \otimes \chi_s$ of rank $s$ is one dimensional.*

*Proof.* The generalized theta function of rank $s$ of course provides one nontrivial complex analytic relatively automorphic function for each factor of automorphy $\rho_z \cdot \chi_s$, so it is only necessary to show that the dimension of the space of complex analytic relatively automorphic functions is at most one. Actually it is only necessary to prove the last assertion for the special case $s = g$, since it is an easy matter to show that if it is true for some rank $s \geq g$ it is necessarily true for the rank $s+1$. Indeed note that as a consequence of Theorem 11 there is an exact sequence of complex analytic vector bundles over $J(M)$ of the form

$$0 \longrightarrow \rho_{z_1}^{-1} \longrightarrow \chi_{s+1} \longrightarrow T_{t_1}\chi_s \longrightarrow 0$$

for any point $z_1 \in \tilde{M}$, where $t_1 = \tilde{\phi}(z_1 - z)$. Upon tensoring this exact sequence with the line bundle $\rho_z$ for another point $z \in \tilde{M}$ and noting that $\rho_z \cdot T_{t_1}\chi_s = T_{t_1}(\rho_z \cdot \chi_s)$ there results the exact sequence of complex analytic vector bundles

$$0 \longrightarrow \rho_t \longrightarrow \rho_z \cdot \chi_{s+1} \longrightarrow T_{t_1}(\rho_z \cdot \chi_s) \longrightarrow 0\,, \tag{47}$$

where $t = \tilde{\phi}(z - z_1)$. Now for any fixed point $z$ choose $z_1$ so that the factor of automorphy $\rho_t$ is not analytically trivial; that that is possible is evident from the discussion of the structure of scalar factors of automorphy over complex tori, as contained in the appendix. Since the factor of automorphy $\rho_t$ is then topologically trivial but not analytically trivial it follows that $\rho_t$ admits no complex analytic relatively automorphic functions; and therefore the space of complex analytic relatively automorphic functions for the factor of automorphy $\rho_z \cdot \chi_{s+1}$ is mapped injectively into the space of complex analytic relatively automorphic functions for the factor of automorphy $\rho_z \cdot \chi_s$ by the composition of the homomorphisms induced by the exact sequence of vector bundles (47) and by translation by $t_1$, from which the desired inductive step clearly follows.

Considering then the special case $s=g$, suppose that $f(t)$ is a complex analytic relatively automorphic function for the factor of automorphy $\rho_z \cdot \chi_g$ of rank $g$. The proof of the theorem will be completed by showing that $f(t)=c\theta(t,z)$ for some complex constant $c$, where $\theta(t,z)$ is the generalized theta function of rank $g$ for the theta factor $\chi_g$. Select points $z_2, \dots, z_g$ of $\tilde{M}$ such that $z, z_2, \dots, z_g$ represent distinct points of $M$ and none of the factors of automorphy $\rho_z, \rho_{z_2}, \dots, \rho_{z_g}$ are analytically equivalent, which is readily seen possible upon referring to the discussion in the appendix of scalar factors of automorphy for complex tori; and introduce the $g \times g$ matrix-valued functions

$$F(t)=(f(t), \theta(t,z_2), \dots, \theta(t,z_g)) \quad \text{and}$$
$$H(t)=(\theta(t,z), \theta(t,z_2), \dots, \theta(t,z_g))$$

of $t \in \mathbb{C}^g$. The functions $\det F(t)$ and $\det H(t)$ are both relatively automorphic for the factor of automorphy $\rho_z \cdot \rho_{z_2} \cdot \cdots \cdot \rho_{z_g} \cdot \det \chi_g = \rho_{t_0} \cdot \det \chi_g$, where $t_0 = \tilde{\phi}(z + z_2 + \cdots + z_g - g z_0)$; and from Corollary 1 to Theorem 9 it follows that $\det H(t)=0$ precisely when $t \in -W_{g-1}+k+t_0$. It is a consequence of Theorem 12 that the factor of automorphy $\rho_{t_0} \cdot \det \chi_g$ is analytically equivalent to $\rho_{t_0} \cdot \xi = T_{t_0} \xi$, where $\xi$ is the Abelian theta factor of automorphy; and since the space of complex analytic relatively automorphic functions for the factor of automorphy $\xi$ is one dimensional, as observed in the appendix, necessarily $\det F(t) = c \det H(t)$ for some complex constant $c$. Now using Cramer's rule note that

$$\text{(48)} \qquad f(t)=c_1(t)\theta(t,z)+\textstyle\sum_{j=2}^{g} c_j(t)\theta(t,z_j)$$

for some uniquely determined meromorphic functions $c_j(t)$ having poles at most along the subvariety $-W_{g-1}+k+t_0$, since $\det H(t)$ vanishes only along that subvariety. Upon substituting (48) into the matrix $F(t)$ note that $\det F(t) = c_1(t) \det H(t)$, and consequently that $c_1(t)=c$ is necessarily a constant. The vectors $f(t)-c\theta(t,z)$, $\theta(t,z_2), \dots, \theta(t,z_g)$ are thus linearly dependent whenever $t \notin -W_{g-1}+k+t_0$, and hence by continuity of the determinant are linearly dependent for all $t \in \mathbb{C}^g$. Again by Corollary 1 to Theorem 9 the vectors $\theta(t,z_2), \dots, \theta(t,z_g)$ are linearly independent outside the subvariety $-W_{g-2}+k+\tilde{\phi}(z_2+\cdots+z_g-(g-1)z_0)$; the functions $c_2(t), \dots, c_g(t)$ of (48) are consequently analytic outside that subvariety, and since it is of codimension 2 it follows that the functions $c_2(t), \dots, c_g(t)$ must actually be complex analytic throughout $\mathbb{C}^g$. Now upon replacing $t$ by $t+\lambda$ for some $\lambda \in \mathscr{L}$ and recalling that all the functions involved are relatively automorphic (48) becomes

$$\rho_z(\lambda)\chi_g(\lambda,t)f(t)=c\rho_z(\lambda)\chi_g(\lambda,t)\theta(t,z)+\textstyle\sum_{j=2}^{g} c_j(t+\lambda)\rho_{z_j}(\lambda)\chi_g(\lambda,t)\theta(t,z_j),$$

and therefore

$$f(t)=c\theta(t,z)+\textstyle\sum_{j=2}^{g} \rho_z(\lambda)^{-1} c_j(t+\lambda)\rho_{z_j}(\lambda)\theta(t,z_j);$$

but since the representation (48) is unique necessarily

$$c_j(t+\lambda)=\rho_z(\lambda)\,\rho_{z_j}(\lambda)^{-1}c_j(t)\quad\text{for all}\quad \lambda\in\mathscr{L}\,.$$

However since the functions $c_j(t)$ are complex analytic throughout $\mathbb{C}^g$ and the factors of automorphy $\rho_z$ and $\rho_{z_j}$ are topologically but not analytically equivalent this can only happen when $c_j(t)=0$; therefore (48) reduces to $f(t)=c\theta(t,z)$, and that suffices to conclude the proof of the theorem.

It is a relatively easy matter to extend the preceding theorem to the complex analytic vector bundles $\chi_s$ defined over suitable subsets of the Jacobi variety $J(M)$ for indices $s\leq g-1$, or to the corresponding sheaves $\tilde{\chi}_s$ defined over all of $J(M)$, assuming some familiarity with the results and techniques of sheaf theory in complex analysis. However to be consistent with the assumption of prerequisites in the preceding discussion the presentation of these extensions will be organized so as to minimize the required knowledge of sheaf theory, at the expense of course of a somewhat lengthier exposition.

Considering again the homomorphism of complex analytic vector bundles

$$T_{t_0}\theta_g(z_1,\ldots,z_n)\colon R^{-1}\longrightarrow T_{t_0}\chi_g\,,$$

where $z_1,\ldots,z_n\in\tilde{M}$ represent $n$ distinct points $p_1,\ldots,p_n\in M$ for $1\leq n\leq g-1$, $t_0=-\tilde{\phi}(z_1+\cdots+z_n-nz_0)$, and $R$ is the flat complex vector bundle $R=\rho_{z_1}\oplus\cdots\oplus\rho_{z_n}$, recall that that homomorphism can be viewed as determining the exact sequence (45) of analytic sheaves over $J(M)$ and hence as defining the analytic sheaf $\tilde{\chi}_{g-n}$. In some detail, if $U\subseteq J(M)$ is an open neighborhood of a point of $J(M)$ and $\tilde{U}$ is the complete inverse image of $U$ under the natural projection $\mathbb{C}^g\longrightarrow J(M)=\mathbb{C}^g/\mathscr{L}$ then $\tilde{U}$ is an $\mathscr{L}$-invariant subset of $\mathbb{C}^g$. A complex analytic section of the vector bundle $T_{t_0}\chi_g$ over $U$ is a complex analytic mapping $f\colon\tilde{U}\longrightarrow\mathbb{C}^g$ such that $f(t+\lambda)=\chi_g(\lambda,t-t_0)f(t)$ for all $t\in\tilde{U}$ and $\lambda\in\mathscr{L}$; and the set of all sections form a complex vector space $\Gamma(U,\mathcal{O}(T_{t_0}\chi_g))$. Similarly the set of all complex analytic sections of the vector bundle $R^{-1}$ over $U$ form a complex vector space $\Gamma(U,\mathcal{O}(R^{-1}))$; and multiplication by the function $T_{t_0}\theta_g(z_1,\ldots,z_n)$ determines an injective linear mapping

$$\theta\colon\Gamma(U,\mathcal{O}(R^{-1}))\longrightarrow\Gamma(U,\mathcal{O}(T_{t_0}\chi_g))\,.$$

The set of quotient spaces

$$\Gamma(U,\tilde{\chi}_{g-n})=\Gamma(U,\mathcal{O}(T_{t_0}\chi_g))/\theta\Gamma(U,\mathcal{O}(R^{-1}))$$

for the family of all open subsets $U\subseteq J(M)$ comprise the sheaf $\tilde{\chi}_{g-n}$; and as in § 9 the space $\Gamma(U,\tilde{\chi}_{g-n})$ can be identified with the space of complex analytic cross sections of the vector bundle $\chi_{g-n}$ over any open subset $U\subseteq X_{n-1}\subseteq J(M)$. A *section* of the sheaf $\tilde{\chi}_{g-n}$ over $J(M)$ is described by a collection of elements $f_\alpha\in\Gamma(U_\alpha,\tilde{\chi}_{g-n})$ for some open covering $\{U_\alpha\}$ of $J(M)$ such that these sections are compatible in the intersections $U_\alpha\cap U_\beta$, in the sense that $f_\alpha|U_\alpha\cap U_\beta=$

$f_\beta | U_\alpha \cap U_\beta$ with the natural restriction mappings; sections described by different coverings are of course identified if the restrictions to a common refinement of the two coverings coincide. Equivalently a section of the sheaf $\tilde{\chi}_{g-n}$ can be described by a collection of sections $f_\alpha \in \Gamma(U_\alpha, \mathcal{O}(T_{t_0}\chi_g))$ for some open covering $\{U_\alpha\}$ of $J(M)$ such that

$$f_\alpha(t) - f_\beta(t) = \theta_g(t - t_0; z_1, \ldots, z_n) g_{\alpha\beta}(t) \tag{49}$$

whenever $t \in \tilde{U}_\alpha \cap \tilde{U}_\beta$, for some sections $g_{\alpha\beta} \in \Gamma(U_\alpha \cap U_\beta, \mathcal{O}(R^{-1}))$. Somewhat more generally the matrix function $T_{t_0}\theta_g(z_1, \ldots, z_n)$ also describes a homomorphism of complex analytic vector bundles

$$T_{t_0}\theta_g(z_1, \ldots, z_n)\colon \rho_z \cdot R^{-1} \longrightarrow \rho_z \cdot T_{t_0}\chi_g \tag{50}$$

for any line bundle $\rho_z$; and the preceding constructions then lead to a sheaf $\mathcal{O}(\rho_z) \cdot \tilde{\chi}_{g-n} = \mathcal{O}(\rho_z) \otimes \tilde{\chi}_{g-n}$ and its space of sections.

**Corollary 1 to Theorem 13.** *For any integer* $1 \leq n \leq g-1$ *and any point* $z \in \tilde{M}$ *the space of sections of the sheaf* $\mathcal{O}(\rho_z) \cdot \tilde{\chi}_{g-n}$ *over* $J(M)$ *is one-dimensional. For any integer* $1 \leq n \leq g-2$ *and any point* $z \in \tilde{M}$ *the space of complex analytic cross-sections of the complex analytic vector bundle* $\rho_z \cdot \chi_{g-n}$ *over the open subset* $X_{n-1} \subset J(M)$ *is also one-dimensional.*

*Proof.* Given the point $z \in \tilde{M}$ choose points $z_1, \ldots, z_n \in \tilde{M}$ such that $z, z_1, \ldots, z_n$ represent distinct points of $M$; and note that none of the flat factors of automorphy $\rho_z \rho_{z_i}^{-1}, 1 \leq i \leq n$, are analytically trivial since by assumption $g \geq 1$. There is then the exact sequence of sheaves

$$0 \longrightarrow \mathcal{O}(\rho_z \cdot R^{-1}) \longrightarrow \mathcal{O}(\rho_z \cdot T_{t_0}\chi_g) \longrightarrow \mathcal{O}(\rho_z) \cdot \tilde{\chi}_{g-n} \longrightarrow 0 \tag{51}$$

over $J(M)$ induced by the homomorphism (50) of complex analytic vector bundles. The basic step in the proof of the first assertion of the Corollary is the demonstration that there is an associated exact sequence of vector spaces of the form

$$\begin{aligned} &\Gamma(J(M), \mathcal{O}(\rho_z \cdot R^{-1})) \longrightarrow \Gamma(J(M), \mathcal{O}(\rho_z \cdot T_{t_0}\chi_g)) \\ &\qquad \longrightarrow \Gamma(J(M), \mathcal{O}(\rho_z) \cdot \tilde{\chi}_{g-n}) \longrightarrow H^1(\mathcal{L}, \rho_z \cdot R^{-1}), \end{aligned} \tag{52}$$

where the cohomology groups $H^1(\mathcal{L}, \rho_z \cdot R^{-1}) = \sum_{i=1}^n H^1(\mathcal{L}, \rho_z \rho_{z_i}^{-1})$ are as discussed in § 19 of the Appendix. The first of these vector spaces is trivial, since $\rho_z \cdot R^{-1}$ is a direct sum of complex analytic line bundles which are topologically trivial but not analytically trivial; the second is one-dimensional by Theorem 13; the third is the unknown space of current interest; and the fourth is trivial, as demonstrated in § 19 of the Appendix. The desired result, that $\Gamma(J(M), \mathcal{O}(\rho_z) \cdot \tilde{\chi}_{g-n})$ is one-dimensional, is then an immediate consequence.

The exactness of the sequence (52) at the second of these spaces is quite obvious, since the homomorphism of complex analytic vector bundles (50) clearly induces an injective homomorphism between the spaces of sections. As a useful preliminary observation for proving the exactness of the sequence (52) at the third of these spaces, note that the matrix function $T_{t_0}\theta_g(z_1, \ldots, z_n)$ can be viewed as describing a bundle homomorphism

$$T_{t_0}\theta_g(z_1, \ldots, z_n)\colon I_n \longrightarrow I_g$$

between the identity bundles of ranks $n$ and $g$ over all of $\mathbb{C}^g$; and that this bundle homomorphism leads to an exact sequence of sheaves of the form

$$(53) \qquad 0 \longrightarrow \mathcal{O}(I_n) \longrightarrow \mathcal{O}(I_g) \longrightarrow \mathcal{S} \longrightarrow 0$$

over all of $\mathbb{C}^g$, where the construction of the sheaf $\mathcal{S}$ is the analogue of the construction of the sheaf $\tilde{\chi}_{g-n}$. Since $\mathbb{C}^g$ is a Stein manifold it follows from Cartan's Theorem B that the associated sequence of sections

$$(54) \qquad 0 \longrightarrow \Gamma(\mathbb{C}^g, \mathcal{O}(I_n)) \longrightarrow \Gamma(\mathbb{C}^g, \mathcal{O}(I_g)) \longrightarrow \Gamma(\mathbb{C}^g, \mathcal{S}) \longrightarrow 0$$

is also an exact sequence of vector spaces [22]. Now to return to the exactness argument consider a section $\tilde{f} \in \Gamma(J(M), \mathcal{O}(\rho_z)\cdot\tilde{\chi}_{g-n})$, described by a collection of sections $f_\alpha \in \Gamma(U_\alpha, \mathcal{O}(\rho_z \cdot T_{t_0}\chi_g))$ satisfying (49) for some open covering $\{U_\alpha\}$ of $J(M)$. Ignoring the conditions that the functions $f_\alpha, g_{\alpha\beta}$ be relatively automorphic for the appropriate factors of automorphy, these functions can also be viewed as describing a section in $\Gamma(\mathbb{C}^g, \mathcal{S})$; and it follows from the exactness of the sequence (54) that this section is the image of some section $f \in \Gamma(\mathbb{C}^g, \mathcal{O}(I_g))$. Thus there exist complex analytic mappings $f\colon \mathbb{C}^g \longrightarrow \mathbb{C}^g$ and $g_\alpha\colon U_\alpha \longrightarrow \mathbb{C}^n$ such that

$$f(t) = f_\alpha(t) + \theta_g(t - t_0; z_1, \ldots, z_n)\cdot g_\alpha(t)$$

whenever $t \in U_\alpha$, after refinement of the covering $\{U_\alpha\}$ if necessary; and it is evident that the most general mapping $f$ so associated to the given section $\tilde{f}$ is of the form $f + T_{t_0}\theta_g(z_1, \ldots, z_n)\cdot g$ for some complex analytic mapping $g\colon \mathbb{C}^g \longrightarrow \mathbb{C}^n$. Note that for any $t \in U_\alpha$ and $\lambda \in \mathcal{L}$

$$\begin{aligned}
f(t+\lambda) &+ \theta_g(t+\lambda-t_0; z_1, \ldots, z_n)g(t+\lambda)\\
&= f_\alpha(t+\lambda) + \theta_g(t+\lambda-t_0; z_1, \ldots, z_n)[g_\alpha(t+\lambda) + g(t+\lambda)]\\
&= \rho_z(\lambda)\chi_g(\lambda, t-t_0)f_\alpha(t)\\
&\quad + \chi_g(\lambda, t-t_0)\theta_g(t-t_0; z_1, \ldots, z_n)R(\lambda)[g_\alpha(t+\lambda) + g(t+\lambda)]\\
&= \rho_z(\lambda)\chi_g(\lambda, t-t_0)[f(t) + \theta_g(t-t_0; z_1, \ldots, z_n)g(t)]\\
&\quad + \rho_z(\lambda)\chi_g(\lambda, t-t_0)\theta_g(t-t_0; z_1, \ldots, z_n)\sigma(\lambda, t),
\end{aligned}$$

where

$$\sigma(\lambda, t) = \rho_z(\lambda)^{-1}R(\lambda)[g_\alpha(t+\lambda) + g(t+\lambda)] - [g_\alpha(t) + g(t)].$$

The expression $\sigma(\lambda, t)$, which is clearly independent of $\alpha$, is a one-cocycle $\sigma$ in $Z^1(\mathcal{L}, \rho_z R^{-1}) = \sum_{i=1}^n Z^1(\mathcal{L}, \rho_z\rho_{z_i}^{-1})$, as is evident upon examining the definition given in § 19 of the Appendix; and since a change in $g$ has the effect of changing $\sigma$ by a one-coboundary the cohomology class in $H^1(\mathcal{L},\rho_z R^{-1})=\sum_{i=1}^{n} H^1(\mathcal{L},\rho_z\rho_{z_i}^{-1})$ represented by $\sigma$ is a uniquely determined linear function of the original section $\tilde{f}\in\Gamma(J(M),\mathcal{O}(\rho_z)\cdot\tilde{\chi}_{g-n})$. Moreover this cohomology class is clearly trivial precisely when there exists a complex analytic mapping $g\colon\mathbb{C}^g\longrightarrow\mathbb{C}^n$ such that the function $f+T_{t_0}\theta_g(z_1,\dots,z_n)\cdot g$ is relatively automorphic for the factor of automorphy $\rho_z\cdot T_{t_0}\chi_g$, hence precisely when $\tilde{f}$ is the image of a section

$$f+T_{t_0}\theta_g(z_1,\dots,z_n)\cdot g\in\Gamma(J(M),\mathcal{O}(\rho_z\cdot T_{t_0}\chi_g)).$$

That completes the proof of the exactness of the sequence (52) and hence of the first part of the Corollary.

For the second part of the Corollary note that the complex analytic cross-sections of the vector bundle $\rho_z\cdot\chi_{g-n}$ over the open subset $X_{n-1}\subset J(M)$ can be identified with the sections of the sheaf $\mathcal{O}(\rho_z)\cdot\tilde{\chi}_{g-n}$ over $X_{n-1}$. Again the basic step in the proof is the demonstration that associated to the restriction of the exact sequence of sheaves (51) to the open subset $X_{n-1}\subset J(M)$ there is an exact sequence of vector spaces of the form

$$\text{(55)}\qquad \Gamma(X_{n-1},\mathcal{O}(\rho_z\cdot R^{-1}))\longrightarrow\Gamma(X_{n-1},\mathcal{O}(\rho_z\cdot T_{t_0}\chi_g))$$
$$\longrightarrow\Gamma(X_{n-1},\mathcal{O}(\rho_z)\cdot\tilde{\chi}_{g-n})\longrightarrow H^1(\mathcal{L},\rho_z\cdot R^{-1}|\tilde{X}_{n-1}).$$

The preceding proof of the exactness of the sequence (52) carries over immediately to give the exactness of the sequence (55), if there is an analogue of the exact sequence (54) over the subset $\tilde{X}_{n-1}\subset\mathbb{C}^g$ covering $X_{n-1}\subset J(M)$. There is always an exact sequence of the form

$$0\longrightarrow\Gamma(\tilde{X}_{n-1},\mathcal{O}(I_n))\longrightarrow\Gamma(\tilde{X}_{n-1},\mathcal{O}(I_g))\longrightarrow\Gamma(\tilde{X}_{n-1},\mathcal{S})\longrightarrow H^1(\tilde{X}_{n-1},\mathcal{O}(I_n)),$$

[22]; and if $1\leq n\leq g-2$ then $\tilde{X}_{n-1}$ is the complement of a complex analytic subvariety of codimension $d=g-(n-1)\geq 3$ in $\mathbb{C}^g$, and it is a known result of the theory of functions of several complex variables that in such a case $H^1(\tilde{X}_{n-1},\mathcal{O}(I_n))=\sum_{i=1}^{n} H^1(\tilde{X}_{n-1},\mathcal{O})=0$, [14, Theorem 22]. Now in the exact sequence (55) an element of $\Gamma(X_{n-1},\mathcal{O}(\rho_z\cdot R^{-1}))$ is a complex analytic relatively automorphic function for the factor of automorphy $\rho_z R^{-1}$ in the complement of a complex analytic subvariety of codimension $d=g-(n-1)\geq 2$ in $\mathbb{C}^g$; but it follows immediately from the extended Riemann removable singularities theorem that any such function automatically extends to all of $\mathbb{C}^g$, and hence $\Gamma(X_{n-1},\mathcal{O}(\rho_z\cdot R^{-1}))\cong\Gamma(J(M),\mathcal{O}(\rho_z\cdot R^{-1}))$ and similarly $\Gamma(X_{n-1},\mathcal{O}(\rho_z\cdot T_{t_0}\chi_g))\cong\Gamma(J(M),\mathcal{O}(\rho_z\cdot T_{t_0}\chi_g))$. For the same reason the one-cocycles and one-coboundaries for the group $\mathcal{L}$ acting on $\tilde{X}_{n-1}$ extend to those for the group $\mathcal{L}$ acting on $\mathbb{C}^g$, so $H^1(\mathcal{L},\rho_z\cdot R^{-1}|\tilde{X}_{n-1})\cong H^1(\mathcal{L},\rho_z\cdot R^{-1})$. Thus except for the unknown third space the exact sequences (52) and (55) are isomorphic; and it then follows from the observations made in the first part of the proof that $\Gamma(X_{n-1},\mathcal{O}(\rho_z)\cdot\tilde{\chi}_{g-n})$ is one-dimensional, to conclude the proof of the Corollary.

A further simple consequence of Theorem 13 is also worth pointing out here. An *endomorphism* for a factor of automorphy $\chi$ of rank $s$ for the action of a lattice subgroup $\mathscr{L}$ on $\mathbb{C}^g$ is a complex analytic mapping $C: \mathbb{C}^g \longrightarrow \mathbb{C}^{s\times s}$ from $\mathbb{C}^g$ into the space of all $s\times s$ complex matrices such that $\chi(\lambda, t)C(t)=C(t+\lambda)\chi(\lambda, t)$ for all $\lambda\in\mathscr{L}$. The automorphisms for a factor of automorphy, as considered earlier, are those endomorphisms such that the matrix $C(t)$ is nonsingular for all $t\in\mathbb{C}^g$. Any constant scalar matrix is obviously an endomorphism, even an automorphism, for an arbitrary factor of automorphy; but these will be considered as trivial endomorphisms or automorphisms.

**Corollary 2 to Theorem 13.** *A theta factor of automorphy of rank $s\geq g$ for a Jacobi variety admits only trivial endomorphisms.*

*Proof.* If $C$ is any endomorphism for a theta factor of automorphy $\chi_s$ of rank $s$ and $\theta$ is a generalized theta function of rank $s$ associated to the theta factor $\chi_s$ then clearly both $\theta(t, z)$ and $C(t)\theta(t, z)$ are complex analytic relatively automorphic functions for the factor of automorphy $\rho_z\cdot\chi_s$ for any point $z\in\tilde{M}$. It is of course an immediate consequence of Theorem 13 that $C(t)\theta(t, z)=c(z)\theta(t, z)$ for some scalar $c(z)$ depending perhaps on $z$; but for any fixed point $t\in\mathbb{C}^g$ the components of the function $\theta(t, z)$ are linearly independent functions of $z$, so necessarily $C(t)=c(z)\cdot I$. That means that $C(t)$ is a constant scalar matrix and hence a trivial endomorphism, as was to be proved.

## § 11. Induced Theta Factors and Theta Functions on Riemann Surfaces

The Jacobi mapping $\tilde{\phi}: \tilde{M} \longrightarrow \mathbb{C}^g$ has the property that $\tilde{\phi}(Tz)=\tilde{\phi}(z)+\phi(T)$ for every element $T\in\Gamma$, where $\phi: \Gamma \longrightarrow \mathscr{L}$ is the homomorphism which maps an element $T\in\Gamma$ to the vector $\phi(T)=(\omega_1(T), \ldots, \omega_g(T))\in\mathscr{L}$ consisting of the periods of the canonical Abelian differentials. It is therefore evident that any factor of automorphy $\chi$ of rank $s$ for the action of the lattice subgroup $\mathscr{L}$ on $\mathbb{C}^g$ induces a factor of automorphy $\phi^*\chi$ of rank $s$ for the action of the group $\Gamma$ on $\tilde{M}$, where

$$\phi^*\chi(T, z)=\chi(\phi(T), \tilde{\phi}(z)-\tilde{\phi}(z_0)).$$

Correspondingly any complex analytic relatively automorphic function $\theta$ for the factor of automorphy $\chi$ induces a complex analytic relatively automorphic function $\phi^*\theta$ for the factor of automorphy $\phi^*\chi$, where

$$\phi^*\theta(z)=\theta(\tilde{\phi}(z)-\tilde{\phi}(z_0));$$

but the function $\phi^*\theta$ may of course be identically zero. This is a very familiar construction in the classical theory of scalar theta functions, where the Abelian theta factor of automorphy for $\mathscr{L}$ induces the Riemannian theta factor of automorphy for $\Gamma$ and correspondingly the Abelian theta function on $\mathbb{C}^g$ induces the Riemannian theta function on $\tilde{M}$; and an indication of its usefulness can be found in § 18 of the Appendix. It is of some interest to examine this construction for the theta factors of higher rank as well, particularly for the standard theta factor of rank $g$, and for the associated generalized theta functions; and as in the classical case it is useful also to consider arbitrary translates of these theta factors and theta functions.

Considering the translate $T_t\chi$ of a theta factor of automorphy $\chi$ of rank $g$ for the action of the group $\mathscr{L}$, the induced factor of automorphy of rank $g$ for the action of the group $\Gamma$ will be denoted by $\chi_t$ to simplify the notation; thus

$$\chi_t(T, z) = (\phi^* T_t\chi)(T, z) = \chi(\phi(T), \tilde{\phi}(z) - \tilde{\phi}(z_0) - t)$$

for all elements $T \in \Gamma$. Correspondingly for the associated generalized theta function $\theta(t_1, z_1)$ of rank $g$ the induced function on $\tilde{M}$ will be denoted by $\theta_t(z, z_1)$; thus

$$\theta_t(z, z_1) = (\phi^* T_t\theta)(z, z_1) = \theta(\tilde{\phi}(z) - \tilde{\phi}(z_0) - t, z_1).$$

This induced theta function is therefore a complex analytic mapping $\theta_t\colon \tilde{M} \times \tilde{M} \longrightarrow \mathbb{C}^g$ with the properties that

(56) $$\theta_t(Tz, z_1) = \rho_{z_1}(\phi(T))\,\chi_t(T, z)\,\theta_t(z, z_1) \quad \text{and}$$

(57) $$\theta_t(z, Tz_1) = \rho_{\tilde{\phi}(z) - \tilde{\phi}(z_0) - t}(T)\,\zeta(T, z)^{2g-1}\,\theta_t(z, z_1)$$

for all $T \in \Gamma$, as consequences of (6) and (7) respectively; recall that

(10) $$\rho_z(\phi(T)) = \rho_{\tilde{\phi}(z) - \tilde{\phi}(z_0)}(T),$$

with the conventions used before. In the present discussion though the most useful property of this induced theta function is that, as an immediate consequence of Corollary 1 to Theorem 9, for any index $n$ in the range $1 \leq n \leq g$ and any points $z_1, \ldots, z_n$ of $\tilde{M}$ which represent distinct points $p_1, \ldots, p_n$ of $M$,

$$\operatorname{rank}(\theta_t(z, z_1), \ldots, \theta_t(z, z_n)) < n$$

if and only if $\tilde{\phi}(z) - \tilde{\phi}(z_0) - t$ represents a point of the subvariety $-W_{n-1} + k + \phi(p_1 + \cdots + p_n - np_0) \subset J(M)$, or equivalently, if and only if

$$k + t + \tilde{\phi}(z_1 + \cdots + z_n - nz_0) - \tilde{\phi}(z - z_0)$$

represents a point of the subvariety $W_{n-1} \subset J(M)$. In order to state a first additional consequence of this property consider the subset

$$W_1 - W_1 = \{\phi(p_1) - \phi(p_2) \mid p_1, p_2 \in M\} \subseteq J(M),$$

noting that it is a two-dimensional complex analytic subvariety of $J(M)$ whenever $g \geq 2$, [12]; and as usual let $W_1 - W_1$ also denote the corresponding subvariety of $\mathbb{C}^g$.

**Theorem 14a.** (i) *If* $t+k \notin W_1 - W_1$ *then* $\theta_t(z, z_1) \neq 0$ *for all points* $(z, z_1) \in \tilde{M} \times \tilde{M}$. (ii) *If* $t+k \in W_1 - W_1$ *but* $t+k \notin \mathscr{L}$, *then when* $M$ *is not hyperelliptic there is a unique point* $(p, p_1) \in M \times M$ *such that* $\theta_t(z, z_1) = 0$ *precisely when* $(z, z_1) \in \tilde{M} \times \tilde{M}$ *represents* $(p, p_1) \in M \times M$; *while if* $M$ *is hyperelliptic with hyperelliptic involution* $\eta$ *there is a unique pair of points* $(p, p_1)$, $(\eta p_1, \eta p)$ *of* $M \times M$ *such that* $\theta_t(z, z_1) = 0$ *precisely when* $(z, z_1) \in \tilde{M} \times \tilde{M}$ *represents either* $(p, p_1)$ *or* $(\eta p_1, \eta p)$. (iii) *If* $t+k \in \mathscr{L}$ *then* $\theta_t(z, z_1) = 0$ *precisely when* $z \in \Gamma z_1$.

*Proof*. For the special case $n=1$ the last observation in the preceding paragraph shows that $\theta_t(z, z_1) = 0$, or equivalently rank $\theta_t(z, z_1) < 1$, if and only if $k + t + \tilde{\phi}(z_1 - z_0) - \tilde{\phi}(z - z_0)$ represents the point $W_0$, hence if and only if $t + k \equiv \tilde{\phi}(z - z_1) \bmod \mathscr{L}$. Thus if $t+k \notin W_1 - W_1$ there are no such points $z, z_1$, and consequently $\theta_t(z, z_1) \neq 0$ for all points $(z, z_1) \in \tilde{M} \times \tilde{M}$. On the other hand if $t+k \in W_1 - W_1$ there are points $p, p_1$ of $M$ such that $t+k$ represents the point $\phi(p - p_1)$, and hence $\theta_t(z, z_1) = 0$ whenever $(z, z_1) \in \tilde{M} \times \tilde{M}$ represents $(p, p_1) \in M \times M$. If there are other points $p', p'_1$ of $M$ such that $t+k$ also represents the point $\phi(p' - p'_1)$ then of course $\phi(p + p'_1) = \phi(p_1 + p')$, and conversely. If $M$ is not hyperelliptic an equality of this last form can only arise when $p + p'_1 = p_1 + p'$ as divisors on $M$, [10]. Thus either $p = p'$ and $p_1 = p'_1$, or $p = p_1$ and $p' = p'_1$; and in the latter case $t+k$ represents the point $\phi(p - p_1) = \phi(p) - \phi(p) = W_0$, so that $t + k \in \mathscr{L}$. If $M$ is hyperelliptic with hyperelliptic involution $\eta: M \longrightarrow M$ there is the additional possibility that $p'_1 = \eta(p)$, $p' = \eta(p_1)$, [10]. Thus the only possibilities are just those listed in the theorem and the proof is thereby concluded.

In particular of course the induced theta function $\theta_t(z, z_1)$ is a nontrivial complex analytic vector-valued function of $z$ for any fixed points $z_1 \in \tilde{M}$ and $t \in \mathbb{C}^g$. If $t+k \notin \mathscr{L}$ then provided that $z_1$ is chosen so as to avoid a proper complex analytic subvariety of $M$ (consisting indeed of at most two points of $M$) the function $\theta_t(z, z_1)$ is a nowhere vanishing complex analytic vector field on $\tilde{M}$. If $t+k \in \mathscr{L}$ the function $\theta_t(z, z_1)$ vanishes precisely at the points $\Gamma z_1$.

These results can be extended to yield a useful further consequence of the property discussed above. As a preliminary observation note that whenever $t+k \notin \mathscr{L}$ the space of complex analytic relatively automorphic functions for the factor of automorphy $\kappa \rho_{t+k}^{-1}$ has dimension $g-1$; and if $h_1, \ldots, h_{g-1}$ is any basis for that space of functions consider the complex analytic subvarieties $Y_n \subset \tilde{M}^n$ defined by

$$Y_n = \{(z_1, \ldots, z_n) \in \tilde{M}^n \,|\, \operatorname{rank}(h_i(z_j)) < n\}$$

for $1 \leq n \leq g-1$. These are evidently proper complex analytic subvarieties of $\tilde{M}^n$, since $n \leq g-1$, and are clearly independent of the choice of the basis $h_i$. They are also invariant under the natural action of $\Gamma^n$, and hence define proper complex analytic subvarieties of $M^n$ as well. The complement $\tilde{M}^n - Y_n \subset \tilde{M}^n$ is a dense open subset of $\tilde{M}^n$.

**Theorem 14b.** *If* $t+k\notin\mathscr{L}$ *and* $(z_1,\ldots,z_n)\in\tilde{M}^n-Y_n$ *for some index n in the range* $1\leq n\leq g-1$ *then*

$$\operatorname{rank}(\theta_t(z,z_1),\ldots,\theta_t(z,z_n))=n$$

*for all points* $z\in\tilde{M}$.

*Proof.* As observed in the paragraph preceding the statement of Theorem 14a, $\operatorname{rank}(\theta_t(z,z_1),\ldots,\theta_t(z,z_n))<n$ precisely when $t+k+t_1-\tilde{\phi}(z-z_0)$ represents a point of the subvariety $W_{n-1}\subset J(M)$, where $t_1=\tilde{\phi}(z_1+\cdots+z_n-nz_0)$. Thus for any given points $z_1,\ldots,z_n$ of $\tilde{M}$ there exists a point $z$ at which $\operatorname{rank}(\theta_t(z,z_1),\ldots,\theta_t(z,z_n))<n$ precisely when $t+k+t_1$ represents a point of the subvariety $W_n\subset J(M)$; and that is in turn equivalent to the condition that

$$\gamma(\rho_{t+k}\zeta_{z_1}\cdots\zeta_{z_n})=\gamma(\rho_{t+k+t_1}\zeta^n)\geq 1\,,$$

or by the Riemann-Roch theorem to the condition that

$$\gamma(\kappa\rho_{t+k}^{-1}\zeta_{z_1}^{-1}\ldots\zeta_{z_n}^{-1})\geq g-n\,.$$

Now $\gamma(\kappa\rho_{t+k}^{-1}\zeta_{z_1}^{-1}\ldots\zeta_{z_n}^{-1})$ is also equal to the dimension of the space of those complex analytic relatively automorphic functions for the factor of automorphy $\kappa\rho_{t+k}^{-1}$ which vanish at the points $z_1,\ldots,z_n$, hence is equal to $g-1-\operatorname{rank}(h_i(z_j))$. Therefore there exists a point $z$ at which $\operatorname{rank}(\theta_t(z,z_1),\ldots,\theta_t(z,z_n))<n$ precisely when $\operatorname{rank}(h_i(z_j))\leq n-1$, hence when $(z_1,\ldots,z_n)\in Y_n$; and since that is clearly equivalent to the desired result, it serves to conclude the proof of the theorem.

Note incidentally that the restriction that $(z_1,\ldots,z_n)\notin Y_n$ implies that the points $z_i$ represent distinct points of $M$. Note further that the preceding assertion fails if $t+k\in\mathscr{L}$, since then the column vector $\theta_t(z,z_i)$ vanishes at the points $\Gamma z_i$; and it also fails if $n=g$, since $W_g=J(M)$. However at least something can be said about the case $n=g$, as follows.

**Theorem 14c.** *If* $t+\tilde{\phi}(z_1+\cdots+z_g-gz_0)$ *does not represent a point of the subvariety* $-W_{g-2}\subset J(M)$ *then there are points* $z\in\tilde{M}$ *at which*

$$\operatorname{rank}(\theta_t(z,z_1),\ldots,\theta_t(z,z_g))=g\,.$$

*Proof.* As in the proof of Theorem 14b note that $\operatorname{rank}(\theta_t(z,z_1),\ldots,\theta_t(z,z_g))<g$ for all points $z\in\tilde{M}$ precisely when $t+k+t_1-\tilde{\phi}(z-z_0)$ represents a point of the subvariety $W_{g-1}$ for all points $z\in\tilde{M}$, where $t_1=\tilde{\phi}(z_1+\cdots+z_g-gz_0)$, hence precisely when $p-W_1\subseteq W_{g-1}$ where $p$ is the point of $J(M)$ represented by $t+k+t_1$. However using the notation of [12, p. 42] and the results noted in [12, pp. 47 and 49], that is equivalent to the condition that

$$p\in W_{g-1}\ominus(-W_1)=W_g^2=k-W_{g-2}\,,$$

hence that $t+t_1$ represents a point of the subvariety $-W_{g-2}\subset J(M)$, as desired.

In a sense Theorem 14b is the analogue for the induced theta function of rank $g$ over $M$ of Corollary 2 to Theorem 9 for the generalized theta functions of ranks $s \geq g+1$ over $J(M)$; and there is an interpretation of Theorem 14b paralleling the interpretation of Corollary 2 to Theorem 9 discussed in § 9. Indeed it follows from (56) that for any points $z_1, \ldots, z_n$ of $\tilde{M}$

$$(58) \qquad \begin{aligned} &(\theta_t(Tz, z_1), \ldots, \theta_t(Tz, z_n)) \cdot \operatorname{diag}(\rho_{z_1}(\phi(T))^{-1}, \ldots, \rho_{z_n}(\phi(T))^{-1}) \\ &\quad = \chi_t(T, z) \cdot (\theta_t(z, z_1), \ldots, \theta_t(z, z_n)) \end{aligned}$$

for all $T \in \Gamma$, where $\operatorname{diag}(\rho_{z_1}(\phi(T))^{-1}, \ldots, \rho_{z_n}(\phi(T))^{-1})$ denotes the diagonal $n \times n$ matrix with the listed entries along the diagonal. As before an equation of the form (58) can be interpreted as the assertion that the complex analytic mapping $(\theta_t(z, z_1), \ldots, \theta_t(z, z_n))$ determines a bundle homomorphism

$$\theta_t(z_1, \ldots, z_n) \colon \phi^* \rho_{z_1}^{-1} \oplus \cdots \oplus \phi^* \rho_{z_n}^{-1} \longrightarrow \chi_t$$

from the direct sum of the complex analytic line bundles over $M$ represented by the factors of automorphy $\phi^* \rho_{z_1}^{-1}, \ldots, \phi^* \rho_{z_n}^{-1}$ into the complex analytic vector bundle of rank $g$ over $M$ represented by the factor of automorphy $\chi_t$. Furthermore if $1 \leq n \leq g-1$, $t+k \notin \mathscr{L}$, and $(z_1, \ldots, z_n) \notin Y_n \subset \tilde{M}^n$ then it follows from Theorem 14b that this homomorphism is injective and exhibits the complex analytic vector bundle $\phi^* \rho_{z_1}^{-1} \oplus \cdots \oplus \phi^* \rho_{z_n}^{-1}$ as a subbundle of the complex analytic vector bundle $\chi_t$. This assertion is particularly interesting for the case $n = g-1$, when it can be completed as follows.

**Corollary 1 to Theorem 14.** *If $t+k \notin \mathscr{L}$ and $(z_1, \ldots, z_{g-1}) \notin Y_{g-1} \subset \tilde{M}^{g-1}$ then there is an exact sequence of complex analytic vector bundles over $M$ of the form*

$$(59) \qquad 0 \longrightarrow \phi^* \rho_{z_1}^{-1} \oplus \cdots \oplus \phi^* \rho_{z_{g-1}}^{-1} \xrightarrow{\theta_t(z_1, \ldots, z_{g-1})} \chi_t \longrightarrow \rho_{k+t+t_1} \zeta^g \longrightarrow 0,$$

*where* $t_1 = \tilde{\phi}(z_1 + \cdots + z_{g-1} - (g-1) z_0)$.

*Proof.* As noted above it follows immediately from Theorem 14b that there is an exact sequence of complex analytic vector bundles over $M$ of the form

$$0 \longrightarrow \phi^* \rho_{z_1}^{-1} \oplus \cdots \oplus \phi^* \rho_{z_{g-1}}^{-1} \longrightarrow \chi_t \longrightarrow \alpha \longrightarrow 0$$

for some complex analytic line bundle $\alpha$; or equivalently, the factor of automorphy $\chi_t$ is analytically equivalent to one of the form

$$(60) \qquad \begin{pmatrix} R^{-1} & S \\ 0 & \alpha \end{pmatrix}$$

where $R(T, z) = \rho_{z_1}(\phi(T)) \oplus \cdots \oplus \rho_{z_{g-1}}(\phi(T))$ is a flat factor of automorphy of rank $g-1$ and $\alpha(T, z)$ is a scalar factor of automorphy. Thus up to complex analytic equivalence

$$\begin{aligned}\det \chi_t(T, z) &= \det R(T, z)^{-1} \cdot \alpha(T, z)\\ &= \rho_{z_1}(\phi(T))^{-1} \cdot \dots \cdot \rho_{z_{g-1}}(\phi(T))^{-1} \cdot \alpha(T, z)\\ &= \rho_{t_1}(T)^{-1} \cdot \alpha(T, z).\end{aligned}$$

Now for the theta factor of automorphy $\chi$ of rank $g$ over $J(M)$ it follows from Theorem 12 that $\det \chi$ is analytically equivalent to the Abelian theta factor of automorphy $\xi$, hence that the translate $T_t \det \chi = \det T_t \chi$ is analytically equivalent to the translate $T_t \xi$. However with the notation used in § 18 of the Appendix $T_t \xi = T_t \xi_{k-r} = \xi_{k-r+t}$; and as demonstrated in the Appendix the induced factor of automorphy $\phi^* \xi_{k-r+t}$, the Riemannian theta factor of automorphy associated to the parameter $k-r+t$, is analytically equivalent to the factor of automorphy $\rho_{k+t} \zeta^g$. Thus up to analytic equivalence $\alpha = \rho_{t_1} \cdot \phi^* T_t \det \chi = \rho_{t_1} \cdot \phi^* \xi_{k-r+t} = \rho_{k+t+t_1} \zeta^g$; and that suffices to conclude the proof.

For any point $z_1 \in \tilde{M}$ the induced theta function $\theta_t(z, z_1)$ as a function of $z$ is a nontrivial complex analytic relatively automorphic function for the factor of automorphy $\phi^* \rho_{z_1} \otimes \chi_t$. It is interesting to note that the constant multiples of $\theta_t(z, z_1)$ are the only complex analytic relatively automorphic functions for that factor of automorphy, and hence that any complex analytic relatively automorphic function for the induced factor of automorphy $\phi^* \rho_z \otimes \chi_t = \phi^*(\rho_z \otimes T_t \chi)$ over $M$ is induced by a complex analytic relatively automorphic function for the factor of automorphy $\rho_z \otimes T_t \chi$ over $J(M)$, at least when $t+k \notin \mathscr{L}$.

**Corollary 2 to Theorem 14.** *If $t+k \notin \mathscr{L}$ then for any point $z \in \tilde{M}$ the space of complex analytic relatively automorphic functions for the factor of automorphy $\phi^* \rho_z \otimes \chi_t$ is one dimensional.*

*Proof.* Choose points $z_1, \dots, z_{g-1} \in \tilde{M}$ such that $z, z_1, \dots, z_{g-1}$ represent distinct points of $M$ and such that $(z_1, \dots, z_{g-1}) \notin Y_{g-1} \subset \tilde{M}^{g-1}$; so by Corollary 1 to Theorem 14 there is an exact sequence of complex analytic vector bundles over $M$ of the form (59), or equivalently the factor of automorphy $\chi_t$ is analytically equivalent to one of the form (60) where $\alpha = \rho_{k+t+t_1} \cdot \zeta^g$ for the point $t_1 = \tilde{\phi}(z_1 + \dots + z_{g-1} - (g-1) z_0)$. If $f = {}^t(f_1, \dots, f_g)$ is any complex analytic relatively automorphic function for the factor of automorphy $\phi^* \rho_z \otimes \chi_t$ when $\chi_t$ has this form (60) then the function $f_g$ is a scalar relatively automorphic function for the factor of automorphy $\phi^* \rho_z \otimes \rho_{k+t+t_1} \cdot \zeta^g$. Now note that up to analytic equivalence

$$\phi^* \rho_z \otimes \rho_{k+t+t_1} \cdot \zeta^g = \rho_{k+t+t_1+\tilde{\phi}(z-z_0)} \zeta^g = \rho_{k+t} \zeta_{z_1} \dots \zeta_{z_{g-1}} \zeta_z;$$

hence by the Riemann-Roch theorem

$$\gamma(\phi^* \rho_z \otimes \rho_{k+t+t_1} \cdot \zeta^g) = \gamma(\kappa \rho_{k+t}^{-1} \zeta_{z_1}^{-1} \dots \zeta_{z_{g-1}}^{-1} \zeta_z^{-1}) + 1;$$

but $\gamma(\kappa\rho_{k+t}^{-1}\zeta_{z_1}^{-1}\ldots\zeta_{z_{g-1}}^{-1}\zeta_z^{-1})$ is the dimension of the space of those complex analytic relatively automorphic functions for the factor of automorphy $\kappa\rho_{k+t}^{-1}$ which vanish at the points $z_1,\ldots,z_{g-1},z$, and that dimension is zero since there are not even any nontrivial such functions vanishing at the points $z_1,\ldots,z_{g-1}$ for $(z_1,\ldots,z_{g-1})\notin Y_{g-1}$. Consequently $\gamma(\phi^*\rho_z\otimes\rho_{k+t+t_1}\cdot\zeta^g)=1$, from which it follows that $f_g$ is necessarily a constant multiple $c$ of the last component of the induced theta function $\theta_t(*,z)$; and therefore after replacing the function $f$ by $f-c\theta_t(*,z)$ it can be assumed that $f_g\equiv 0$. The components $f_i$ for $1\leq i\leq g-1$ are then necessarily scalar relatively automorphic functions for the factors of automorphy $\phi^*\rho_z\otimes\phi^*\rho_{z_i}^{-1}=\rho_{\tilde{\phi}(z-z_i)}$; but since $z$ and $z_i$ represent distinct points of $M$ these factors of automorphy are topologically trivial but not analytically trivial, recalling Theorem 4, and hence $f_i\equiv 0$ as well. The function $f$ must therefore be a constant multiple of the induced theta function, and that suffices to conclude the proof.

This last observation can be used in turn to show that the extension of vector bundles in the exact sequence (59) is necessarily nontrivial, or slightly more generally, to show that when the factor of automorphy $\chi_t$ for $k+t\notin\mathscr{L}$ is written in the form (60) then none of the components of the matrix $S$ can vanish identically. Indeed if one of the components of the matrix $S$ does vanish identically then the factor of automorphy is decomposable, in the sense that it can be written as a nontrivial direct sum of factors of automorphy of smaller ranks; the desired result is therefore an immediate consequence of the following yet more general one.

**Corollary 3 to Theorem 14.** *If $t+k\notin\mathscr{L}$ then the factor of automorphy $\chi_t$ is indecomposable.*

*Proof.* Suppose contrarily that the factor of automorphy $\chi_t$ is decomposable, so that up to analytic equivalence $\chi_t=\chi_t'\oplus\chi_t''$ for some factors of automorphy $\chi_t';\chi_t''$ of ranks $r',r''$ with $1\leq r',r''<g$. Note that any complex analytic relatively automorphic function $f$ for the factor of automorphy $\phi^*\rho_z\otimes\chi_t$ can be decomposed correspondingly into a direct sum $f=f'\oplus f''$ of complex analytic relatively automorphic functions for the summands $\phi^*\rho_z\otimes\chi_t'$ and $\phi^*\rho_z\otimes\chi_t''$; but then $f'\oplus 0$ and $0\oplus f''$ are both complex analytic relatively automorphic functions for the factor of automorphy $\phi^*\rho_z\otimes\chi_t$, and since the space of such functions is one-dimensional as a consequence of Corollary 2 to Theorem 14 it follows that either $f'\equiv 0$ or $f''\equiv 0$. However it is evident from Theorem 14c that there are points $z\in\tilde{M}$ such that the induced theta function is a complex analytic relatively automorphic function $f$ for the factor of automorphy $\phi^*\rho_z\otimes\chi_t$ and that neither $f'\equiv 0$ nor $f''\equiv 0$; and that contradiction suffices to conclude the proof.

These results, particularly Corollary 1 to Theorem 14, do convey a good deal of information about the complex analytic vector bundles over $M$ described by the induced theta factors of automorphy $\chi_t$, but of course do not suffice to give a complete characterization of these bundles. Unfortunately there is not as yet a fully satisfactory classification theory for complex analytic vector bundles over compact Riemann surfaces, although there are various ways in which such bundles can be completely specified. One way is to describe the precise extensions involved in the exact sequences (59); another way is to investigate the subbundles of these vector bundles and then to describe the appropriate normal forms; and yet another way is to describe some of the matrix divisors associated to these bundles. There are disadvantages to each of these separate descriptions; and there are considerable gaps in the present state of knowledge of the relationships between these various descriptions. However it is perhaps of some interest briefly to examine the induced theta factors of automorphy from each of these points of view.

First if $t+k\notin\mathscr{L}$ and $(z_1,\dots,z_{g-1})\notin Y_{g-1}$ then the factor of automorphy $\chi_t$ is analytically equivalent to one of the form (60), where $R(T,z)=\rho_{z_1}(\phi(T))\oplus\cdots\oplus\rho_{z_{g-1}}(\phi(T))$ and $\alpha(T,z)=\rho_{k+t+t_1}(T)\cdot\zeta(T,z)^g$ for $t_1=\tilde{\phi}(z_1+\cdots+z_{g-1}-(g-1)z_0)$; and the components $s_i(T,z)$ of the column vector $S$ in (60) describe the particular extensions (59). It is an easy consequence of the condition that $\chi_t$ be a factor of automorphy that these functions $s_i(T,z)$ satisfy

$$s_i(ST,z)=\rho_{z_i}(\phi(S))^{-1}s_i(T,z)+s_i(S,Tz)\alpha(T,z)$$

for any $S,T\in\Gamma$, hence that the products $\sigma_i(T,z)=\rho_{z_i}(\phi(T))s_i(T,z)$ satisfy

$$\sigma_i(ST,z)=\rho_{z_i}(\phi(T))\alpha(T,z)\cdot\sigma_i(S,Tz)+\sigma_i(T,z)$$

for any $S,T\in\Gamma$; thus the mapping $\sigma_i$ can be viewed as a one-cocycle $\sigma_i\in Z^1(\Gamma,\rho_{z_i}^{-1}\alpha^{-1})$, using the notation introduced in § 19 of the Appendix. The equivalence class of the particular extension described by the cocycle $\sigma_i$ depends only on the cohomology class in $H^1(\Gamma,\rho_{z_i}^{-1}\alpha^{-1})$ represented by $\sigma_i$; that is a quite familiar construction, discussed in [1] and [21] among other places. Furthermore the cohomology group $H^1(\Gamma,\rho_{z_i}^{-1}\alpha^{-1})$ can be identified with the dual to the space of complex analytic relatively automorphic functions for the factor of automorphy $\kappa\rho_{z_i}\alpha$, by an application of the Serre duality theorem [12]; and that leads quite easily to the following explicit form for the linear functional associated to the equivalence class of the extension described by the mapping $\sigma_i$. The induced theta function $\theta_t(z,z_0)=(f_i(z))$ is by Theorem 14a and Eq. (56) a nontrivial complex analytic relatively automorphic function for the factor of automorphy $\chi_t$; thus

$$\begin{aligned}f_g(Tz)&=\alpha(T,z)f_g(z),\\ f_i(Tz)&=\rho_{z_i}(\phi(T))^{-1}f_i(z)+s_i(T,z)f_g(z),\qquad 1\le i\le g-1\end{aligned}$$

for any $T\in\Gamma$. Assuming further that $\Gamma z_0\neq\Gamma z_i$ for $1\le i\le g-1$ then the function $f_g$ is itself not identically zero; for otherwise the functions $f_i$ would be complex analytic relatively automorphic functions for topologically but not analytically trivial factors of automorphy, and hence would also all be zero. The functions $g_i(z)=f_i(z)f_g(z)^{-1}$ for $1 \le i \le g-1$ are then well defined meromorphic functions on $M$ and satisfy

$$g_i(Tz)=\rho_{z_i}(\phi(T))^{-1}\alpha(T,z)^{-1}g_i(z)+\alpha(T,z)^{-1}s_i(T,z)$$

for any $T\in\Gamma$. Now let $\phi_j(z)$ for $1\le j\le 2g$ be a basis for the space of complex analytic relatively automorphic functions for the factor of automorphy $\kappa \rho . \chi$: and view each $\phi_j(z)$ as a complex analytic differential form of type (1, 0) on $\tilde{M}$ satisfying

$$\phi_j(Tz)=\rho_{z_i}(\phi(T))\,\alpha(T,z)\,\phi_j(z)$$

for any $T\in\Gamma$. The products $g_i(z)\phi_j(z)$ are consequently meromorphic differential forms on $\tilde{M}$ satisfying

$$g_i(Tz)\phi_j(Tz)=g_i(z)\phi_j(z)+\sigma_i(T,z)\phi_j(z)$$

for any $T\in\Gamma$. Since the differential forms $\sigma_i(T,z)\phi_j(z)$ are complex analytic the singularities of the meromorphic differential forms $g_i(z)\phi_j(z)$ are necessarily invariant under $\Gamma$; therefore these meromorphic differential forms have well defined $\Gamma$-invariant residues, which will be denoted by $\mathrm{res}_M(g_i\phi_j)$. The mapping

$$\phi_j \longrightarrow \mathrm{res}_M(g_i\phi_j)$$

is then the linear mapping associated to the equivalence class of the extension described by the mapping $\sigma_i$. This description of the extensions, and hence of the factor of automorphy $\chi_t$, is really not very useful though.

In a sense the reduction of the factor of automorphy $\chi_t$ to the form (60) is from the very beginning a somewhat unsatisfactory method of obtaining a useful characterization of the vector bundle it represents, since the subbundle $R^{-1}$ is far from being unique. Roughly speaking, analytic vector bundles over compact Riemann surfaces fall into two classes: the unstable bundles, those for which there are reasonably well determined subbundles which provide convenient means for characterizing the bundles, and the stable bundles, those for which there are no reasonably uniquely determined subbundles but for which there are other effective characterizations. The dichotomy is particularly clear for bundles of rank 2, [11]. That suggests a closer examination of the subbundles of the vector bundle represented by $\chi_t$; but such an examination will only be carried out here for subbundles of rank one.

It is clear that a scalar factor of automorphy $\xi$ represents a subbundle of $\chi_t$, written $\xi\subseteq\chi_t$, precisely when there is a complex analytic nowhere vanishing relatively automorphic function for the factor of automorphy $\xi^{-1}\otimes\chi_t$. More generally of course if there is a complex analytic relatively automorphic function $f$ for the factor of automorphy $\xi^{-1}\otimes\chi_t$ and if all the components of the function $f$ vanish at some points $p_i\in M$ then it is possible to write $f=g_1^{-\nu_1}\dots g_n^{-\nu_n}f_0$, where $g_i$ are complex analytic relatively automorphic functions for the scalar factors of automorphy $\zeta_{p_i}$, the function $g_i$ vanishes to first order at $p_i$ and is otherwise nonzero, and $f_0$ is a complex analytic nowhere vanishing relatively automorphic function for the factor of automorphy $\zeta_{p_1}^{-\nu_1}\dots\zeta_{p_n}^{-\nu_n}\xi^{-1}\otimes\chi_t$; and consequently

$$\zeta_{p_1}^{\nu_1}\dots\zeta_{p_n}^{\nu_n}\xi\subseteq\chi_t\,.$$

The characteristic classes of the factors of automorphy which represent subbundles of $\chi_t$ are bounded from above; indeed it is clear from the representation (60) that $c(\xi)\leq g$ whenever $\xi\subseteq\chi_t$, since $\gamma(\xi^{-1}\otimes\chi_t)=0$ if $c(\xi)>g$. The *divisor order* of the bundle $\chi_t$, written $\operatorname{div}\chi_t$, is defined to be the maximum of the characteristic classes of all the factors of automorphy which represent subbundles of $\chi_t$; thus the preceding observation can be restated as $\operatorname{div}\chi_t\leq g$. Somewhat more can be asserted, as follows.

**Corollary 4 to Theorem 14.** *If* $t+k\notin\mathscr{L}$ *then* $\operatorname{div}\chi_t\leq 1$. *Moreover if* $\xi\subseteq\chi_t$ *for a factor of automorphy* $\xi$ *with* $c(\xi)=1$ *then necessarily* $\xi=\rho_{k+t}\zeta$.

*Proof.* Any factor of automorphy $\xi$ with $c(\xi)=\nu$ is analytically equivalent to one of the form $\rho_x\zeta^\nu$ for some point $x\in J(M)$; and if $\xi\subseteq\chi_t$ then there is a nontrivial complex analytic relatively automorphic function $h$ for the factor of automorphy $\xi^{-1}\otimes\chi_t=\rho_x^{-1}\zeta^{-\nu}\otimes\chi_t$. As before the factor of automorphy $\chi_t$ can be written in the form (60) for any points $(z_1,\dots,z_{g-1})\notin Y_{g-1}\subset\tilde{M}^{g-1}$; and the components of the function $h$ then satisfy

$$h_g(Tz)=\rho_{k+t+t_1-x}(T)\zeta(T,z)^{g-\nu}h_g(z),$$
$$h_i(Tz)=\rho_x(T)^{-1}\zeta(T,z)^{-\nu}\left[\rho_{z_i}(\phi(T))^{-1}h_i(z)+s_i(T,z)h_g(z)\right],\quad 1\leq i\leq g-1,$$

for any $T\in\Gamma$. If $\nu\geq 1$ and $h_g(z)\equiv 0$ then necessarily $h_i(z)\equiv 0$ for all $i$, which is impossible since $h$ was assumed to be nontrivial; thus $h_g$ is a nontrivial complex analytic relatively automorphic function for the factor of automorphy $\rho_{k+t+t_1-x}\zeta^{g-\nu}$ if $\nu\geq 1$, so that $\gamma(\rho_{k+t+t_1-x}\zeta^{g-\nu})\geq 1$ or equivalently

$$k+t+t_1-x\in W_{g-\nu}\,.$$

However since the point $(z_1,\dots,z_{g-1})$ can be taken to be an arbitrary point of the dense open subset $\tilde{M}^{g-1}-Y_{g-1}\subseteq\tilde{M}^{g-1}$ this containment must hold for all points $t_1$ in a dense open subset of $W_{g-1}$; and consequently

$$k+t-x+W_{g-1}\subseteq W_{g-\nu}\,.$$

It follows immediately from this that $\nu\leq 1$, and that if $\nu=1$ then $x=k+t$, [12, p. 45]; and that suffices to conclude the proof.

If $t+k\in W_1-W_1\subseteq J(M)$ but $t+k\notin\mathscr{L}$, so that $t+k=\tilde{\phi}(z_2-z_1)$ for some points $z_1,z_2\in\tilde{M}$ which represent distinct points of $M$, then by Eq. (56) and Theorem 14a the function $\theta_t(z,z_1)$ is a complex analytic relatively automorphic function for the factor of automorphy $\phi^*\rho_{z_1}\otimes\chi_t$ and vanishes at the point $z_2$.

The function $\theta_t(z, z_1)$ must vanish to the first order at the point $z_2$, in the sense that in an open neighborhood of $z_2$ the function $\theta_t(z, z_1)$ can be written as the product of an analytic scalar function vanishing to the first order at $z_2$ and a nowhere vanishing analytic vector function, as a consequence of Corollary 4 to Theorem 14. Thus the function $\theta_t(z, z_1)$ leads to a nowhere vanishing complex analytic relatively automorphic function for the factor of automorphy $\zeta_{z_2}^{-1} \cdot \phi^* \rho_{z_1} \otimes \chi_t = \rho_{k+t}^{-1} \zeta^{-1} \otimes \chi_t$, so that in this case $\rho_{k+t}\zeta$ is indeed a subbundle of $\chi_t$ and consequently the bundle $\chi_t$ is not stable. In the special case of a Riemann surface $M$ of genus $g=2$ note that $W_1 - W_1 = J(M)$, since $W_1 - W_1$ is a two-dimensional complex analytic subvariety of the two-dimensional complex manifold $J(M)$. Thus $\rho_{k+t}\zeta$ is in this case a subbundle of $\chi_t$ whenever $k+t \notin \mathscr{L}$; and it is further clear that the factor of automorphy $\chi_t$ can be written in the canonical form

$$\begin{pmatrix} \rho_{k+t}\zeta & s \\ 0 & \zeta \end{pmatrix},$$

since $\det \chi_t = \rho_{k+t}\zeta^2$. This is a convenient canonical form, and is uniquely determined except of course that the cocycle $s$ can be replaced by any cohomologous cocycle to describe the equivalence class of this bundle extension. Indeed the cocycle $s$ can be replaced by one of the form $a(T)\zeta(T, z)$ for some constants $a(T)$; and that leads to the still more interesting canonical form

$$\zeta(T, z) \cdot \begin{pmatrix} \rho_{k+t}(T) & a(T) \\ 0 & 1 \end{pmatrix}$$

for the factor of automorphy $\chi_t$, as the product of the scalar factor of automorphy $\zeta$ and a rather explicitly given flat matrix factor of automorphy. For an arbitrary genus $g$ the factor of automorphy $\det \chi_t$ is analytically equivalent to $\rho_{k+t}\zeta^g$, as in Corollary 1 to Theorem 14, so the determinant of the factor of automorphy $\zeta^{-1}\chi_t$ has zero characteristic class; and since this factor of automorphy $\zeta^{-1}\chi_t$ is also indecomposable if $k+t \notin \mathscr{L}$, by Corollary 3 to Theorem 14, it follows from Weil's theorem that it is analytically equivalent to a flat factor of automorphy, as in [27] or [11]. Thus whenever $k+t \notin \mathscr{L}$ the factor of automorphy $\chi_t$ is analytically equivalent to one of the form $\zeta \cdot \Xi$ for some flat factor of automorphy $\Xi$ of rank $g$; in the general case though it seems rather hard to say much more about the representation $\Xi \in \mathrm{Hom}(\Gamma, GL(g, \mathbb{C}))$.

This is a convenient point at which to insert a discussion of the factor of automorphy $\chi_t$ in the exceptional case that $t+k \in \mathscr{L}$; this is actually a very easy case, in which the bundle $\chi_t$ can be described completely as follows.

**Corollary 5 to Theorem 14.** *If* $k+k \in \mathscr{L}$ *then the factor of automorphy* $\chi_t$ *is analytically equivalent to the diagonal factor of automorphy* $\zeta \oplus \cdots \oplus \zeta$, *where* $\zeta$ *is the factor of automorphy describing the point bundle* $\zeta_{p_0}$.

*Proof.* Again as a consequence of Eq. (56) and Theorem 14a, for any point $z_1 \in \tilde{M}$ the function $\theta_t(z, z_1)$ is a complex analytic relatively automorphic function for the factor of automorphy $\phi^*\rho_{z_1}\otimes\chi_t$ and vanishes at the point $z_1$; so by Corollary 4 to Theorem 14 this function necessarily vanishes to the first order at $z_1$, hence leads to a complex analytic nowhere vanishing relatively automorphic function $\theta_t^*(z, z_1)$ for the factor of automorphy $\zeta_{z_1}^{-1}\cdot\phi^*\rho_{z_1}\otimes\chi_t=\zeta^{-1}\otimes\chi_t$. It follows from Theorem 14c that there exist points $z_1, \dots, z_g$ such that the $g\times g$ matrix-valued function $(\theta_t(z, z_1), \dots, \theta_t(z, z_g))$ is not identically singular; and since $\theta_t^*(z, z_i)$ is a scalar multiple of $\theta_t(z, z_i)$ for each $i$ the $g\times g$ matrix-valued function $\Theta^*(z)=(\theta_t^*(z, z_1), \dots, \theta_t^*(z, z_g))$ is also not identically singular. The relative automorphy of each column of this matrix implies that the matrix itself satisfies $\Theta^*(Tz)\cdot\zeta(T, z)=\chi_t(T, z)\cdot\Theta^*(z)$ for each $T\in\Gamma$, where $\zeta(T, z)$ is a scalar factor of automorphy, or equivalently is viewed as a scalar factor of automorphy times the $g\times g$ identity matrix; hence $\Theta^*(z)$ can be considered as defining a bundle homomorphism

$$\Theta^*: \zeta\oplus\cdots\oplus\zeta \longrightarrow \chi_t\,.$$

Now the function $\theta^*(z)=\det\Theta^*(z)$ is a nontrivial complex analytic function such that

$$\theta^*(Tz)=\zeta(T, z)^{-g}\det\chi_t(T, z)\cdot\theta^*(z)$$

for each $T\in\Gamma$; and since it is evident from Corollary 1 to Theorem 14 that the factor of automorphy $\det\chi_t(T, z)$ is analytically equivalent to $\rho_{t+k}\zeta^g$, while $\rho_{t+k}\zeta^g$ is analytically equivalent to $\zeta^g$ for $t+k\in\mathscr{L}$, it follows necessarily that the function $\theta^*(z)$ is nonvanishing. The homomorphism $\Theta^*$ is consequently an isomorphism, and that suffices to conclude the proof.

Finally, supposing again that $t+k\notin\mathscr{L}$, it follows from Theorems 14b and 14c that there exists points $z_1, \dots, z_g$ such that the complex analytic $g\times g$ matrix-valued function $\Theta(z)=(\theta_t(z, z_1), \dots, \theta_t(z, z_g))$ has its first $g-1$ columns linearly independent and is not identically singular; and it follows from Eq. (56) that this matrix satisfies

$$\Theta(Tz)=\chi_t(T, z)\,\Theta(z)\,R(T)$$

for all $T\in\Gamma$, where $R(T)=\rho_{z_1}(\phi(T))\oplus\cdots\oplus\rho_{z_g}(\phi(T))$. Now for each point $z_i$ choose a simple path from $z_0$ to $z_i$ and let $p_i(z)$ be the local prime function associated to that path; thus $p_i(z)$ is a meromorphic function on $\tilde{M}$, has simple zeros at the points $\Gamma z_i$, simple poles at the points $\Gamma z_0$, and is otherwise regular, and satisfies $p_i(Tz)=\rho_{z_i}(\phi(T))\,p_i(z)$ for all $T\in\Gamma$. The $g\times g$ diagonal matrix $P(z)=p_1(z)\ominus\cdots\oplus p_g(z)$ is an invertible meromorphic matrix-valued function which satisfies

$$P(Tz)=R(T)\,P(z)=P(z)\,R(T)$$

for all $T\in\Gamma$; and the product $F(z)=\Theta(z)\,P(z)^{-1}$ is also an invertible matrix-valued function, and it satisfies

$$F(Tz)=\chi_t(T, z)\,F(z)$$

for all $T \in \Gamma$. This meromorphic function $F$ can then be used to describe the factor of automorphy $\chi_t$; and the divisor of this function $F$ can be used to describe the analytic equivalence class of the factor of automorphy $\chi_t$.

The divisor of a meromorphic matrix-valued function is quite analogous to the divisor of a meromorphic scalar-valued function. Two meromorphic matrix-valued functions $F_1, F_2$ defined in an open neighborhood of a point $p$ are called equivalent at $p$ if there exists a complex analytic mapping $F$ from an open neighborhood of $p$ into $GL(n, \mathbb{C})$ such that $F_1(z) = F(z) F_2(z)$ for all points $z$ in an open neighborhood of $p$; it is clear that this is indeed an equivalence relation. The divisor of a meromorphic matrix-valued function $F$ on $\tilde{M}$ is the mapping which associated to each point of $\tilde{M}$ the equivalence class at that point containing $F$. In the scalar case this amounts to the usual notion of the divisor of a function, since the equivalence class at a point is uniquely determined by the order at that point for any meromorphic function; in the matrix case the equivalence class is somewhat more complicated, but is in principle equally useful. For the function $F$ at hand it is evident from the functional equation that the divisor of $F$ is invariant under $\Gamma$, hence can be viewed as a divisor on $M$; and it is an easy matter to see that this divisor determines the analytic equivalence class of the factor of automorphy $\chi_t$. On the other hand the divisor of the function $F$ is relatively easy to describe; for the function $\Theta(z)$ is by construction equivalent at each point to a function of the form

$$(\delta_1, \delta_2, \ldots, \delta_{g-1}, f(z))$$

where $\delta_i$ are the columns of the $g \times g$ identity matrix, and the matrix $P(z)$ is quite simple. A more detailed investigation would lead too far afield at this point, so will not be pursued further. It will suffice to note in conclusion that the complete description of analytic equivalence classes of factors of automorphy or of vector bundles requires the introduction of the obvious equivalence relation among matrix divisors paralleling linear equivalence of scalar divisors.

# Chapter IV. Prym Differentials

## § 12. Prym Differentials and Generalized Theta Functions

A *meromorphic Prym differential* on the Riemann surface $M$ associated to a representation $\rho \in \mathrm{Hom}(\Gamma, \mathbb{C}^*)$ is a meromorphic differential form $\psi(z)$ on $\tilde{M}$ such that $\psi(Tz) = \rho(T)\psi(z)$ for all $T \in \Gamma$; the classical Prym differentials are just the complex analytic or everywhere regular meromorphic Prym differentials. It is of course clear that the divisor of a meromorphic Prym differential $\psi$ is invariant under the action of the group $\Gamma$; hence such a divisor can be viewed as a divisor on $M$, and will then be denoted by $\mathfrak{d}(\psi)$. For any fixed representation $\rho$ the space of associated meromorphic Prym differentials can be identified with the space of meromorphic relatively automorphic functions for the factor of automorphy $\rho\kappa$; similarly for any fixed divisor $\mathfrak{d} = \sum_{i=1}^{n} \nu_i \cdot p_i$ on $M$ the space of meromorphic Prym differentials $\psi$ on $M$ associated to $\rho$ and satisfying $\mathfrak{d}(\psi) + \mathfrak{d} \geq 0$ can be identified with the space of complex analytic relatively automorphic functions for the factor of automorphy $\rho\kappa\zeta_{p_1}^{\nu_1} \dots \zeta_{p_n}^{\nu_n}$, and is consequently finite dimensional. Indeed if the order $r = |\mathfrak{d}|$ of the divisor $\mathfrak{d}$ is strictly positive then by the Riemann-Roch theorem the dimension of this space of Prym differentials is equal to $r + g - 1$, and is hence independent of the particular divisor $\mathfrak{d}$ and representation $\rho$ chosen. It can then be asked whether there is a parametrization for the spaces of these Prym differentials analogous to the parametrization of the spaces of relatively automorphic functions by generalized theta functions; and it is clear that the generalized theta functions themselves should readily provide such a parametrization. This is particularly interesting for positive divisors $\mathfrak{d}$ of a fixed order $r \geq 1$, so only that case will be considered here.

First of all in order conveniently to describe the universal covering manifold of the set $\mathrm{Hom}(\Gamma, \mathbb{C}^*)$ of representations of the group $\Gamma$, which is the natural parameter space, introduce the homomorphism

$$\rho : \mathbb{C}^g \times \mathbb{C}^g \longrightarrow \mathrm{Hom}(\Gamma, \mathbb{C}^*),$$

which associates to a pair of vectors $(s, t) \in \mathbb{C}^g \times \mathbb{C}^g$ the representation $\rho_{s,t} \in \mathrm{Hom}(\Gamma, \mathbb{C}^*)$ for which

$$\rho_{s,t}(A_j) = \exp 2\pi i s_j,$$

$$\rho_{s,t}(B_j) = \exp 2\pi i (t_j + \textstyle\sum_{k=1}^{g} \omega_{jk} s_k);$$

and note that any representation of the group $\Gamma$ is of the form $\rho_{s,t}$ for some parameters $s, t$. Note further that $\rho_{0,t}=\rho_t$, and that the representations $\rho_{s,t}$ for a fixed parameter $t$ are precisely the representations which are analytically equivalent to $\rho_t$; for the complex analytic function

$$h_s(z)=\exp 2\pi i \sum_{k=1}^{g} s_k[w_k(z)-w_k(z_0)] \tag{61}$$

is nowhere vanishing on $\tilde{M}$ and satisfies $h_s(Tz)=\rho_{s,0}(T)h_s(z)$ for all $T\in\Gamma$. Next it is also convenient to identify the universal covering space $\tilde{M}$ of the Riemann surface $M$ with the unit disc, so that the canonical factor of automorphy $\kappa$ can be taken to be of the form $\kappa(T,z)=T'(z)^{-1}$. This factor of automorphy is analytically equivalent to one of the form $\rho_k\zeta^{2g-2}$; and upon selecting and henceforth holding fixed a particular point $k\in\mathbb{C}^g$ representing the canonical point of $J(M)$ the factor of automorphy $\zeta$ can be so chosen that $\rho_k(T)\zeta(T,z)^{2g-2}=T'(z)^{-1}$. Thus $f(z)$ is a meromorphic relatively automorphic function for the factor of automorphy $\rho_{s,t+k}\zeta^{2g-2}$ precisely when $\psi(z)=f(z)dz$ is a meromorphic Prym differential associated to the representation $\rho_{s,t}$. Finally, to construct a useful auxiliary function, select two fixed points $z', z''$ of $\tilde{M}$ which represent distinct points of $M$ and select nontrivial complex analytic relatively automorphic functions $h', h''$ for the respective factors of automorphy $\rho_{\tilde{\phi}(z'-z_0)}\zeta$, $\rho_{\tilde{\phi}(z''-z_0)}\zeta$; thus $\mathfrak{d}(h')=\Gamma z'$ and $\mathfrak{d}(h'')=\Gamma z''$. The function

$$q(z_1,z_2)=h'(z_2)h''(z_1)p(z_1,z',z_2,z''), \tag{62}$$

where $p(z_1,z',z_2,z'')$ is the prime function of the Riemann surface $M$, clearly is a complex analytic function on $\tilde{M}\times\tilde{M}$, has simple zeros at the points $z_1=Tz_2$ for all $T\in\Gamma$ but is otherwise nonvanishing, and satisfies

$$\begin{aligned} q(Tz_1,z_2)&=\rho_{z_2}(\phi(T))\zeta(T,z_1)q(z_1,z_2), \\ q(z_1,Tz_2)&=\rho_{z_1}(\phi(T))\zeta(T,z_2)q(z_1,z_2) \end{aligned} \tag{63}$$

for all $T\in\Gamma$. It is obvious from these properties that the function $q(z_1,z_2)$ is skew symmetric in the variables $z_1,z_2$, so $q(z_1,z_2)=-q(z_2,z_1)$. With this notation and these conventions the basic result about parametrized Prym differentials can be stated as follows.

**Theorem 15.** *There are $r+g-1$ meromorphic functions $f_i(z_1,\ldots,z_r,s,t;z)$ on $\tilde{M}^r\times\mathbb{C}^{2g}\times\tilde{M}$, having as singularities at most simple poles along the subvarieties $z=Tz_j$ for $1\le j\le r$ and for all $T\in\Gamma$, such that for any fixed point $(z_1,\ldots,z_r,s,t)\in\tilde{M}^r\times\mathbb{C}^{2g}$ the differential forms $\psi_i(z_1,\ldots,z_r,s,t;z)=f_i(z_1,\ldots,z_r,s,t;z)dz$ are a basis for the space of those Prym differentials $\psi$ associated to the representation $\rho_{s,t}$ and satisfying $\mathfrak{d}(\phi)+z_1+\cdots+z_r\ge 0$. Moreover these functions can be so* *chosen that the vector-valued function* $f(z_1,\dots,z_r,s,t;z)=(f_i(z_1,\dots,z_r,s,t;z))$ *satisfies*

(64) $$f(z_1,\dots,z_r,s,t;z)=h_s(z)f(z_1,\dots,z_r,0,t;z),$$

(65) $$\begin{aligned}f(z,\dots,z_r,s,t+\lambda;z)&=\rho_z(\lambda)\chi_{r+g-1}(\lambda,t+k+\tilde{\phi}(z_1+\cdots+z_r-rz_0))f(z_1,\dots,z_r,s,t;z)\\&=\chi_{r+g-1}(\lambda,t+k+\tilde{\phi}(z_1+\cdots+z_r-rz_0))f(z_1,\dots,z_r,s+n,t;z)\end{aligned}$$

*for all* $\lambda=\sum_{j=1}^{g}(m_j\omega_j+n_j\omega_{j+g})\in\mathscr{L}$, *and*

(66) $$\begin{aligned}&f(z_1,\dots,z_{j-1},Tz_j,z_{j+1},\dots,z_r,s,t;z)\\&=\zeta(T,z_j)^{-1}\chi_{r+g-1}(\phi(T),t+k+\tilde{\phi}(z_1+\cdots+z_r-rz_0))f(z_1,\dots,z_r,s,t;z)\end{aligned}$$

*for all* $T\in\Gamma$ *and* $1\leq j\leq g$; *and when so chosen this function* $f(z_1,\dots,z_r,s,t;z)$ *is unique up to a nonzero constant scalar factor.*

*Proof*. Letting $\theta(t,z)=(\theta_i(t,z))$ be the generalized theta function of rank $r+g-1$, it is a straightforward calculation to show that the functions

(67) $$\begin{aligned}&f_i(z_1,\dots,z_r,s,t;z)\\&=\theta_i(t+k+\tilde{\phi}(z_1+\cdots+z_r-rz_0),z)h_s(z)\prod_{j=1}^{r}q(z_j,z)^{-1}\end{aligned}$$

satisfy all the desired conditions. That these are indeed meromorphic functions having the required singularities is completely obvious from the definition. For any fixed point $(z_1,\dots,z_r,s,t)\in\tilde{M}^r\times\mathbb{C}^g$ it is an immediate consequence of the properties of the generalized theta function that

$$f_i(z_1,\dots,z_r,s,t;Tz)=\rho_{s,t+k}(T)\zeta(T,z)^{2g-2}f_i(z_1,\dots,z_r,s,t;z)$$

for all $T\in\Gamma$, so that

$$\psi_i(z_1,\dots,z_r,s,t;z)=f_i(z_1,\dots,z_r,s,t;z)dz$$

are linearly independent meromorphic Prym differentials associated to the representation $\rho_{s,t}$ and having the desired singularities; and (64), (65) and (66) follow equally readily from the definitions and the properties of the generalized theta function, noting for the second equality in (65) that $\rho_z(\lambda)=h_n(z)$ when $\lambda$ has the indicated form and hence that $\rho_z(\lambda)h_s(z)=h_{s+n}(z)$. On the other hand in order to demonstrate the uniqueness suppose that $f(z_1,\dots,z_r,s,t;z)=(f_i(z_1,\dots,z_r,s,t;z))$ is any meromorphic vector-valued function which satisfies all of the conditions of the theorem. Then for any fixed point $(z_1,\dots,z_r,z)\in\tilde{M}^{r+1}$ such that $z\notin\bigcup_{j=1}^{r}\Gamma z_j$ the function $f(z_1,\dots,z_r,0,t;z)$ is a complex analytic function of the variable $t$ which as a consequence of (65) is relatively automorphic for the factor of automorphy $\rho_z \otimes T_{-k-t_1}\chi_{r+g-1}$, where $t_1 = \tilde{\phi}(z_1 + \cdots + z_r - rz_0)$ and $T_{-k-t_1}\chi_{r+g-1}$ is the translate of the theta factor of automorphy of rank $r+g-1$; but the translate $T_{-k-t_1}\theta(t,z) = \theta(t+k+t_1, z)$ of the generalized theta function of rank $r+g-1$ is also relatively automorphic for this factor of automorphy, so it follows from Theorem 13 that

$$f(z_1, \ldots, z_r, 0, t; z) = c(z_1, \ldots, z_r; z)\theta(t+k+t_1, z) \tag{68}$$

for some scalar $c(z_1, \ldots, z_r; z)$ which is independent of $t$. If $r=1$ then $\theta(t+k+t_1, z)=0$ only when $t \in \mathscr{L}$, by Corollary 1 to Theorem 9, while if $r>1$ then $\theta(t+k+t_1, z) \neq 0$ for all $t \in \mathbb{C}^g$ by Corollary 2 to Theorem 9; then it follows from (68) that $c(z_1, \ldots, z_r; z)$ can be extended to a meromorphic function on $\tilde{M}^{r+1}$ having as singularities at most simple poles along the subvarieties $z = Tz_j$ for $1 \leq j \leq r$ and for all $T \in \Gamma$. Upon substituting (68) into (66) note that for any $T \in \Gamma$

$$\begin{aligned} c(z_1, \ldots, Tz_j, \ldots, z_r; z)\theta(t+k+t_1+\phi(T), z) \\ = \zeta(T, z_j)^{-1}\chi_{r+g-1}(\phi(T), t+k+t_1)c(z_1, \ldots, z_r; z)\theta(t+k+t_1, z); \end{aligned}$$

and since

$$\theta(t+k+t_1+\phi(T), z) = \rho_z(\phi(T))\chi_{r+g-1}(\phi(T), t+k+t_1)\theta(t+k+t_1, z)$$

it follows that

$$c(z_1, \ldots, Tz_j, \ldots, z_r; z) = \rho_z(\phi(T))^{-1}\zeta(T, z_j)^{-1}c(z_1, \ldots, z_j, \ldots, z_r; z).$$

The product

$$c(z_1, \ldots, z_r; z)\textstyle\prod_{j=1}^{r} q(z_j, z) = c(z) \tag{69}$$

is therefore a complex analytic function on $\tilde{M}^{r+1}$ which is invariant under the action of the group $\Gamma$ on each of the variables $z_1, \ldots, z_r$, and which is consequently independent of the variables $z_1, \ldots, z_r$. Finally recalling that for any $T \in \Gamma$

$$f(z_1, \ldots, z_r, 0, t; Tz) = \rho_{t+k}(T)\zeta(T, z)^{2g-2} f(z_1, \ldots, z_r, 0, t; z),$$

and substituting (68) and (69) into this equation it follows readily that $c(Tz) = c(z)$ for all $T \in \Gamma$, and hence that $c(z) = c$ is a scalar constant. Altogether then the function $f(z_1, \ldots, z_r, s, t, z)$ must be the scalar constant $c$ times the function having the components (67); and that establishes the uniqueness and concludes the proof of the theorem.

The differential forms

$$\psi_i(z_1, \ldots, z_r, s, t; z) = f_i(z_1, \ldots, z_r, s, t; z)dz$$

of the preceding theorem are given quite explicitly in terms of the generalized theta functions by Eq. (67), so can be viewed as known functions; they will be called the *canonical meromorphic Prym differentials having singularities of order r*, and are uniquely determined up to a common scalar factor by the choice of a normal form for the theta factor of automorphy of rank $r+g-1$ or for the generalized theta function of rank $r+g-1$. For many applications the characterization of these forms given by the preceding theorem is more useful than the explicit formula (67). However as an instance in which that is not the case note that formula (67) is completely symmetric in the variables $z_1, \ldots, z_r$; thus *the canonical Prym differentials* $\psi_i(z_1, \ldots, z_r, s, t; z)$ *are symmetric functions of the variables* $z_1, \ldots, z_r$, so the parameter $(z_1, \ldots, z_r)$ can be viewed either as a point of $\tilde{M}^r$ or as a point of the quotient space of $\tilde{M}^r$ under the action of the symmetric group on $r$ elements by permutation of the factors.

It is interesting to examine these differential forms in somewhat more detail at the identity representation $\rho_{0,0}$, at least for some values of the index $r$. For $r=1$ the differential forms $\psi_i(z_1, 0, 0; z)$ are $g$ linearly independent meromorphic Abelian differentials on $M$ having as singularities at most simple poles at the point $p_1 \in M$ represented by $z_1 \in \tilde{M}$. Since the total residue of any meromorphic Abelian differential on a compact Riemann surface is necessarily zero, however, the differential forms $\psi_i(z_1, 0, 0; z)$ must really be nonsingular; hence

$$\psi_i(z_1, 0, 0; z) = \textstyle\sum_{j=1}^{g} c_{ij}(z_1)\, \omega_j(z) \tag{70}$$

for a uniquely determined complex analytic mapping

$$C: \tilde{M} \longrightarrow GL(g, \mathbb{C}) .$$

It is tempting to try to reduce this mapping $C$ to the identity mapping by choosing a suitable normal form for the theta factor of automorphy of rank $g$; but unfortunately that cannot be done. Indeed it is clear that achieving such a reduction amounts to finding a complex analytic mapping

$$C^*: \mathbb{C}^g \longrightarrow GL(g, \mathbb{C})$$

such that $C(z_1) = C^*(\tilde{\phi}(z_1 - z_0))$. If there were such a mapping $C^*$ then necessarily $C(Tz_1) = C(z_1)$ for all $T \in \Gamma$ for which $\phi(T) = 0$, hence for all $T$ in the commutator subgroup $[\Gamma, \Gamma]$ of $\Gamma$; but since

$$C(Tz_1) = \zeta(T, z_1)^{-1} \chi_g(\phi(T), k + \tilde{\phi}(z_1 - z_0))\, C(z_1)$$

as a consequence of (66) and (70) it would then follow that $\zeta(T, z_1) = 1$ for all $T \in [\Gamma, \Gamma]$ and all $z_1 \in \tilde{M}$, which is a contradiction. Note that the differential forms $\psi_i(z_1, s, t; z)$ are also nonsingular for all $s \in \mathbb{C}^g$, $t \in \mathscr{L}$, since the representations $\rho_{s,t}$ are then analytically equivalent to the identity representation; but for other values of the parameters $s, t$ at least some of the differential forms must be singular, since $\gamma(\rho_{s,t}\kappa) = g-1$. For $r=2$ the differential forms $\psi_i(z_1, z_2, 0, 0; z)$ are $g+1$ linearly independent meromorphic Abelian differentials on $M$ having as singularities at most simple poles at the points $p_1, p_2 \in M$ represented by $z_1, z_2 \in \tilde{M}$ if $p_1 \neq p_2$ and at most double poles at the single point $p_1 = p_2$ otherwise. Thus if $p_1 \neq p_2$ then

(71) $$\psi_i(z_1, z_2, 0, 0; z) = c_i(z_1, z_2)\,\omega_\delta(z) + \sum_{j=1}^{g} c_{ij}(z_1, z_2)\,\omega_j(z),$$

where $c_i(z_1, z_2)$, $c_{ij}(z_1, z_2)$ for $1 \leq i \leq g+1$, $1 \leq j \leq g$ are complex analytic functions on the complement of the analytic subvariety

$$\{(z_1, z_2) \mid z_1 \in \Gamma z_2\} \subset \tilde{M}^2\,;$$

and $\omega_\delta(z)$ is the canonical Abelian differential of the third kind associated to any simple path $\delta$ from $z_2$ to $z_1$; and the $(g+1) \times (g+1)$ matrix composed of these functions $c_i(z_1, z_2)$, $c_{ij}(z_1, z_2)$ is nonsingular. It is only natural to ask what happens in this formula as the points $p_1, p_2$ approach coincidence. The coefficients $c_{ij}(z_1, z_2)$ are merely the periods of the differential forms $\psi_i(z_1, z_2; 0, 0; z)$ on the canonical generators $A_j$ of the group $\Gamma$, hence are evidently complex analytic functions on the entire complex manifold $\tilde{M}^2$; therefore the differential forms $c_i(z_1, z_2)\,\omega_\delta(z)$ viewed as functions of $z_1, z_2$ extend to complex analytic functions on the entire parameter manifold $\tilde{M}^2$ as well, and have a well defined limiting value as $z_2$ approaches $z_1$. Choosing a simple closed path $\gamma$ surrounding the arc $\delta$, note that for any function $f(z)$ defined and complex analytic in an open neighborhood of the path $\gamma$ and its interior

$$\begin{aligned}\int_\gamma f(z)\,\psi_i(z_1, z_2, 0, 0; z) &= \int_\gamma f(z)\left[\psi_i(z_1, z_2, 0, 0; z) - \textstyle\sum_j c_{ij}(z_1, z_2)\,\omega_j(z)\right] \\ &= \int_\gamma f(z)\,c_i(z_1, z_2)\,\omega_\delta(z) = 2\pi i\, c_i(z_1, z_2)\left[f(z_1) - f(z_2)\right];\end{aligned}$$

and if $\psi_i(z_1, z_1, 0, 0; z)$ has a Laurent expansion at the point $z_1$ with principal part $a_i(z_1)(z - z_1)^{-2}dz$ in terms of the standard coordinate $z$ in the unit disc $\tilde{M}$ then as $z_2$ approaches $z_1$

$$\begin{aligned}\lim_{z_2 \to z_1} \int_\gamma f(z)\,\psi_i(z_1, z_2, 0, 0; z) &= \int_\gamma f(z)\,\psi_i(z_1, z_1, 0, 0; z) \\ &= 2\pi i\, a_i(z_1)\, f'(z_1),\end{aligned}$$

so that

$$\lim_{z_2 \to z_1} c_i(z_1, z_2)\left[f(z_1) - f(z_2)\right] = a_i(z_1)\, f'(z_1)$$

or equivalently

$$\lim_{z_2 \to z_1} (z_1 - z_2)\, c_i(z_1, z_2) = a_i(z_1)\,.$$

Thus $c_i(z_1, z_2)$ extends to a meromorphic function with at most simple poles along the subvarieties $z_1 = T z_2$ for all $T \in \Gamma$. The canonical Abelian differential of the third kind $\omega_\delta(z)$ approaches zero as $p_2$ approaches $p_1$, as is evident either from the preceding discussion or from the expression of $\omega_\delta(z)$ in terms of the prime function.

It is also interesting to examine the residues of the canonical Prym differentials. For this purpose let

$$(72) \qquad R_j\psi_i(z_1, \ldots, z_r, s, t) = \mathrm{residue}_{z=z_j}\psi_i(z_1, \ldots, z_r, s, t; z)\,;$$

and let $R_j\psi(z_1, \ldots, z_r, s, t) = (R_j\psi_i(z_1, \ldots, z_r, s, t))$ be the column vector of length $r+g-1$ having the residues (72) as components. Note that since $q(z_1, z)$ has a simple zero at $z=z_1$ then in terms of the standard coordinate $z$ in the unit disc $\tilde{M}$

$$q(z_1) = \left.\frac{\partial q(z_1, z)}{\partial z}\right|_{z=z_1} = \lim_{z\to z_1}\frac{q(z_1, z)}{z-z_1}$$

is a nowhere vanishing complex analytic function in $\tilde{M}$. Using this observation and the explicit formula (67) for the canonical Prym differentials, *whenever* $z_j \notin \bigcup_{k\neq j}\Gamma z_k$ *the differential form* $\psi_i(z_1, \ldots, z_r, s, t; z)$ *has a simple pole at* $z=z_j$ *and*

$$(73) \qquad \begin{aligned} R_j\psi_i(z_1, \ldots, z_r, s, t) &= \lim_{z\to z_j}(z-z_j)\, f_i(z_1, \ldots, z_r, s, t; z) \\ &= \theta_i(t+k+t_1, z_j)\, h_s(z_j)\, q(z_j)^{-1} \textstyle\prod_{k\neq j} q(z_k, z_j)^{-1}, \end{aligned}$$

where $t_1 = \tilde{\phi}(z_1 + \cdots + z_r - rz_0)$ as usual. Thus the residue function $R_j\psi_i(z_1, \ldots, z_r, s, t)$ is a complex analytic function on the complement of the analytic subvariety

$$X_j = \{(z_1, \ldots, z_r, s, t) \,|\, z_j \in \textstyle\bigcup_{k\neq j}\Gamma z_k\} \subset \tilde{M}^r \times \mathbb{C}^{2g},$$

and extends to a meromorphic function on the entire manifold $\tilde{M}^r \times \mathbb{C}^{2g}$ having as singularities at most simple poles along the subvariety $X_j$. Of course the residue function $R_j\psi_i(z_1, \ldots, z_r, s, t)$ has a well defined finite value even at points at which $z_j = Tz_k$ for some $T\in\Gamma$ and $k\neq j$, but is not continuous at such points. To examine this exceptional behavior consider for example the residue $R_1\psi_i(z_1, z_1, z_3, \ldots, z_r, s, t)$ at a point at which $z_1 \notin \bigcup_{k=3}^{r}\Gamma z_k$. If $\gamma$ is a small circle around $z_1$ and contains none of the other points $\Gamma z_k$, $3\leq k\leq g$, in its interior then whenever $z_1, z_2$ are inequivalent points inside $\gamma$

$$\begin{aligned} &R_1\psi_i(z_1, z_2, \ldots, z_r, s, t) + R_2\psi_i(z_1, z_2, \ldots, z_r, s, t) \\ &\quad = \frac{1}{2\pi i}\int_{z\in\gamma}\psi_i(z_1, \ldots, z_r, s, t; z)\,. \end{aligned}$$

The integral is clearly a complex analytic function of the variables $z_1, z_2$ in $\gamma$ and remains analytic even when $z_2$ approaches $z_1$; thus the sum $R_1\psi_i + R_2\psi_i$ remains analytic when $z_2$ approaches $z_1$, and

$$\begin{aligned} &R_1\psi_i(z_1, z_1, z_3, \ldots, z_r, s, t) \\ &\quad = \lim_{z_2\to z_1}[R_1\psi_i(z_1, z_2, \ldots, z_r, s, t) + R_2\psi_i(z_1, z_2, \ldots, z_r, s, t)]\,. \end{aligned}$$

Substituting into this formula the expression (73), which is valid so long as $z_2 \neq z_1$, and recalling that $q(z_1, z_2) = -q(z_2, z_1)$ it follows that

$$\begin{aligned} R_1\psi_i(z_1, z_1, z_3, \ldots, z_r, s, t) &= \lim_{z_2 \to z_1} q(z_1, z_2)^{-1}[\theta_i(t+k+t_1, z_2)h_s(z_2)q(z_2)^{-1}\textstyle\prod_{k=3}^r q(z_k, z_2)^{-1} \\ &\quad -\theta_i(t+k+t_1, z_1)h_s(z_1)^{-1}q(z_1)^{-1}\textstyle\prod_{k=3}^r q(z_k, z_1)^{-1}] \\ &= q(z_1)^{-1}\frac{\partial}{\partial z}(\theta_i(t+k+t_1, z)h_s(z)q(z)^{-1}\textstyle\prod_{k=3}^r q(z_k, z)^{-1})\big|_{z=z_1}. \end{aligned}$$

Note that this last expression also extends meromorphically to the entire parameter manifold, but fails to represent the residue if there are further coincidences among the points of $M$ represented by $z_1, z_3, \ldots, z_r$. Similar formulas can readily be obtained for any other pattern of coincidences among these points, although the limits leading to the appropriate derivatives involve symmetric expressions of a rather more general form than the usual difference quotient. The most interesting cases of the preceding formulas however are

$$(74) \qquad R_1\psi_i(z_1, s, t) = \theta_i(t+k+\tilde{\phi}(z_1 - z_0), z_1)h_s(z_1)q(z_1)^{-1}$$

and

$$(75) \qquad R_1\psi_i(z_1, z_1, s, t) = q(z_1)^{-1}\frac{\partial}{\partial z}\ \theta_i(t+k+2\tilde{\phi}(z_1 - z_0), z)h_s(z)q(z)^{-1}\big|_{z=z_1};$$

these formulas are valid for all points $z_1, s, t$ and express the residues $R_1\psi_i(z_1, s, t)$ and $R_1\psi_i(z_1, z_1, s, t)$ as complex analytic functions on $\tilde{M} \times \mathbb{C}^{2g}$. In the second case it is also easy to determine the entire principal part of the double pole of the Prym differential $\psi_i(z_1, z_1, s, t; z)$; for using the explicit expression (67) again

$$(76) \qquad \lim_{z \to z_1}(z - z_1)^2 f_i(z_1, z_1, s, t; z) = \theta_i(t+k+2\tilde{\phi}(z_1 - z_0), z_1)h_s(z_1)q(z_1)^{-2}.$$

Although the auxiliary function $q(z)$ in the preceding discussion was defined rather explicitly in terms of the prime function of the Riemann surface $M$ and can thus be considered as a known function, it is perhaps interesting as an incidental observation to derive an alternative characterization of this function. By definition $q(z) = q_2(z, z)$ where $q_2(z_1, z_2) = dq(z_1, z_2)/dz_2$. Now using (63) note that

$$\begin{aligned} q_2(Tz, z_2) &= \frac{d}{dz_2}[\rho_{z_2}(\phi(T))\zeta(T, z)q(z, z_2)] \\ &= d\rho_{z_1}(\phi(T))/dz_2 \cdot \zeta(T, z)q(z, z_2) + \rho_{z_2}(\phi(T))\zeta(T, z)q_2(z, z_2); \end{aligned}$$

and setting $z_2 = Tz$ and recalling that $q(z, Tz) = 0$ it follows that

$$(77) \qquad q(Tz) = q_2(Tz, Tz) = \rho_{Tz}(\phi(T))\zeta(T, z)q_2(z, Tz).$$

On the other hand though, again using (63), note that

$$\begin{aligned} q_2(z_1, Tz)\, T'(z) &= \frac{d}{dz} q_2(z_1, Tz) \\ &= \frac{d}{dz}\left[\rho_{z_1}(\phi(T))\zeta(T, z) q(z_1, z)\right] \\ &= \rho_{z_1}(\phi(T))\zeta'(T, z) q(z_1, z) + \rho_{z_1}(\phi(T))\zeta(T, z) q_2(z_1, z)\,; \end{aligned}$$

and setting $z_1 = z$ it follows that

$$q_2(z, Tz) = T'(z)^{-1}\, \rho_z(\phi(T))\zeta(T, z) q(z)\,. \tag{78}$$

Combining Eqs. (77) and (78) and recalling that $T'(z)^{-1} = \rho_k(T)\zeta(T, z)^{2g-2}$ it then follows that

$$q(Tz) = \mu(T, z)\zeta(T, z)^{2g} q(z) \tag{79}$$

for all $T \in \Gamma$, where

$$\mu(T, z) = \rho_{k+\phi(T)+2\tilde{\phi}(z-z_0)}(T)\,;$$

hence $q(z)$ is a complex analytic relatively automorphic function for the factor of automorphy $\mu\zeta^{2g}$. Observe in particular that

$$\begin{aligned} &\mu(A_i, z) = 1\,, \\ &\mu(B_i, z) = \exp 2\pi i\left[2w_i(z) - 2w_i(z_0) + k_i + \omega_{i,i+g}\right]\,; \end{aligned}$$

so referring to Eq. (119) of the Appendix it is apparent that $\mu = \xi_\tau^{-2}$ where $\xi_\tau$ is the Riemannian theta factor of automorphy associated to the parameter $\tau = -\frac{1}{2}k - \varepsilon$, with $\varepsilon \in \mathbb{C}^g$ being the vector having components $\varepsilon_i = \frac{1}{2}\omega_{i,i+g}$. Since $\xi_\tau$ is analytically equivalent to $\rho_{\tau+r}\zeta^g$ where $r \in \mathbb{C}^g$ is the Riemann vector with components given in Eq. (120) of the Appendix, and since $2(r - \varepsilon) = k$ by Eq. (123) of the Appendix, it follows that the factor of automorphy $\mu$ is analytically equivalent to $\zeta^{-2g}$. Thus (79) implies that $q(z)$ is relatively automorphic for an analytically trivial factor of automorphy, and that equation therefore determines the function $q(z)$ uniquely up to a constant factor. Of course since $q(z)$ is complex analytic and nowhere vanishing that already shows that the factor of automorphy $\mu\zeta^{2g}$ is analytically trivial; but the alternative proof is interesting in that the Riemannian theta factors of automorphy appear so naturally.

The preceding formulas for the residues of the canonical Prym differentials can be used in turn to derive some relations between these differentials. For example recall first from Theorem 11 that there is an exact sequence of complex analytic vector bundles over $J(M)$ of the form

$$0 \longrightarrow \rho_{z_2}^{-1} \xrightarrow{\theta(z_2)} \chi_{g+1} \xrightarrow{C} T_{t_2}\chi_g \longrightarrow 0$$

for any fixed point $z_2 \in \tilde{M}$, where $t_2 = \tilde{\phi}(z_2 - z_0) \in \tilde{M}$; hence equivalently there is a complex analytic mapping

$$C: \mathbb{C}^g \longrightarrow \mathbb{C}^{g \times (g+1)}$$

such that $C(t)$ is a $g \times (g+1)$ matrix of rank $g$ for every $t \in \mathbb{C}^g$, that $C(t) \cdot \theta(t, z_2) = 0$ for all $t \in \mathbb{C}^g$ where $\theta(t, z)$ is the generalized theta function of rank $g+1$, and that

$$C(t+\lambda) \cdot \chi_{g+1}(\lambda, t) = \chi_g(\lambda, t - t_2) \cdot C(t)$$

for all $\lambda \in \mathscr{L}$. If $z_1 \in \tilde{M}$ is another fixed point and if $z_1 \notin \Gamma z_2$ it follows from these observations and (73) that

$$C(t+k+t_1+t_2) \cdot R_2 \psi(z_1, z_2, s, t) = 0,$$

where $t_j = \tilde{\phi}(z_j - z_0)$ for $j = 1, 2$, hence that

$$\text{residue}_{z=z_2} C(t+k+t_1+t_2) \cdot \psi(z_1, z_2, s, t; z) = 0;$$

the components of the vector $C(t+k+t_1+t_2) \cdot \psi(z_1, z_2, s, t; z)$ are therefore $g$ linearly independent meromorphic Prym differentials associated to the representation $\rho_{s,t}$ which have as singularities at most simple poles at the points $\Gamma z_1$, and consequently

$$C(t+k+t_1+t_2) \cdot \psi(z_1, z_2, s, t; z) = X(z_1, z_2, s, t) \cdot \psi(z_1, s, t; z)$$

for a uniquely determined matrix $X(z_1, z_2, s, t) \in GL(g, \mathbb{C})$. The matrix $X(z_1, z_2, s, t)$ is clearly a complex analytic function of the variables $s, t$; and it follows readily from the properties of the canonical Prym differentials listed in Theorem 15 that the matrix $X(z_1, z_2, s, t)$ is actually a scalar and independent of the parameters $s, t$. Indeed since $\psi(z_1, z_2, s, t; z)$ and $\psi(z_1, s, t; z)$ both involve the parameter $s$ only in the common factor $h_s(z)$ it is obvious that $X(z_1, z_2, s, t)$ is independent of $s$. Using (65) it is a simple calculation to verify that

$$X(z_1, z_2, t+\lambda) \chi_g(\lambda, t+k+t_1) = \chi_g(\lambda, t+k+t_1) X(z_1, z_2, t)$$

for all $\lambda \in \mathscr{L}$; thus $X(z_1, z_2, t)$ is an endomorphism of the factor of automorphy $\chi_g$, so by Corollary 2 to Theorem 13 the matrix $X(z_1, z_2, t)$ is a scalar and is independent of $t$. Thus altogether

$$C(t+k+t_1+t_2) \cdot \psi(z_1, z_2, s, t; z) = x(z_1, z_2) \cdot \psi(z_1, s, t; z)$$

for a nonzero scalar factor $x(z_1, z_2)$.

There are of course similar results for the canonical meromorphic Prym differentials with higher order singularities; but more interesting than that, there are also somewhat similar results involving the everywhere regular Prym differentials and leading to canonical forms for these differentials. Recall that

Theorem 11 can be extended to the vector bundles which are the analogues of generalized theta factors of automorphy of ranks less than $g$, leading to exact sequences of complex analytic vector bundles of the form (41); thus for example over the complement $X_0 \subset J(M)$ of the point $k \in J(M)$ there is an exact sequence of complex analytic vector bundles of the form

$$0 \longrightarrow \rho_{z_1}^{-1} | X_0 \xrightarrow{T_{-t_1}\theta_g(z_1)} T_{-t_1}\chi_g | X_0 \longrightarrow \chi_{g-1} \longrightarrow 0$$

for any fixed point $z_1 \in \tilde{M}$, where $t_1 = \tilde{\phi}(z_1 - z_0) \in \tilde{M}$, hence equivalently over the complement $X \subset J(M)$ of the point $W_0 \subset J(M)$ represented by $0 \in \mathbb{C}^g$ there is an exact sequence of complex analytic vector bundles of the form

$$0 \longrightarrow \rho_{z_1}^{-1} | X \xrightarrow{T_{-t_1-k}\theta_g(z_1)} T_{-t_1-k}\chi_g | X \xrightarrow{C} T_{-k}\chi_{g-1} \longrightarrow 0 . \tag{80}$$

Let $\{U_\alpha\}$ be a covering of the complement of the subset $\mathscr{L} \subset \mathbb{C}^g$ by open sets projecting homeomorphically into $\mathbb{C}^g/\mathscr{L}$ such that the vector bundles in the exact sequence (80) can be described by coordinate bundles in terms of this covering when it is viewed as a covering of $X \subset J(M)$. Of course for the bundles $\rho_{z_1}^{-1} | X$ and $T_{-t_1-k}\chi_g | X$ the coordinate transformations are the identity mappings in $U_\alpha \cap U_\beta$ and the appropriate factors of automorphy in $U_\alpha \cap (U_\beta + \lambda)$ for any $\lambda \in \mathscr{L}$; and for the bundle $\chi_{g-1}$ the coordinate transformations are some complex analytic mappings

$$\chi_{g-1}^{\alpha\beta} : (U_\alpha/\mathscr{L}) \cap (U_\beta/\mathscr{L}) \longrightarrow GL(g+1, \mathbb{C}) .$$

The bundle homomorphism $C$ in the exact sequence (80) is then described by some complex analytic mappings

$$C^\alpha : U_\alpha \longrightarrow \mathbb{C}^{(g-1)\times g}$$

such that $C^\alpha(t)$ is a $(g-1)\times g$ matrix of rank $g-1$ for every $t \in U_\alpha$ and that

$$C^\alpha(t) \cdot \theta_g(t + t_1 + k, z_1) = 0$$

for all $t \in U_\alpha$, where $\theta_g(t, z)$ is the generalized theta function of rank $g$. Then recalling (74) it follows that $C^\alpha(t) \cdot R_1 \psi(z_1, s, t) = 0$ for all $t \in U_\alpha$, hence that

$$\text{residue}_{z=z_1} C^\alpha(t) \cdot \psi(z_1, s, t; z) = 0$$

for all $t \in U_\alpha$; the components of the vector

$$\psi^\alpha(s, t; z) = C^\alpha(t) \cdot \psi(z_1, s, t; z)$$

are therefore $g-1$ linearly independent complex analytic Prym differentials associated to the representation $\rho_{s,t}$ for any $s \in \mathbb{C}^g$ and $t \in U_\alpha$. These components

$$\psi_i^\alpha(s, t; z) = f_i^\alpha(s, t; z)\, dz$$

will be called the *canonical analytic Prym differentials*. The functions $f_i^\alpha(s,t;z)$ are complex analytic in $\mathbb{C}^g \times U_\alpha \times \tilde{M}$, when $U_\alpha$ is viewed as a subset of $J(M)$. It follows from (64) that

$$(81) \qquad f^\alpha(s,t;z) = h_s(z) f^\alpha(0,t;z);$$

and since $f(z_1,s,t;z)$ describes a cross-section of the vector bundle $\rho_z \otimes T_{-t_1-k}\chi_g$ by (65) it follows from the sequence (80) that $\{f^\alpha(s,t;z)\}$ describes a cross-section of the vector bundle $\rho_z \otimes T_{-k}\chi_{g-1}$, hence that

$$(82) \qquad f^\alpha(s,t;z) = \rho_z^{\alpha\beta}(t) \otimes T_{-k}\chi_{g-1}^{\alpha\beta}(t) \cdot f^\beta(s,t;z)$$

whenever $t/\mathscr{L} \in (U_\alpha/\mathscr{L}) \cap (U_\beta/\mathscr{L})$. There remains the question of the extent to which these canonical Prym differentials depend on the choice of the auxiliary point $z_1$ used in this construction. As demonstrated at the end of § 9 the bundle $\chi_{g-1}$ is independent of the choice of the point $z_1$ in the exact sequence (80); hence the coordinate transformations $\chi_{g-1}^{\alpha\beta}$ can be assumed to be fixed and given. It is an immediate consequence of Corollary 1 to Theorem 13 that the functions $f^\alpha(s,t;z)$ are determined uniquely up to a scalar factor $x(z)$ by properties (81) and (82); and the Prym differential condition then ensures that the function $x(z)$ is necessarily independent of $z$. Thus the canonical Prym differentials are determined uniquely up to a scalar constant factor, and to that extent are independent of the choice of the point $z_1$ in this construction.

## § 13. Periods and the Period Matrix for Prym Differentials

The canonical meromorphic Prym differentials having specified singularities can be expressed quite explicitly in terms of the generalized theta functions as in Eq. (67), hence are really just the functions considered before but in another guise. However the introduction of the periods of these differentials leads to a quite different and considerably more extensive class of functions, but a class of functions still similar in many ways to the generalized theta functions. The *period* of a meromorphic Prym differential $\psi_i(z_1,\ldots,z_r,s,t;z)$ along a path $\delta \subset \tilde{M}$ containing none of the points $\bigcup_j \Gamma z_j$ is by definition the integral

$$\psi_i(z_1,\ldots,z_r,s,t;\delta) = \int_{z\in\delta} \psi_i(z_1,\ldots,z_r,s,t;z).$$

It is evident that this period is a complex analytic function on $(\tilde{M} - \Gamma\delta)^r \times \mathbb{C}^{2g}$; and the set

$$\psi(z_1,\ldots,z_r,s,t;\delta) = (\psi_i(z_1,\ldots,z_r,s,t;\delta))$$

of these periods can thus be viewed as a complex analytic mapping

$$\psi(\delta): (\tilde{M} - \Gamma\delta)^r \times \mathbb{C}^{2g} \longrightarrow \mathbb{C}^{r+g-1}.$$

As functions of the variables $s, t$ it is obvious from (65) that the Prym periods satisfy

$$(83)\qquad \begin{aligned}&\psi(z_1,\dots,z_r,s-n,t+\lambda;\delta)\\&\quad=\chi_{r+g-1}(\lambda,t+k+\tilde{\phi}(z_1+\cdots+z_r-rz_0))\psi(z_1,\dots,z_r,s,t;\delta)\end{aligned}$$

for all $\lambda=\sum_{j=1}^{g}(m_j\omega_j+n_j\omega_{j+g})\in\mathscr{L}$; thus letting the lattice $\mathscr{L}$ act on the space $\mathbb{C}^{2g}$ by setting $\lambda\cdot(s,t)=(s-n,t+\lambda)$ where $\lambda\in\mathscr{L}$ is as above, Eq. (83) can be interpreted as the assertion that the Prym periods are complex analytic relatively automorphic functions for this group action, with respect to the factor of automorphy induced by the appropriate translate of the theta factor of automorphy of rank $r+g-1$. The variables $s$ and $t$ can be separated further for the Prym differentials than for the Prym periods; indeed it follows from (64) and the first equality in (65) that the Prym differentials satisfy the rather more extensive functional equations characterizing the appropriate generalized theta functions, but these additional relations do not carry over upon integration. Such a situation is of course only to be expected since there are obviously vastly more Prym period functions than Prym differentials, corresponding to all the various choices of paths of integration. However upon using (64) and recalling the explicit form (61) for the auxiliary function $h_s(z)$ it does follow immediately that

$$(84)\qquad \frac{\partial}{\partial s_j}\psi(z_1,\dots,z_r,s,t;\delta)=2\pi i\int_{z\in\delta}[w_j(z)-w_j(z_0)]\,\psi(z_1,\dots,z_r,s,t;z)\,.$$

Quadratic integral expressions of this sort are a particularly interesting but little explored additional set of invariants. Furthermore, again using (64) and (61), for any fixed $z_j, t$ note that as a function of $s$ alone

$$(85)\qquad \begin{aligned}&|\psi_i(z_1,\dots,z_r,s,t;\delta)|\\&\quad=\Big|\int_{z\in\delta}\psi_i(z_1,\dots,z_r,0,t;z)\exp 2\pi i\textstyle\sum_k s_k[w_k(z)-w_k(z_0)]\Big|\le Me^{m|s|}\end{aligned}$$

for some positive constants $M, m$, where $|s|^2=\sum_k|s_k|^2$; thus any Prym period $\psi_i(z_1,\dots,z_r,s,t;\delta)$ is an entire function of order 1 in each variable $s_j$ separately.

As functions of the variables $z_j$ the Prym periods have a rather more complicated analytic behavior. If $\delta$ is a simple path from $z_-$ to $z_+$ in $\tilde{M}$ and contains no $\Gamma$-equivalent points then it is possible to choose a simple closed path $\gamma$ surrounding $\delta$ and also containing no $\Gamma$-equivalent points either on itself or in its interior. If none of the points $\Gamma z_j$ are contained within or on $\gamma$ then the differential form $\psi(z_1,\dots,z_r,s,t;z)$ is a closed complex analytic form inside $\gamma$ and can hence be written as $\psi(z_1,\dots,z_r,s,t;z)=dh(z)$ for some complex analytic function $h(z)$ inside $\gamma$; and then in addition

$$\begin{aligned}\psi(z_1,\dots,z_r,s,t;\delta)&=h(z_+)-h(z_-)\\&=\frac{1}{2\pi i}\int_{z\in\gamma}h(z)\omega_\delta(z)\\&=-\frac{1}{2\pi i}\int_{z\in\gamma}w_\delta(z)\psi(z_1,\dots,z_r,s,t;z)\,.\end{aligned}$$

If the point $z_1$ is near the path $\gamma$ then $\gamma$ can be deformed to a closed path $\gamma_1$ containing $z_1$ but none of the other points of $\bigcup_j \Gamma z_j$ in its interior by adding to $\gamma$ a small closed loop $\gamma'_1$ containing $z_1$ in its interior; and then

$$\begin{aligned}\frac{1}{2\pi i}&\int_{z\in\gamma_1} w_\delta(z)\psi(z_1,\ldots,z_r,s,t;z)\\&=\frac{1}{2\pi i}\int_{z\in\gamma+\gamma'_1} w_\delta(z)\psi(z_1,\ldots,z_r,s,t;z)\\&=-\psi(z_1,\ldots,z_r,s,t;\delta)+w_\delta(z_1)\cdot R_1\psi(z_1,\ldots,z_r,s,t)\,.\end{aligned}$$

Now since $z_1$ is in the interior of $\gamma_1$ the first integral above clearly remains a complex analytic function of $z_1$ even as $z_1$ approaches the path $\delta$; and consequently as $z_1$ approaches $\delta$ the singularity of the Prym period $\psi(z_1,\ldots,z_r,s,t;\delta)$ is the same as the singularity of the function $w_\delta(z_1)\cdot R_1\psi(z_1,\ldots,z_r,s,t)$. The canonical Abelian integral $w_\delta(z_1)$ has an analytic continuation across the interior points of the path $\delta$ but has logarithmic branch points at the end points $z_-$ and $z_+$; and the residue function $R_1\psi(z_1,\ldots,z_r,s,t)$ remains complex analytic since none of the points $\Gamma z_2,\ldots,\Gamma z_r$ are contained inside $\gamma_1$. Thus as a function of each variable $z_j$ separately the Prym period $\psi(z_1,\ldots,z_r,s,t;\delta)$ can be continued analytically across the interior points of the path $\delta$ but has logarithmic branch points at the end points $z_-$ and $z_+$. The behavior at the translates $T\delta$ is similar, since it follows readily from (66) that

$$\begin{aligned}(86)\qquad &\psi(z_1,\ldots,z_{j-1},Tz_j,z_{j+1},\ldots,z_r,s,t;\delta)\\&=\zeta(T,z_j)^{-1}\chi_{r+g-1}(\phi(T),t+k+\tilde{\phi}(z_1+\cdots+z_r-rz_0))\psi(z_1,\ldots,z_r,s,t;\delta)\end{aligned}$$

for all $T\in\Gamma$. A similar analysis of the behavior of the Prym periods as functions of the variables $z_j$ can be carried out in the rather more interesting case that $\delta$ is a path from $z_0$ to $Sz_0$ for some $S\in\Gamma$ and represents a simple closed curve on the surface $M$. Alternatively though the path $\delta$ can be written as a sum $\delta=\sum_j\delta_j$ of segments each of which contains no $\Gamma$-equivalent points, and the Prym period can be written correspondingly as a sum

$$\psi(z_1,\ldots,z_r,s,t;\delta)=\sum_j\psi(z_1,\ldots,z_r,s,t;\delta_j)$$

of periods of the sort already considered; and since all the logarithmic branch points cancel except perhaps those at the points $\Gamma z_0$ it follows again that as a function of each variable $z_j$ separately the Prym period $\psi(z_1,\ldots,z_r,s,t;\delta)$ can be continued analytically across the interior points of the paths $\Gamma\delta$ but has as singularities at most logarithmic branch points at the points $\Gamma z_0$. To examine in somewhat more detail these possible singularities let $\delta_1$ and $\delta_n$ be the first and last segments of the path $\delta$, so that $\delta_1$ begins at $z_0$, $\delta_n$ ends at $Sz_0$, and $S^{-1}\delta_n$ ends at $z_0$. For fixed points $z_2,\ldots,z_r$ inside $\tilde{\Delta}$ the period $\psi(z_1,\ldots,z_r,s,t;\delta_1)$ viewed as a function of $z_1$ in an open neighborhood of $z_0$ differs from $w_{\delta_1}(z_1)R_1\Psi(z_1,\ldots,z_r,s,t)$ by a complex analytic function of $z_1$ in that neighborhood, as noted above; and similarly of course the period $\psi(z_1, \ldots, z_r, s, t; \delta_n) = \rho_{s,t}(S)\psi(z_1, \ldots, z_r, s, t; S^{-1}\delta_n)$ differs from $\rho_{s,t}(S) w_{S^{-1}\delta_n}(z_1) R_1 \psi(z_1, \ldots, z_r, s, t)$ by a complex analytic function of $z_1$ in that neighborhood. Since the periods over the remaining segments $\delta_j$ are all complex analytic functions of $z_1$ near $z_0$ it follows that the period $\psi(z_1, \ldots, z_r, s, t; \delta)$ differs from

$$[w_{\delta_1}(z_1) + \rho_{s,t}(S) w_{S^{-1}\delta_n}(z_1)] R_1 \psi(z_1, \ldots, z_r, s, t)$$

by a complex analytic function of $z_1$ in an open neighborhood of $z_0$; hence if $\rho_{s,t}(S) = 1$ the period $\psi(z_1, \ldots, z_r, s, t; \delta)$ continues analytically across $\delta$ even at $z_0$ but has a jump discontinuity along that path, while otherwise the period has a nontrivial logarithmic branch point at $z_0$.

The most interesting of the Prym periods are those associated to the canonical liftings $\tilde{\alpha}_1, \ldots, \tilde{\alpha}_g, \tilde{\beta}_1, \ldots, \tilde{\beta}_g$ to $\tilde{M}$ of the paths of the marking for the Riemann surface $M$; they are called the *basic Prym periods* for that marked surface, and can be grouped into the $(r+g-1) \times 2g$ *Prym period matrix*

$$\Psi(z_1, \ldots, z_r, s, t) = (\psi(z_1, \ldots, z_r, s, t; \tilde{\alpha}_1), \ldots, \psi(z_1, \ldots, z_r, s, t; \tilde{\beta}_g)) .$$

This is of course only defined when the points $z_i$ are disjoint from all of the paths $\Gamma\tilde{\alpha}_j$ and $\Gamma\tilde{\beta}_j$; so in particular it is well defined when all of the points $z_i$ are contained in the canonical fundamental polygon $\tilde{\Delta}$.

As an illustrative example consider the basic Prym period $\psi(z_1, 0, t; \tilde{\alpha}_j)$ associated to the path $\tilde{\alpha}_j$. When viewed as a function of $z_1$ for any fixed $t$ this period is analytic in $\tilde{M} - \Gamma\tilde{\alpha}_j$; and since $\rho_{0,t}(A_j) = \rho_t(A_j) = 1$ this function has an analytic continuation across all of $\Gamma\tilde{\alpha}_j$. Let $\psi^0(z_1, 0, t; \tilde{\alpha}_j)$ denote the maximal analytic continuation of $\psi(z_1, 0, t; \tilde{\alpha}_j)$ from the canonical fundamental polygon $\tilde{\Delta}$; thus $\psi^0(z_1, 0, t; \tilde{\alpha}_j) = \psi(z_1, 0, t; \tilde{\alpha}_j)$ whenever $z_1 \in \tilde{\Delta}$, and is otherwise obtained by analytic continuation.

**Theorem 16 a.** *The analytic continuations $\psi^0(z_1, 0, t; \tilde{\alpha}_j)$ of the Prym periods $\psi(z_1, 0, t; \tilde{\alpha}_j)$ as functions of $z_1 \in \tilde{\Delta}$ are complex analytic functions on $\tilde{M} \times \mathbb{C}^g$; and for any $T \in \Gamma$*

$$\psi^0(Tz_1, 0, t; \tilde{\alpha}_j) = \zeta(T, z_1)^{-1} \chi_g(\phi(T), t + k + \tilde{\phi}(z_1 - z_0)) \psi^0(z_1, 0, t; \tilde{\alpha}_j) + \sigma_j(T; z_1, t)$$

*for some complex analytic functions $\sigma_j(T; z_1, t)$ on $\tilde{M} \times \mathbb{C}^g$.*

*Proof.* The period $\psi(z_1, 0, t; \tilde{\alpha}_j)$ is a complex analytic function of $z_1$ in the connected component of $\tilde{M} - \Gamma\tilde{\alpha}_j$ which contains $\tilde{\Delta}$, and coincides with the function $\psi^0(z_1, 0, t; \tilde{\alpha}_j)$ there; consequently the latter function has an analytic continuation across each boundary arc of the connected component of $\tilde{M} - \Gamma\tilde{\alpha}_j$ which contains $\tilde{\Delta}$, hence to an open neighborhood of the closure of $\tilde{\Delta}$ in $\tilde{M}$. If it is demonstrated that the function $\psi^0(z_1, 0, t; \tilde{\alpha}_j)$ satisfies functional equations of the desired form for all points $z_1$ in the closure of $\tilde{\Delta}$ and all transformations $T \in \Gamma$ such that $\tilde{\Delta}$ and $T\tilde{\Delta}$ have a common boundary arc, then since such transformations generate $\Gamma$ and the transforms of $\tilde{\Delta}$ by $\Gamma$ cover $\tilde{M}$ it follows readily that

$\psi^0(z_1, 0, t; \tilde{\alpha}_j)$ has an analytic continuation to all of $\tilde{M}$ and satisfies the desired functional equations. The boundary of $\tilde{\Delta}$ consists of the segments $C_1 \ldots C_{k-1}\tilde{\alpha}_k$, $-C_1 \ldots C_k B_k \tilde{\alpha}_k$, $C_1 \ldots C_{k-1} A_k \tilde{\beta}_k$, $-C_1 \ldots C_k \tilde{\beta}_k$ for $k=1, \ldots, g$, as noted in § 4; and

$$C_1 \ldots C_k B_k \tilde{\alpha}_k = S_k(C_1 \ldots C_{k-1}\tilde{\alpha}_k) \quad \text{and} \quad C_1 \ldots C_k \tilde{\beta}_k = T_k(C_1 \ldots C_{k-1} A_k \tilde{\beta}_k)$$

where $S_k$, $T_k$ are the elements of $\Gamma$ given by

$$S_k = C_1 \ldots C_k B_k (C_1 \ldots C_{k-1})^{-1} \quad \text{and} \quad T_k = C_1 \ldots C_k (C_1 \ldots C_{k-1} A_k)^{-1}.$$

The elements $S_k$, $T_k$ are also generators of $\Gamma$; and the transforms $S_k\tilde{\Delta}$ for $k \neq j$ and $T_k\tilde{\Delta}$ for all $k$ are contained in the same connected component of $\tilde{M} - \Gamma\tilde{\alpha}_j$ as $\tilde{\Delta}$. It therefore follows that the function $\psi^0(z_1, 0, t; \tilde{\alpha}_j)$ satisfies Eq. (86) if $T = S_k$ for $k \neq j$ or $T = T_k$ for any $k$; and that is an equation of the desired form with $\sigma_j(T; z_1, t) = 0$. To consider the behavior of this function under the remaining transformation $S_j$ consider a point $z_1 \in \tilde{\Delta}$ near the arc $C_1 \ldots C_{j-1}\tilde{\alpha}_j$, so that $S_j z_1$ will be near the arc $C_1 \ldots C_j B_j \tilde{\alpha}_j$, outside $\tilde{\Delta}$ but within the domain to which the function $\psi^0(z_1, 0, t; \tilde{\alpha}_j)$ has been analytically continued; and let $\tilde{z}_1$ be a point inside $\mathring{\tilde{\Delta}}$ near $S_j z_1$, and $\sigma$ be a simple closed curve enclosing $\tilde{z}_1$ as in Fig. 4. Then

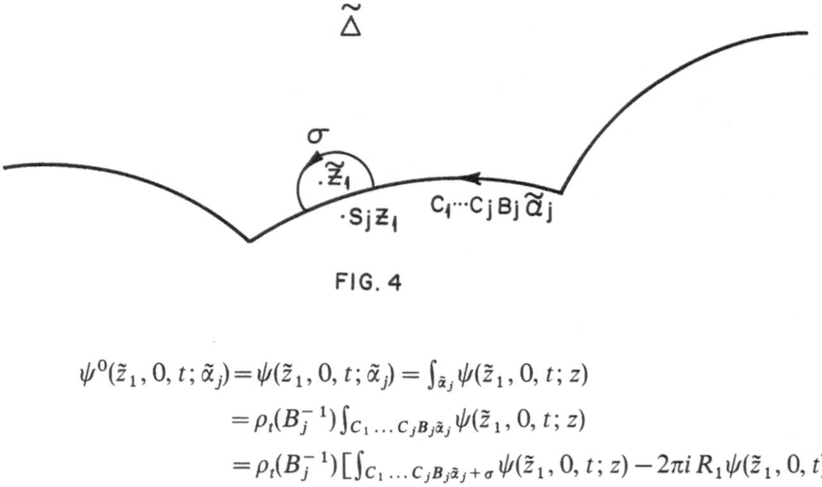


FIG. 4

$$\begin{aligned}\psi^0(\tilde{z}_1, 0, t; \tilde{\alpha}_j) = \psi(\tilde{z}_1, 0, t; \tilde{\alpha}_j) &= \textstyle\int_{\tilde{\alpha}_j} \psi(\tilde{z}_1, 0, t; z) \\ &= \rho_t(B_j^{-1}) \textstyle\int_{C_1 \ldots C_j B_j \tilde{\alpha}_j} \psi(\tilde{z}_1, 0, t; z) \\ &= \rho_t(B_j^{-1}) \left[\textstyle\int_{C_1 \ldots C_j B_j \tilde{\alpha}_j + \sigma} \psi(\tilde{z}_1, 0, t; z) - 2\pi i R_1 \psi(\tilde{z}_1, 0, t)\right].\end{aligned}$$

The common parts of the paths $C_1 \ldots C_j B_j \tilde{\alpha}_j$ and $\sigma$ can be cancelled in the last integral above; and letting $\tilde{z}_1$ approach $S_j z_1$ avoiding the remainder of the path of integration that integral approaches

$$\begin{aligned}&\rho_t(B_j^{-1}) \textstyle\int_{C_1 \ldots C_j B_j \tilde{\alpha}_j + \sigma} \psi(S_j z_1, 0, t; z) \\ &= \rho_t(B_j^{-1}) \textstyle\int_{C_1 \ldots C_j B_j \tilde{\alpha}_j} \psi(S_j z_1, 0, t; z) \\ &= \textstyle\int_{\tilde{\alpha}_j} \psi(S_j z_1, 0, t; z) = \psi(S_j z_1, 0, t; \tilde{\alpha}_j),\end{aligned}$$

which by (86) is in turn equal to

$$\zeta(S_j, z_1)^{-1}\chi_g(\phi(S_j), t+k+\tilde{\phi}(z_1-z_0))\psi(z_1, 0, t; \tilde{\alpha}_j)$$
$$=\zeta(S_j, z_1)^{-1}\chi_g(\phi(S_j), t+k+\tilde{\phi}(z_1-z_0))\psi^0(z_1, 0, t; \tilde{\alpha}_j).$$

Since $\psi^0(\tilde{z}_1, 0, t; \tilde{\alpha}_j)$ approaches $\psi^0(S_j z_1, 0, t; \tilde{\alpha}_j)$ as $\tilde{z}_1$ approaches $S_j z_1$ it follows that

$$\psi^0(S_j z_1, 0, t; \tilde{\alpha}_j)=\zeta(S_j, z_1)^{-1}\chi_g(\phi(S_j), t+k+\tilde{\phi}(z_1-z_0))\psi^0(z_1, 0, t; \tilde{\alpha}_j)$$
$$-2\pi i\,\rho_t(B_j^{-1})R_1\psi(S_j z_1, 0, t);$$

that is a transformation of the desired form, with $\sigma_j(S_j, z_1, t) = -2\pi i\,\rho_t(B_j^{-1})R_1\psi(S_j z_1, 0, t)$, and that then suffices to conclude the proof of the theorem.

In the course of the proof of the preceding theorem it was also demonstrated that

$$\sigma_j(S_k; z_1, t)= -2\pi i\,\delta_k^j\rho_t(B_j^{-1})R_1\psi(S_j z_1, 0, t),$$
$$\sigma_j(T_k; z_1, t)=0$$

for $k=1, \dots, g$; since the elements $S_k$, $T_k$ generate $\Gamma$ and since it is easily seen that

$$\sigma_j(ST; z_1, t)=\sigma_j(S; Tz_1, t)+\zeta(S, Tz_1)^{-1}\chi_g(\phi(S), t+k+\tilde{\phi}(Tz_1-z_0))\sigma_j(T, z_1)$$

for any $S, T\in\Gamma$ it follows that the functions $\sigma_j(T; z_1, t)$ can be expressed in terms of known functions for all $T\in\Gamma$. Now *the functional equation of Theorem* 16a *actually completely determines the function* $\psi^0(z_1, 0, t; \tilde{\alpha}_j)$ *and hence the Prym period* $\psi(z_1, 0, t; \tilde{\alpha}_j)$; so that that period can be viewed as a given well determined, if perhaps not well known, function of the variables $z_1$ and $t$. Indeed for any fixed $t\in\mathbb{C}^g$ the function $\psi^0(z_1, 0, t; \tilde{\alpha}_j)$ is clearly determined uniquely up to a relatively automorphic function for the induced factor of automorphy $\zeta(T, z_1)^{-1}\chi_{-t-k}(T, z_1)$ of rank $g$ for the action of the group $\Gamma$ on $\tilde{M}$. The product of any such relatively automorphic function by a nontrivial relatively automorphic function for the scalar factor of automorphy $\zeta(T, z_1)$ will be a relatively automorphic function $f(z_1)$ for the factor of automorphy $\chi_{-t-k}(T, z_1)$, with the property that $f(z_0)=0$. If $t\notin\mathscr{L}$ the space of complex analytic relatively automorphic functions for that factor of automorphy $\chi_{-t-k}(T, z_1)$ is one-dimensional by Corollary 2 to Theorem 14, and consequently $f(z_1)=c\theta_{-t-k}(z_1, z_0)= c\theta(\tilde{\phi}(z_1-z_0)+t+k, z_0)$ for some constant $c$; but since $\theta(t+k, z_0)\neq 0$ when $t\notin\mathscr{L}$ it is necessary that $c=0$, hence $f(z_1)$ and the original relatively automorphic function are both trivial. That demonstrates the desired uniqueness when $t\notin\mathscr{L}$, and the remainder follows by continuity.

The analysis of the periods $\psi(z_1, 0, t; \tilde{\beta}_j)$ or more generally of the periods $\psi(z_1, s, t; \tilde{\alpha}_j)$ or $\psi(z_1, s, t; \tilde{\beta}_j)$ is somewhat more complicated, since in addition to the jump discontinuities along the paths of integration there are logarithmic singularities at the points $\Gamma z_0$. These singularities can be eliminated by considering suitable linear combinations of the periods, and such linear combinations can be analyzed and determined much as above; but it seems somewhat pointless to carry that analysis out much further just as present. Instead some further properties of the Prym periods will be examined, as follows.

**Theorem 16b.** *If the points $z_j$ are contained inside the canonical fundamental polygon $\tilde{\Delta}\subset\tilde{M}$ then the total residues of the canonical meromorphic Prym differentials in $\tilde{\Delta}$ are given by*

$$\text{residue}_{z\in\tilde{\Delta}}\psi_i(z_1,\ldots,z_r,s,t;z)$$
$$=\frac{1}{2\pi i}\sum_{j=1}^{g}[(1-\rho_{s,t}(B_j))\psi_i(z_1,\ldots,z_r,s,t;\tilde{\alpha}_j)-(1-\rho_{s,t}(A_j))\psi_i(z_1,\ldots,z_r,s,t;\tilde{\beta}_j)].$$

*Proof.* If the singularities $z_j$ are contained in $\tilde{\Delta}$ then by the residue theorem

$$\text{residue}_{z\in\tilde{\Delta}}\psi_i(z_1,\ldots,z_r,s,t;z)=\frac{1}{2\pi i}\int_{z\in\partial\tilde{\Delta}}\psi_i(z_1,\ldots,z_r,s,t;z);$$

and recalling from § 4 the explicit form of the boundary of $\tilde{\Delta}$ and using the transformational properties of the Prym differentials it follows that

$$\begin{aligned}&\int_{\partial\tilde{\Delta}}\psi_i(z_1,\ldots,z_r,s,t;z)\\&\quad=\sum_{j=1}^{g}\Big[\int_{\tilde{\alpha}_j}(1-\rho_{s,t}(B_j))\psi_i(z_1,\ldots,z_r,s,t;z)\\&\qquad+\int_{\tilde{\beta}_j}(\rho_{s,t}(A_j)-1)\psi_i(z_1,\ldots,z_r,s,t;z)\Big],\end{aligned}$$

which leads immediately to the desired result and therewith concludes the proof of the theorem.

**Theorem 16c.** *For any point $z_1\in\tilde{\Delta}$ and any index $r\geq 1$ the rank of the Prym period matrix $\Psi(rz_1,s,t)$ is independent of s and*

$$W_{r-2}^{r+g-d}-(r-1)\tilde{\phi}(z_1-z_0)=\{t\,\mathbb{C}^g\,|\,\text{rank}\,\Psi(rz_1,s,t)<d\}/\mathscr{L}.$$

*Proof.* If $h(z)$ is a meromorphic function on $\tilde{M}$ such that $h(Tz)=\rho_{s,t}(T)h(z)$ for all $T\in\Gamma$ and $\mathfrak{d}(h)+(r-1)\cdot\Gamma z_1-\Gamma z_0\geq 0$ then $\psi(z)=dh(z)$ is a meromorphic Prym differential associated to the representation $\rho_{s,t}$, has a divisor such that $\mathfrak{d}(\psi)+r\cdot\Gamma z_1\geq 0$, and has periods

$$\psi(\tilde{\alpha}_j)=h(A_jz_0)-h(z_0)=0,\qquad \psi(\tilde{\beta}_j)=h(B_jz_0)-h(z_0)=0.$$

Conversely if $\psi(z)$ is a meromorphic Prym differential associated to the representation $\rho_{s,t}$, has a divisor such that $\mathfrak{d}(\psi)+r\cdot\Gamma z_1\geq 0$, and has zero periods along the paths $\tilde{\alpha}_j,\tilde{\beta}_j$ then it follows from Theorem 16b that

$$\text{residue}_{z=z_1}\psi(z)=\text{residue}_{z\in\tilde{\Delta}}\psi(z)=0;$$

therefore $\psi(z)=dh(z)$ for some meromorphic function $h$ on $\tilde{M}$, and that function is uniquely determined if it is also required that $h(z_0)=0$. It is clear that $h(Tz)=\rho_{s,t}(T)h(z)+\psi(T)$ for some constant $\psi(T)$ whenever $T\in\Gamma$; and since $\psi(A_j)=h(A_j z_0)=\int_{z\in\tilde{\alpha}_j}\psi(z)=0$ and $\psi(B_j)=h(B_j z_0)=\int_{z\in\tilde{\beta}}\psi(z)=0$ it follows directly that $h(Tz)=\rho_{s,t}(T)h(z)$ for all $T\in\Gamma$. The divisor of this function $h$ is then $\Gamma$-invariant, so $\mathfrak{d}(h)+(r-1)\cdot\Gamma z_1-\Gamma z_0\geq 0$. Thus differentiation establishes an isomorphism between the spaces of such functions $h$ and such differentials $\psi$. Setting $t_1=(r-1)\tilde{\phi}(z_1-z_0)$ note that the dimension of the space of such functions $h$ is equal to $\gamma(\rho_{s,t}\zeta_{z_1}^{r-1}\zeta^{-1})=\gamma(\rho_{t+t_1}\zeta^{r-2})$, hence is independent of $s$. On the other hand note that these differentials $\psi$ are exactly those meromorphic Prym differentials which can be written in the form $\psi(z)=\sum_{i=1}^{r+g-1}c_i\psi_i(rz_1,s,t;z)$ for some vector $c=(c_i)\in\mathbb{C}^{r+g-1}$ such that ${}^t c\cdot\Psi(rz_1,s,t)=0$; hence the dimension of the space of such differentials is equal to $r+g-1-\operatorname{rank}\Psi(rz_1,s,t)$. Consequently

$$\gamma(\rho_{t+t_1}\zeta^{r-2})=r+g-1-\operatorname{rank}\Psi(rz_1,s,t),$$

so that $\operatorname{rank}\Psi(rz_1,s,t)<d$ precisely when $\gamma(\rho_{t+t_1}\zeta^{r-2})\geq r+g-d$, hence when $t+t_1\in W_{r-2}^{r+g-d}$; and that suffices to conclude the proof of the theorem.

The reason for restricting attention to singular divisors of the form $rz_1$ is that only in that case does the vanishing of the total residue of a meromorphic Prym differential imply that the residue at each singularity is zero, hence that the differential can be written as the exterior derivative of a meromorphic function; otherwise it is necessary to consider functions with logarithmic singularities rather than meromorphic functions, and such relatively automorphic functions have not been considered here. There are several special cases of this theorem which are worth noting particularly.

**Corollary 1 to Theorem 16.** *For any point $z_1\in\tilde{\Delta}$ and for all points $(s,t)\in\mathbb{C}^{2g}$ the $g\times 2g$ Prym period matrix $\Psi(z_1,s,t)$ is of rank $g$.*

*Proof.* Setting $r=1$ and $d=g$, the left hand side of the equality in Theorem 16c is the empty set, hence so is the right hand side; and that is the desired result.

**Corollary 2 to Theorem 16.** *For any point $z_1\in\tilde{\Delta}$ and any index $r\geq 2$*

$$W_{r-2}-(r-1)\tilde{\phi}(z_1-z_0)=\{t\in\mathbb{C}^g\,|\,\operatorname{rank}\psi(rz_1,s,t)<r+g-1\}/\mathscr{L}$$

*Proof.* This is just the statement of Theorem 16c for the special case $d=r+g-1$.

The Prym period matrix $\Psi(rz_1,s,t)$ is an $(r+g-1)\times 2g$ matrix; and for $2\leq r\leq g+1$ the assertion of Corollary 2 to Theorem 16 is that for any fixed points $z_1\in\tilde{\Delta}$ and $s\in\mathbb{C}^g$ the matrix $\Psi(rz_1,s,t)$ has less than maximal rank precisely when $t$ represents a point in the appropriate translate of the subvariety $W_{r-2}\subseteq J(M)$. Thus the Prym period matrices can be used to describe the canonical subvarieties of the Jacobi variety $J(M)$ in a manner quite parallel to the use of the generalized theta functions to describe these subvarieties as in Theorem 9 and its corollaries, in spite of the fact that these matrices are functions of an additional variable $s$ and are not necessarily relatively automorphic functions of the variable $t$ alone for the natural action of the lattice subgroup $\mathscr{L}$ on $\mathbb{C}^g$. It might be expected that this observation could be used to obtain some further properties of the Prym period matrices; and an easy example to illustrate that that is indeed the case is the following.

**Corollary 3 to Theorem 16.** *For any fixed points $z_1 \in \tilde{\Delta}$ and $s \in \mathbb{C}^g$ the determinant of the $2g \times 2g$ Prym period matrix $\Psi((g+1)z_1, s, t)$ as a function of $t$ generates the proper ideal of the analytic subvariety $W_{g-1} - g\tilde{\phi}(z_1 - z_0)$ at each point of $J(M)$. Furthermore*

$$\det \Psi((g+1)z_1, s, t) = \det \Psi((g+1), s, t) \cdot \exp 2\pi i {}^t s \cdot (\tilde{\phi}(z_1 - z_0) + k + \lambda_0)$$

*for some fixed element $\lambda_0 \in \mathscr{L}$, hence that determinant is a relatively automorphic function of $t$ under the action of the lattice subgroup $\mathscr{L}$ on $\mathbb{C}^g$.*

*Proof.* It follows from Corollary 2 to Theorem 16 for the special case $r = g+1$ that

$$W_{g-1} - g\tilde{\phi}(z_1 - z_0) = \{t \in \mathbb{C}^g \mid \det \Psi((g+1)z_1, s, t) = 0\}/\mathscr{L}$$

for any fixed points $z_1 \in \tilde{\Delta}$, $s \in \mathbb{C}^g$. It was proved above that $W_{g-1}$ is an irreducible analytic subvariety of $J(M)$; and the same argument shows that the full inverse image of $W_{g-1}$ under the covering mapping $\mathbb{C}^g \longrightarrow J(M)$ is an irreducible analytic subvariety of $\mathbb{C}^g$, which subvariety will also be denoted by $W_{g-1}$. Indeed the image of the Jacobi mapping $\phi_{g-1}: M^{g-1} \longrightarrow J(M)$ is precisely the subvariety $W_{g-1} \subset J(M)$, so that every point $t \in W_{g-1} \subset \mathbb{C}^g$ can be written as $t = \tilde{\phi}(z_1 + \cdots + z_{g-1} - (g-1)z_0) + \lambda$ for some points $z_i \in \tilde{M}$ and some element $\lambda \in \mathscr{L}$; but since $\lambda = \phi(T)$ for some element $T \in \Gamma$ it follows that $t = \tilde{\phi}(Tz_1 + z_2 + \cdots + z_{g-1} - (g-1)z_0)$, hence the image of the Jacobi mapping $\tilde{\phi}_{g-1}: \tilde{M}^{g-1} \longrightarrow \mathbb{C}^g$ is precisely the subvariety $W_{g-1}$ and that subvariety is therefore irreducible. The function $\det \Psi((g+1)z_1, s, t)$ as a function of $t$ must then vanish to the same order $\nu$ at all points of the subvariety $W_{g-1} - g\tilde{\phi}(z_1 - z_0) \subset \mathbb{C}^g$, and $\nu$ is of course also independent of the choices of the points $z_1 \in \tilde{\Delta}$ and $s \in \mathbb{C}^g$; and since the Abelian theta function $\theta_0(t)$ generates the proper ideal of the subvariety $W_{g-1}$ at each point of $J(M)$, hence at each point of $\mathbb{C}^g$, it follows that

(87) $$\det \Psi((g+1)z_1, s, t) = \theta_0(t + g\tilde{\phi}(z_1 - z_0))^\nu \exp 2\pi i f(z_1, s, t)$$

for some complex analytic function $f(z_1, s, t)$ on $\tilde{\Delta} \times \mathbb{C}^g \times \mathbb{C}^g$. As noted before each component of the matrix $\Psi((g+1)z_1, s, t)$ when viewed as a function of $s$ alone is an entire function of order 1, and the same is of course true for the function $\det \Psi((g+1)z_1, s, t)$ and hence for the function $\exp 2\pi i f(z_1, s, t)$; but since this last function is nonvanishing it follows from the Hadamard factorization theorem that $f(z_1, s, t)$ must be a linear function of $s$, [25]. Thus write

(88) $$f(z_1, s, t) = f_0(z_1, t) + {}^t s \cdot f_1(z_1, t)$$

for some complex analytic mappings

$$f_0 : \tilde{\Delta} \times \mathbb{C}^g \longrightarrow \mathbb{C} \text{ and } f_1 : \tilde{\Delta} \times \mathbb{C}^g \longrightarrow \mathbb{C}^g.$$

Next recalling (83) note that

$$\Psi((g+1)z_1, s-n, t+\lambda) = \chi_{2g}(\lambda, t+k+(g+1)\tilde{\phi}(z_1-z_0)) \cdot \Psi((g+1)z_1, s, t)$$

for all $\lambda = \sum_{j=1}^{g} (m_j \omega_j + n_j \omega_{j+g}) \in \mathcal{L}$. From Theorem 11 and the discussion subsequent thereto it follows that the factors of automorphy $\det \chi_{2g}(\lambda, t)$ and $\det \chi_g(\lambda, t)$ are analytically equivalent and from Theorem 12 it follows that the factor of automorphy $\det \chi_g(\lambda, t)$ is analytically equivalent to the Abelian theta factor of automorphy $\xi(\lambda, t)$, so that

$$\det \chi_{2g}(\lambda, t) = \xi(\lambda, t) \exp 2\pi i [h(t+\lambda) - h(t)]$$

for some entire function $h(t)$; therefore

(89)

$$\det \Psi((g+1)z_1, s-n, t+\lambda) \exp -2\pi i h(t+\lambda+k+(g+1)\tilde{\phi}(z_1-z_0))$$
$$= \xi(\lambda, t+k+(g+1)\tilde{\phi}(z_1-z_0)) \det \Psi((g+1)z_1, s, t) \exp -2\pi i h(t+k+(g+1)\tilde{\phi}(z_1-z_0))$$

for all $\lambda \in \mathcal{L}$. Since $\theta_0(t)$ is relatively automorphic for the factor of automorphy $\xi(\lambda, t)$ and since for any $t_1 \in \mathbb{C}^g$

$$\xi(\lambda, t+t_1) = \rho_{t_1}(\lambda)^{-1} \xi(\lambda, t),$$

where $\rho_{t_1}(\omega_j) = 1$, $\rho_{t_1}(\omega_{j+g}) = \exp 2\pi i t_{1,j}$ as a consequence of the definition (46), then upon combining (87) and (89) it follows that

$$\exp 2\pi i [f(z_1, s-n, t+\lambda) - h(t+\lambda+k+(g+1)\tilde{\phi}(z_1-z_0))]$$
$$= \rho_{k+\tilde{\phi}(z_1-z_0)}(\lambda)^{-1} \xi(\lambda, t+g\tilde{\phi}(z_1-z_0))^{1-\nu}$$
$$\cdot \exp 2\pi i [f(z_1, s, t) - h(t+k+(g+1)\tilde{\phi}(z_1-z_0))]$$

for all $\lambda \in \mathcal{L}$. Upon recalling the more explicit form (88) for the function $f(z_1, s, t)$ the preceding equality can be rewritten

$$\exp 2\pi i [{}^t s \cdot f_1(z_1, t+\lambda) - {}^t n \cdot f_1(z_1, t+\lambda) + f_0(z_1, t+\lambda) - h(t+\lambda+k+(g+1)\tilde{\phi}(z_1-z_0))]$$
$$(90) \quad = \rho_{k+\tilde{\phi}(z_1-z_0)}(\lambda)^{-1} \xi(\lambda, t+g\tilde{\phi}(z_1-z_0))^{1-\nu}$$
$$\cdot \exp 2\pi i [{}^t s \cdot f_1(z_1, t) + f_0(z_1, t) - h(t+k+(g+1)\tilde{\phi}(z_1-z_0))]$$

for all $\lambda \in \mathcal{L}$. Dividing thus last equality by the same one but for the special value $s=0$ note that

$$\exp 2\pi i\, {}^t s \cdot f_1(z_1, t+\lambda) = \exp 2\pi i\, {}^t s \cdot f_1(z_1, t)$$

hence ${}^t s\cdot[f_1(z_1,t+\lambda)-f_1(z_1,t)]\in\mathbb{Z}$ for all $\lambda\in\mathcal{L}$; but that can only be the case when $f_1(z_1,t+\lambda)=f_1(z_1,t)$ for all $\lambda\in\mathcal{L}$, hence when $f_1(z_1,t)$ is independent of $t$. Now if $f_1(z_1,t)=f_1(z_1)$ then Eq. (90) implies that the factor of automorphy

$$\rho_{k+\tilde{\phi}(z_1-z_0)}(\lambda)^{-1}\,\xi(\lambda,t+g\tilde{\phi}(z_1-z_0))^{1-\nu}\exp 2\pi i\,{}^t n\cdot f_1(z_1)$$

is analytically trivial, where $\lambda=\sum_{j=1}^{g}(m_j\omega_j+n_j\omega_{j+g})$; and as demonstrated in the appendix that can only be the case when $\nu=1$ and

$$\rho_{k+\tilde{\phi}(z_1-z_0)}(\lambda)^{-1}\exp 2\pi i\,{}^t n\cdot f_1(z_1)=\exp 2\pi i\,{}^t\lambda\cdot a(z_1)$$

for some complex analytic mapping $a:\tilde{\Delta}\longrightarrow\mathbb{C}^g$. In particular taking $\lambda=\omega_j$ note that $a_j(z_1)\in\mathbb{Z}$, hence that $a(z_1)$ is a constant integral vector; and taking $\lambda=\omega_{j+g}$ note that $f_1(z_1)=\tilde{\phi}(z_1-z_0)+k+\lambda_0$ for some fixed element $\lambda_0\in\mathcal{L}$. Thus

$$\begin{aligned}&\det\Psi((g+1)z_1,s,t)\\&\quad=\theta_0(t+g\tilde{\phi}(z_1-z_0))\exp 2\pi i[f_0(z_1,t)+{}^t s\cdot(\tilde{\phi}(z_1-z_0)+k+\lambda_0)].\end{aligned}$$

That then suffices to conclude the proof of the theorem, although upon considering Eq. (90) again it is easy to see that in addition $f_0(z_1,t)={}^t a\cdot t+h(t+k+(g+1))\tilde{\phi}(z_1-z_0))+g(z_1)$ for some complex analytic function $g$ in $\tilde{\Delta}$.

To continue this string of corollaries next consider indices $r>g+1$, noting then that since $2g<r+g-1$ the $(r+g-1)\times 2g$ Prym period matrix $\Psi(rz_1,s,t)$ always has rank less than $r+g-1$ and Corollary 2 to Theorem 16 is therefore trivially true in this case; however that can be modified as follows, to obtain a nontrivial result.

**Corollary 4 to Theorem 16.** *For any point* $z_1\in\tilde{\Delta}$ *and any index* $r$ *in the range* $g+2\le r\le 2g$

$$k+(r-1)\tilde{\phi}(z_1-z_0)-W_{2g-r}=\{t\in\mathbb{C}^g\,|\,\operatorname{rank}\Psi(rz_1,s,t)<2g\}/\mathcal{L}.$$

*Proof.* Setting $d=2g$ in Theorem 16c it follows that

$$W_{r-2}^{r-g}-(r-1)\tilde{\phi}(z_1-z_0)=\{t\in\mathbb{C}^g\,|\,\operatorname{rank}\Psi(rz_1,s,t)<2g\}/\mathcal{L}.$$

Recalling the definition (16) note that $t\in\mathbb{C}^g$ represents a point of $W_{r-2}^{r-g}$ precisely when $\gamma(\rho_t\zeta^{r-2})\ge r-g$. Now from the Riemann-Roch theorem

$$\gamma(\rho_t\zeta^{r-2})=\gamma(\kappa\rho_t^{-1}\zeta^{2-r})+r-1-g=\gamma(\rho_{k-t}\zeta^{2g-r})+r-1-g;$$

hence $\gamma(\rho_t\zeta^{r-2})\ge r-g$ precisely when $\gamma(\rho_{k-t}\zeta^{2g-r})\ge 1$, which is just the condition that $k-t$ represents a point of $W_{2g-r}$. Thus $W_{r-2}^{r-g}=k-W_{2g-r}$; and on combining this with the first observation the proof of the corollary is concluded.

**Corollary 5 to Theorem 16.** *For any points* $z_1 \in \tilde{\Delta}, s \in \mathbb{C}^g, t \in \mathbb{C}^g$ *and any index* $r > 2g$ *the* $(r+g-1) \times 2g$ *Prym period matrix* $\Psi(rz_1, s, t)$ *is of rank* $2g$

*Proof.* If $r > 2g$ it is a consequence of the Riemann-Roch theorem that $\gamma(\rho_t \zeta^{r-2}) = r-g-1$ for all $t \in \mathbb{C}^g$, hence that $W_{r-2}^{r-g} = \emptyset$; the corollary therefore follows immediately from Theorem 16c upon setting $d=2g$, and the proof is thereby concluded.

The preceding results can be used to describe some interesting and useful alternative bases for most spaces of meromorphic Prym differentials, where such bases are as usual expected to be complex analytic functions of the natural auxiliary parameters. The canonical meromorphic Prym differentials $\psi_i(rz_1, s, t; z)$ are expressed in terms of the generalized theta functions of rank $r+g-1$, and form a basis reflecting the chosen normal form for the theta factor of automorphy of rank $r+g-1$; passing to an equivalent form for that factor of automorphy has the effect of replacing the vector $\psi(rz_1, s, t; z)$ of canonical Prym differentials by $C(t+k+r\tilde{\phi}(z_1-z_0))\,\psi(rz_1, s, t; z)$ for some complex analytic mapping

$$C\colon \mathbb{C}^g \longrightarrow GL(r+g-1, \mathbb{C}).$$

More generally of course, dropping the restriction that the basis for the meromorphic Prym differentials be related to the generalized theta functions as in Theorem 15, the basis $\psi(rz_1, s, t; z)$ can be replaced by $C(z_1, s, t,)\,\psi(rz_1, s, t; z)$ for some complex analytic mapping

$$C\colon \tilde{M} \times \mathbb{C}^g \times \mathbb{C}^g \longrightarrow GL(r+g-1, \mathbb{C}).$$

Note that for any index $r \geq 1$ the Prym differentials $\psi_i(rz_1, s, t; z)$ can be viewed as meromorphic Prym differentials associated to the representation $\rho_{s,t}$ and having as singularities poles at the points $\Gamma z_1$ of orders at most $r+1$, instead of at most $r$; and recalling the dependence of these differentials on the parameter $s$ it follows that

$$\psi(rz_1, s, t; z) = C_1(z_1, t)\,\psi((r+1)\,z_1, s, t; z)$$

for a uniquely determined complex analytic mapping

$$C_1\colon \tilde{M} \times \mathbb{C}^g \longrightarrow \mathbb{C}^{(r+g-1)\times(r+g)}$$

with the property that rank $C_1(z, t) = r+g-1$ at all points $(z_1, t)$. The matrix function $C_1(z, t)$ can be viewed as forming the first $r+g-1$ rows of some complex analytic mapping

$$C\colon \tilde{M} \times \mathbb{C}^g \longrightarrow GL(r+g, \mathbb{C}),$$

since $\tilde{M} \times \mathbb{C}^g$ is a contractible Stein manifold; and for the new basis $\tilde{\psi}((r+1)z_1, s, t; z) = C(z, t)\,\psi((r+1)z_1, s, t; z)$ it then follows that $\tilde{\psi}_i((r+1)z_1, s, t; z) =$

$\psi_i(rz_1, s, t; z)$ for $1 \leq i \leq r+g-1$. The differential $\tilde{\psi}_{r+g}((r+1)z_1, s, t; z)$ must of course have a pole of order precisely $r+1$ at each point $\Gamma z_1$, since the other differentials in the basis have poles of orders at most $r$; and it can even be assumed that the principal part of the differential $\tilde{\psi}_{r+g}((r+1)z_1, s, t; z)$ at the point $z_1$ is just $(z-z_1)^{-r-1}dz$, in terms of the standard coordinate $z$ in the unit disc. Indeed by multiplying the differential $\tilde{\psi}_{r+g}((r+1)z_1, s, t; z)$ by a nowhere vanishing complex analytic function the coefficient of the term $(z-z_1)^{-r-1}dz$ in the principal part can clearly be reduced to one; so it is only necessary to show that by adding to $\tilde{\psi}_{r+g}((r+1)z_1, s, t; z)$ the appropriate linear combination of the differentials $\psi_i(rz_1, s, t; z)$ the lower order terms in the principal part can be reduced to zero. For this purpose note first of all that the residue of any canonical meromorphic Prym differential $\psi_i((r+1)z_1, s, t; z)$ is zero whenever $\rho_{s,t}$ is analytically trivial, that is, whenever $t=\lambda \in \mathcal{L}$; for it follows from Theorem 16a that

$$\text{residue}_{z=z_1} \psi_j((r+1)z_1, 0, 0; z) = 0$$

and from Theorem 15 that

$$\begin{aligned}
&\text{residue}_{z=z_1} \psi_i((r+1)z_1, s, \lambda; z)\\
&\quad= \text{residue}_{z=z_1} \big[\textstyle\sum_j \chi_{ij}(\lambda, t+k+(r+1)\tilde{\phi}(z_1-z_0))\, h_{s+n}(z) \cdot \psi_j((r+1)z_1, 0, 0; z)\big]\\
&\quad= \textstyle\sum_j \chi_{ij}(\lambda, t+k+(r+1)\tilde{\phi}(z_1-z_0))\, h_{s+n}(z_1)\big] \cdot \text{residue}_{z=z_1} \psi_j((r+1)z_1, 0, 0; z)\\
&\quad= 0.
\end{aligned}$$

The same is then true for an arbitrary meromorphic Prym differential, since the canonical meromorphic Prym differentials form a basis. On the other hand the residues $r_i(z_1, s, t) = R_1 \psi_i(z_1, s, t)$ all vanish precisely when $\rho_{s,t}$ is analytically trivial, since only then is $\gamma(\kappa\rho_{s,t}) = g$; indeed recalling (74) and Theorem 10 it follows easily that these residues $r_i(z_1, s, t)$ generate the proper ideal of the complex analytic subvariety $\{(s, t) \in \mathbb{C}^g \times \mathbb{C}^g \mid t \in \lambda\}$. Since $\tilde{M} \times \mathbb{C}^g \times \mathbb{C}^g$ is a Stein manifold it follows from well known results in the theory of functions of several complex variables that

$$R_1 \tilde{\psi}_{r+g}((r+1)z_1, s, t) = \textstyle\sum_{i=1}^{g} f_i(z_1, s, t)\, r_i(z_1, s, t)$$

for some complex analytic function $f_i(z_1, s, t)$; and hence

$$\text{residue}_{z=z_1} \big[\tilde{\psi}_{r+g}((r+1)z_1, s, t; z) - \textstyle\sum_{i=1}^{g} f_i(z_1, s, t)\, \psi_i(z_1, s, t; z)\big] = 0\,.$$

Having taken care of the first order terms in the principal part of the differential $\tilde{\psi}_{r+g}((r+1)z_1, s, t; z)$ the remainder follows from the obvious induction step.

For any index $r \geq 1$ there results a new but not canonical basis $\tilde{\psi}_i(rz_1, s, t; z)$ for the space of meromorphic Prym differentials associated to the representation $\rho_{s,t}$ and having as singularities poles at the points $\Gamma z_1$ of orders at most $r$, such that the differentials $\tilde{\psi}_i(rz_1, s, t; z)$ are complex analytic functions of the parameters $(z_1, s, t) \in \tilde{M} \times \mathbb{C}^g \times \mathbb{C}^g$, that $\tilde{\psi}_i(rz_1, s, t; z) = \psi_i(z_1, s, t; z)$ for $1 \leq i \leq g$, and that the principal part of the differential $\tilde{\psi}_i(rz_1, s, t; z)$ at the point $z_1$ is $(z-z_1)^{g-i-1}dz$ for $g+1 \leq i \leq r$; having made a choice of such a basis these differentials will be called the *canonical meromorphic Prym differentials for singularities*. Note that it can be assumed that $\tilde{\psi}_i(rz_1, s, t; z) = \tilde{\psi}_i((r+1)z_1, s, t; z)$ whenever $r \geq i$, so that these differentials are really independent of the index $r$; hence the notation $\tilde{\psi}_i(*z_1, s, t; z)$ will also be used for this basis. This basis is particularly convenient for analyzing the singularities of meromorphic Prym differentials, but as a function of the parameters $s$ and $t$ it is somewhat more complicated than the usual canonical basis. However it is a straightforward calculation to establish for this basis as an analogue of Theorem 15 that

$$(91) \qquad \tilde{\psi}(rz_1, s, t; z) = \begin{pmatrix} I_g & 0 \\ A & B \end{pmatrix} h_s(z) \tilde{\psi}(rz_1, 0, t; z),$$

where $A(z_1, s, t)$ and $B(z_1, s, t)$ are matrices depending analytically on the parameters $z_1, s, t$ and $B(z_1, s, t)$ is of the form

$$B(z_1, s, t) = h_s(z_1)^{-1} \begin{pmatrix} 1 & 0 & 0 & \dots \\ * & 1 & 0 & \dots \\ * & * & 1 & \dots \\ \dots & \dots & \dots & \end{pmatrix};$$

and that

$$(92) \qquad \tilde{\psi}(rz_1, s, t+\lambda; z) = \begin{pmatrix} \chi_g(\lambda, t+k+\tilde{\phi}(z_1-z_0)) & 0 \\ C & D \end{pmatrix} \rho_z(\lambda) \tilde{\psi}(rz_1, s, t; z)$$

for every $\lambda \in L$, where $C(z_1, s, t)$ and $D(z_1, s, t)$ are matrices depending analytically on the parameters $z_1, s, t$ and $D(z_1, s, t)$ is of the form

$$D(z_1, s, t) = \rho_{z_1}(\lambda)^{-1} \begin{pmatrix} 1 & 0 & 0 & \dots \\ * & 1 & 0 & \dots \\ * & * & 1 & \dots \\ \dots & \dots & \dots & \end{pmatrix}.$$

That there are formulas of the general forms (91) and (92) follows of course from the fact that both sides are bases for the same spaces of meromorphic Prym differentials, and the specific forms of the matrices are easy consequences of the facts that $\tilde{\psi}_i(rz_1, s, t; z) = \psi_i(z_1, s, t; z)$ for $1 \leq i \leq g$ and that $\tilde{\psi}_i(rz_1, s, t; z)$ have the known principal parts for $i \geq g+1$; the details of the verification can be left as an exercise.

To obtain another set of canonical Prym differentials useful in some circumstances write the $g \times 2g$ Prym period matrix $\psi(z_1, s, t)$ in the form

$$\Psi(z_1, s, t) = (\Psi_1(z_1, s, t), \Psi_2(z_1, s, t)),$$

where $\Psi_1(z_1, s, t)$ and $\Psi_2(z_1, s, t)$ are $g \times g$ matrices. Recall from (70) that the Prym differentials $\psi_i(z_1, 0, 0; z)$ are a basis for the space of ordinary Abelian differentials on $M$, and hence that $\Psi_1(z_1, 0, 0)$ is necessarily a nonsingular matrix; therefore $\alpha(z_1, s, t) = \det \Psi_1(z_1, s, t)$ is a not identically vanishing complex analytic function on $\tilde{\Delta} \times \mathbb{C}^g \times \mathbb{C}^g$. The adjugate matrix $\Psi_1^*(z_1, s, t)$, the matrix with entries which are the minors of the corresponding entries in the matrix $\Psi_1(z_1, s, t)$, is also a complex analytic function on $\tilde{\Delta} \times \mathbb{C}^g \times \mathbb{C}^g$; it is nonsingular and is equal to $\alpha(z_1, s, t)\,\Psi_1(z_1, s, t)^{-1}$ whenever $\alpha(z_1, s, t) \neq 0$, while rank $\Psi_1^*(z_1, s, t) = 1$ if rank $\Psi_1(z_1, s, t) = g-1$ and rank $\Psi_1^*(z_1, s, t) = 0$ if rank $\Psi_1(z_1, s, t) < g-1$. Introducing the vector of differentials

$$\psi^*(z_1, s, t; z) = \Psi_1^*(z_1, s, t)\,\psi(z_1, s, t; z)$$

note then that $\psi_i^*(z_1, s, t; z)$ are meromorphic Prym differentials associated to the representation $\rho_{s,t}$ and having as singularities at most simple poles at the points $\Gamma z_1$, that they form a basis for the space of such Prym differentials so long as $\alpha(z_1, s, t) \neq 0$, and that they are complex analytic functions of the parameters $z_1 \in \tilde{\Delta}, s \in \mathbb{C}^g, t \in \mathbb{C}^g$. The period matrix for this set of Prym differentials has the form

$$\Psi^*(z_1, s, t) = (\alpha(z_1, s, t)\, I_g, *).$$

These differentials $\psi_i^*(z_1, s, t; z)$ can also be defined by

$$\psi^*(z_1, s, t; z) = \Psi_1^*(z_1, s, t)\,\tilde{\psi}(z_1, s, t; z)$$

since $\tilde{\psi}(z_1, s, t; z) = \psi(z_1, s, t; z)$; and that suggests the following extension of this definition. For any index $i \geq g+1$ introduce the differential

$$\begin{aligned}\psi_i^*(*z_1, s, t; z) \\ = \alpha(z_1, s, t)\,\tilde{\psi}_i(*z_1, s, t; z) - \textstyle\sum_{j=1}^{g} \tilde{\psi}_i(*z_1, s, t; \tilde{\alpha}_j)\,\psi_j^*(z_1, s, t; z),\end{aligned}$$

noting that this is a meromorphic Prym differential associated to the representation $\rho_{s,t}$ having a singularity at the point $z_1$ with the principal part

$$[\alpha(z_1, s, t)(z - z_1)^{g-i-1} + c_i(z_1, s, t)(z - z_1)^{-1}]\,dz$$

for some analytic function $c_i(z_1, s, t)$ and having the periods $\psi_i^*(*z_1, s, t; \tilde{\alpha}_j) = 0$. Thus for any index $r \geq 1$ the differentials $\psi_i^*(*z_1, s, t; z)$ for $1 \leq i \leq r+g-1$ are meromorphic Prym differentials associated to the representation $\rho_{s,t}$ having as singularities poles at the points $\Gamma z_1$ of orders at most $r$, from a basis for the space of such Prym differentials so long as $\alpha(z_1, s, t) \neq 0$, are complex analytic functions of the parameters $z_1 \in \tilde{\Delta}, s \in \mathbb{C}^g, t \in \mathbb{C}^g$, and have a period matrix of the form

$$\Psi^*(rz_1, s, t) = \begin{pmatrix} \alpha(z_1, s, t)\, I_g & * \\ 0 & * \end{pmatrix};$$

these differentials will be called the *canonical meromorphic Prym differentials for periods*. For emphasis note again that the differentials $\psi_i^*(rz_1, s, t; z)$ form a basis for the appropriate space of meromorphic Prym differentials only when the parameters $(z_1, s, t)$ lie outside the proper analytic subvariety

$$X = \{(z_1, s, t) \in \tilde{\Delta} \times \mathbb{C}^g \times \mathbb{C}^g \mid \alpha(z_1, s, t) = 0\} \subset \tilde{\Delta} \times \mathbb{C}^{2g}.$$

There remains the problem of determining the function $\alpha(z_1, s, t)$, or at least its zero locus $X$. That zero locus $X$ at least is invariant under the action of the lattice group $\mathscr{L}$ on $\mathbb{C}^g \times \mathbb{C}^g$, as an obvious consequence of the functional equation (83); the function $\alpha(z_1, 0, t)$ is essentially determined by Theorem 16a, and G. Kempf has shown in [16] that $X$ is disjoint from the subset $0 \times \mathbb{R}^g \subset \mathbb{C}^g \times \mathbb{C}^g$.

A variant of the preceding construction yields yet another set of canonical Prym differentials. Recall from Corollary 3 to Theorem 16 that the $2g \times 2g$ Prym period matrix $\Psi((g+1)\, z_1, s, t)$ is not identically singular, indeed that $\det \Psi((g+1)\, z_1, s, t) = u(z_1, s, t)\, \theta_0(t + g\tilde{\phi}(z_1 - z_0))$ where $u(z_1, s, t)$ is a nowhere vanishing complex analytic function. The adjugate matrix $\Psi^*((g+1)\, z_1, s, t)$ is then a complex analytic function on $\tilde{\Delta} \times \mathbb{C}^g \times \mathbb{C}^g$ which is nonsingular and equal to $u(z_1, s, t)\, \theta_0(t + g\tilde{\phi}(z_1 - z_0))\, \Psi((g+1)\, z_1, s, t)^{-1}$ whenever $t \notin W_{g-1} - g\tilde{\phi}(z_1 - z_0)$, that is, whenever $\theta_0(t + g\tilde{\phi}(z_1 - z_0)) \neq 0$. Introducing the vector of differentials

$$\psi^{**}((g+1)\, z_1, s, t; z) = \Psi^*((g+1)\, z_1, s, t)\, \psi((g+1)\, z_1, s, t; z),$$

note that $\psi_i^{**}((g+1)\, z_1, s, t; z)$ are meromorphic Prym differentials associated to the representation $\rho_{s,t}$ and having as singularities poles of orders at most $g+1$ at the points $\Gamma z_1$, that they form a basis for the space of such Prym differentials so long as $t \notin W_{g-1} - g\tilde{\phi}(z_1 - z_0)$, that they are complex analytic functions of parameters $z_1 \in \tilde{\Delta}, s \in \mathbb{C}^g, t \in \mathbb{C}^g$, and that the period matrix for this set of Prym differentials is

$$\Psi^{**}((g+1)\, z_1, s, t) = u(z_1, s, t)\, \theta_0(t + g\tilde{\phi}(z_1 - z_0))\, I_{2g}.$$

To extend this construction to meromorphic Prym differentials having higher order singularities, for any $i > 2g$ set

$$\begin{aligned} \psi_i^{**}(*z_1, s, t; z) = {} & u(z_1, s, t)\, \theta_0(t + g\tilde{\phi}(z_1 - z_0))\, \tilde{\psi}(*z_1, s, t; z) \\ & - \textstyle\sum_{j=1}^{g} \tilde{\psi}_i(*z, s, t; \tilde{\alpha}_j)\, \psi_j^{**}((g+1)\, z_1, s, t; z) \\ & - \textstyle\sum_{j=1}^{g} \tilde{\psi}_i(*z, s, t; \tilde{\beta}_j)\, \psi_{j+g}^{**}((g+1)\, z_1, s, t; z). \end{aligned}$$

The principal part of this meromorphic Prym differential at the point $z_1$ is of course $u(z_1, s, t)\, \theta_0(t + g\tilde{\phi}(z_1 - z_0))\, (z - z_1)^{g-i-1} dz +$ pole of order at most $g+1$. Furthermore this differential has by construction zero periods along the canonical paths $\tilde{\alpha}_j, \tilde{\beta}_j$, hence by Theorem 16a has zero residue at each singularity; therefore $\psi_i^{**}(*z_1, s, t; z) = dh_i(*z_1, s, t; z)$ for some meromorphic function $h_i(*z_1, s, t; z)$, and with the normalization $h_i(*z_1, s, t; z_0) = 0$ it is apparent that this function is analytic in the parameters $z_1 \in \tilde{\Delta}, s \in \mathbb{C}^g, t \in \mathbb{C}^g$. Note that for any $T \in \Gamma$ there is a constant $c_i(T)$ such that $h_i(*z_1, s, t; Tz) = \rho_{s,t}(T)\, h_i(*z_1, s, t; z) + c_i(T)$; and setting $z = z_0$ note that $c_i(A_j) = h_i(*z_1, s, t; A_j z_0) = \psi_i^{**}(*z_1, s, t; \tilde{\alpha}_j) = 0$, and similarly

$c_i(B_j)=0$. Thus $c_i(T)=0$ for all $T\in\Gamma$ and $h_i(*z_1, s, t; Tz)=\rho_{s,t}(T)\,h_i(*z_1, s, t; z)$ for all $T\in\Gamma$. The differentials $\psi_i^{**}(*z_1, s, t; z)$ for $1\leq i\leq r+g-1$ are a basis for the space of all meromorphic Prym differentials associated to the representation $\rho_{s,t}$ and having as singularities poles at the points $\Gamma z_1$ of orders at most $r$, provided that $t\notin W_{g-1}-g\tilde{\phi}(z_1-z_0)$; these differentials will be called the *second canonical set of meromorphic Prym differentials for periods.* Incidentally writing

$$\Psi^*((g+1)\,z_1, s, t) = \begin{pmatrix}\alpha(z_1, s, t)\,I_g & B\\ 0 & D\end{pmatrix}$$

for some $g\times g$ matrices $B(z_1, s, t)$, $D(z_1, s, t)$ and recalling that

$$\Psi^*((g+1)\,z_1, s, t) = \begin{pmatrix}\Psi_1^*(z_1, s, t) & 0\\ * & \alpha(z_1, s, t)\,I_g\end{pmatrix}\tilde{\Psi}((g+1)\,z_1, s, t)$$

it follows upon taking determinants that

$$\det D(z_1, s, t)=\tilde{u}(z_1, s, t)\,\alpha(z_1, s, t)^{g-1}\,\theta_0(t+g\,\tilde{\phi}(z_1-z_0))$$

for some nowhere vanishing complex analytic function $\tilde{u}(z_1, s, t)$.

Finally recall from Corollary 5 to Theorem 16 that the $2g\times 2g$ Prym period matrix $\Psi((2g+1)\,z_1, s, t)$ is nonsingular for all points $(z_1, s, t)\in\Delta\times\mathbb{C}^g\times\mathbb{C}^g$; hence the components of the vector of differentials

$$\psi^{***}((2g+1)\,z_1, s, t; z)=\Psi((2g+1)\,z_1, s, t)^{-1}\psi((2g+1)\,z_1, s, t; z)$$

form a basis for the space of meromorphic Prym differentials associated to the representation $\rho_{s,t}$ and having as singularities poles at the points $\Gamma z_1$ of orders at most $2g+1$, are analytic functions of the parameters $(z_1, s, t)$ and have the period matrix

$$\Psi^{***}((2g+1)\,z_1, s, t; z)=I_{2g}.$$

This construction can be extended to yield bases for the spaces of meromorphic Prym differentials having higher order singularities, just as in the case of the second canonical set of meromorphic Prym differentials for periods; the resulting differentials will be called the *third canonical set of meromorphic Prym differentials for periods.*

## § 14. The Riemann Equality for Prym Periods

The Riemann bilinear relations established in Theorem 3 and Theorem 6 proved to be very useful in analyzing the complex analytic and meromorphic Abelian differentials on the Riemann surface $M$; and it might be expected that some analogous relations exist and would be at least of some interest in analyzing the meromorphic Prym differentials for $M$. The simplest analogous relation involves a pair of Prym differentials associated to inverse representations of the group $\Gamma$. For any fixed points $z_1, z_2 \in \tilde{\Delta}$ consider the meromorphic Prym differentials

$$\psi_j(z) = \sum_{i=1}^{r_j+g-1} c_{ji} \psi_i(r_j z_j, s_j, t_j; z)$$

associated to the representations $\rho_j = \rho_{s_j, t_j}$ for $j=1,2$; and suppose that the points $s_j, t_j$ are so chosen that $\rho_1(T) = \rho_2(T^{-1}) = \rho_2(T)^{-1}$ for all $T \in \Gamma$. The basic periods of these differentials will be denoted as usual by $\psi_j(\tilde{\alpha}_k), \psi_j(\tilde{\beta}_k)$; and it is convenient in addition to introduce the linear combinations

$$(93) \qquad \psi_j(\gamma_k) = (1-\rho_j(B_k))\, \psi_j(\tilde{\alpha}_k) - (1-\rho_j(A_k))\, \psi_j(\tilde{\beta}_k)$$

of these basic periods. Let $\tilde{\delta}_j$ be fixed simple paths from $z_0$ to $z_j$ in $\tilde{\Delta}$ for $j=1,2$; and assume that these paths are disjoint except for $z_0$ if $z_1 \neq z_2$ but coincide if $z_1 = z_2$, and that the paths $\tilde{\alpha}_1, \tilde{\delta}_1, \tilde{\delta}_2, \tilde{\beta}_g$ emanate from $z_0$ in that order read counterclockwise, as in Fig. 5, at least in a neighborhood of $z_0$. In the simply

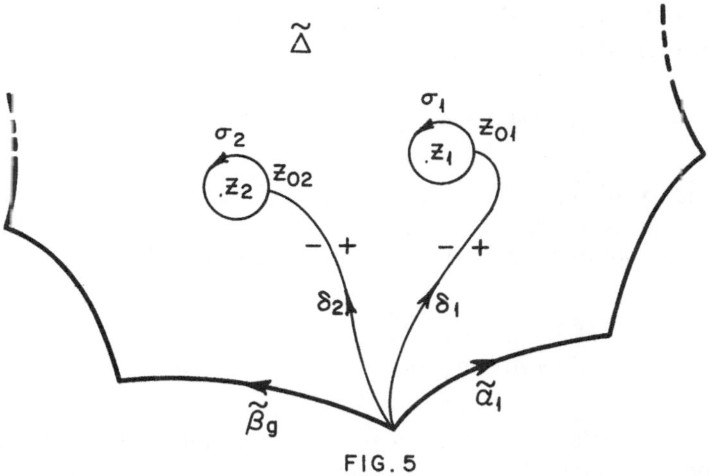

FIG. 5

connected region $\tilde{\Delta} - (\tilde{\delta}_1 \cup \tilde{\delta}_2)$ the differentials $\psi_j(z)$ are complex analytic, so there exist complex analytic functions $h_j(z)$ in that region such that $dh_j(z) = \psi_j(z)$; these functions of course extend analytically across $\partial \tilde{\Delta}$ and across the interior points of the paths $\tilde{\delta}_j$, and are uniquely determined if it is also assumed that $h_j(z)$ approaches zero as $z$ approaches $z_0$ between the paths $\tilde{\delta}_j$ and $\tilde{\alpha}_1$ in $\tilde{\Delta}$. If $z_1 = z_2$ the logarithmic branch points cancel in the linear combination

$$h(z) = h_1(z) \cdot \text{residue}_{z=z_2} \psi_2(z) - h_2(z)\, \text{residue}_{z=z_1} \psi_1(z),$$

so that is is a well defined meromorphic function throughout $\tilde{\Delta}$. Finally write the Laurent expansions of the differentials $\psi_j(z)$ at their singularities $z_j$ explicitly as

$$\psi_j(z)=\sum_\nu c_\nu^j(z-z_j)^\nu dz.$$

**Theorem 17.** *With the notation established above,*

$$\begin{aligned}&\sum_{1\le l<k\le g}\psi_1(\gamma_l)\psi_2(\gamma_k)-\sum_{k=1}^g\psi_1(\gamma_k)[\rho_2(B_k)\psi_2(\tilde\alpha_k)+\psi_2(\tilde\beta_k)]\\&\quad+\sum_{k=1}^g[\rho_2(A_k)\psi_1(\tilde\alpha_k)\psi_2(\tilde\beta_k)-\rho_2(B_k)\psi_2(\tilde\alpha_k)\psi_1(\tilde\beta_k)]\\&=\begin{cases}2\pi i[\text{residue}_{z=z_2}h_1(z)\psi_2(z)-\text{residue}_{z=z_1}h_2(z)\psi_1(z)] & \text{if}\quad z_1\neq z_2,\\ 2\pi i\,\text{residue}_{z=z_1}(z-z_1)^{-1}h(z)-2\pi^2c_{-1}^1c_{-1}^2+2\pi i\sum_{\nu\neq 0}\dfrac{1}{\nu}c_{\nu-1}^1c_{-\nu-1}^2 & \text{if}\quad z_1=z_2.\end{cases}\end{aligned}$$

*Proof.* Let $\sigma_j$ be small oriented circles centered at the points $z_j$ and contained in $\tilde\Delta$, and assume that $\sigma_1$ and $\sigma_2$ are disjoint and have disjoint interiors if $z_1\neq z_2$ but coincide if $z_1=z_2$; and let $\delta_j$ be that portion of the path $\tilde\delta_j$ from $z_0$ to the first point $z_{0j}$ at which $\tilde\delta_j$ meets $\sigma_j$, as sketched in Fig. 5. The differential forms $\psi_j(z)$ and their integrals $h_j(z)$ remain analytic on the boundary of the region $\tilde\Delta_0$ obtained by removing from $\tilde\Delta$ the paths $\delta_j$ and the circles $\sigma_j$ together with their interiors; and since obviously $0=\psi_1(z)\wedge\psi_2(z)=d(h_1(z)\psi_2(z))$ it follows from Stokes' theorem that

$$0=\int_{\tilde\Delta_0}\psi_1(z)\wedge\psi_2(z)=\int_{\partial\tilde\Delta_0}h_1(z)\psi_2(z).$$

The boundary of $\tilde\Delta_0$ consists of the boundary of the canonical fundamental polygon $\tilde\Delta$, the two sides of each path $\delta_j$, and the circles $\sigma_j$; so considering the given orientations and recalling that $\delta_1=\delta_2$ and $\sigma_1=\sigma_2$ if $z_1=z_2$ it follows that

$$\partial\tilde\Delta_0=\begin{cases}\partial\tilde\Delta+\delta_2^- -\sigma_2-\delta_2^+ +\delta_1^- -\sigma_1-\delta_1^+ & \text{if}\quad z_1\neq z_2,\\ \partial\tilde\Delta+\delta_1^- -\sigma_1-\delta_1^+ & \text{if}\quad z_1=z_2,\end{cases}$$

where $\delta_j^+,\delta_j^-$ are the two sides of the path $\delta_j$ as indicated in Fig. 5. It is convenient to consider separately the integrals over various of these segments. First recall from § 4 that $\partial\tilde\Delta=\sum_{k=1}^g\gamma_k$ where

$$\gamma_k=C_1\dots C_{k-1}\tilde\alpha_k+C_1\dots C_{k-1}A_k\tilde\beta_k-C_1\dots C_kB_k\tilde\alpha_k-C_1\dots C_k\tilde\beta_k$$

is the segment of $\partial\tilde\Delta$ from $C_1\dots C_{k-1}z_0$ to $C_1\dots C_kz_0$ as in Fig. 3; so the integral over $\partial\tilde\Delta$ can be written as the sum of the integrals over the segments $\gamma_k$. It is easy to see that the period $\psi_j(\gamma_k)=\int_{\gamma_k}\psi_j(z)$ can be expressed in terms of the basic periods as in Eq. (93), thus justifying that notation. Now when $z\in C_1\dots C_{k-1}\tilde\alpha_k$ then

$$z'=C_1\dots C_kB_k(C_1\dots C_{k-1})^{-1}\cdot z\in C_1\dots C_kB_k\tilde\alpha_k,$$

and $\psi_j(z')=\rho_j(B_k)\psi_j(z)$ so that $h_j(z')=\rho_j(B_k)h_j(z)+c_{jk}$ for some complex constant $c_{jk}$; thus

$$\begin{aligned}\int_{C_1\dots C_{k-1}\tilde{\alpha}_k - C_1\dots C_k B_k\tilde{\alpha}_k} & h_1(z)\psi_2(z)\\ &= \int_{z\in C_1\dots C_{k-1}\tilde{\alpha}_k}[h_1(z)\psi_2(z)-h_1(z')\psi_2(z')]\\ &= \int_{z\in C_1\dots C_{k-1}\tilde{\alpha}_k}[h_1(z)\psi_2(z)-(\rho_1(B_k)h_1(z)+c_{1k})\rho_2(B_k)\psi_2(z)]\\ &= -c_{1k}\rho_2(B_k)\psi_2(\tilde{\alpha}_k)\end{aligned}$$

since $\rho_1(B_k)\rho_2(B_k)=1$. To evaluate the constants $c_{jk}$ set $z=C_1\dots C_{k-1}z_0$ and observe that

$$c_{jk}=h_j(C_1\dots C_k B_k z_0)-\rho_j(B_k)h_j(C_1\dots C_{k-1}z_0).$$

Recalling that $h_j(z)$ was chosen so that $h_j(z)$ approaches zero as $z$ approaches $z_0$ along $\tilde{\alpha}_1$, note that $h_j(C_1\dots C_k B_k z_0)$ is equal to the integral of $\psi_j(z)$ along that segment of $\partial\tilde{\Delta}$ from $z_0$ to $C_1\dots C_k B_k z_0$, and similarly for $h_j(C_1\dots C_{k-1}z_0)$; and calculating these integrals as usual readily yields the formula

$$c_{jk}=(1-\rho_j(B_k))\textstyle\sum_{l=1}^{k-1}\psi_j(\gamma_l)+\psi_j(\gamma_k)+\psi_j(\tilde{\beta}_k).$$

Similarly

$$\int_{C_1\dots C_{k-1}A_k\tilde{\beta}_k - C_1\dots C_k\tilde{\beta}_k} h_1(z)\psi_2(z)=-c'_{1k}\psi_2(\tilde{\beta}_k),$$

where

$$c'_{jk}=(1-\rho_j(A_k^{-1}))\textstyle\sum_{l=1}^{k-1}\psi_j(\gamma_l)+\psi_j(\gamma_k)-\rho_j(A_k^{-1})\psi_j(\tilde{\alpha}_k).$$

Upon combining these results it follows readily that

$$\begin{aligned}\int_{\gamma_k}h_1(z)\psi_2(z)=\textstyle\sum_{l=1}^{k-1}&\psi_1(\gamma_l)\psi_2(\gamma_k)-\psi_1(\gamma_k)[\rho_2(B_k)\psi_2(\tilde{\alpha}_k)+\psi_2(\tilde{\beta}_k)]\\ &+\rho_2(A_k)\psi_1(\tilde{\alpha}_k)\psi_2(\beta_k)-\rho_2(B_k)\psi_2(\tilde{\alpha}_k)\psi_1(\tilde{\beta}_k).\end{aligned}$$

Next if $z_1\neq z_2$ note that $h_1(z)$ extends analytically across the path $\delta_2$ and has the same values on the two sides of that path, so that

$$\int_{\delta_2^- -\sigma_2-\delta_2^+}h_1(z)\psi_2(z)=-2\pi i\,\mathrm{residue}_{z=z_2}h_1(z)\psi_2(z);$$

and since

$$\int_{\delta_1^- -\sigma_1-\delta_1^+}h_1(z)\psi_2(z)=-\int_{\delta_1^- -\sigma_1-\delta_1^+}h_2(z)\psi_1(z)$$

by Stokes' theorem, recalling the normalization of the functions $h_j(z)$, it follows similarly that

$$\int_{\delta_1^- -\sigma_1-\delta_1^+}h_1(z)\psi_2(z)=2\pi i\,\mathrm{residue}_{z=z_1}h_2(z)\psi_1(z).$$

On the other hand if $z_1 = z_2$ then both functions $h_j(z)$ extend analytically across the path $\delta_1 = \delta_2$ but possibly take different values on the two sides of that path; indeed letting $\alpha_j = \text{residue}_{z=z_j} \psi_j(z)$ it is clear that for any point $z \in \delta_1 = \delta_2$ the two boundary values of the function $h_j$ at that point are related by

$$h_j(z^-) - h_j(z^+) = \int_{\sigma_j} \psi_j(z) = 2\pi i \alpha_j,$$

and consequently

$$\begin{aligned}\int_{\delta_1^- - \delta_1^+} h_1(z)\psi_2(z) &= \int_{z\in\delta}[h_1(z^-)\psi_2(z) - h_1(z^+)\psi_2(z)] \\ &= 2\pi i \alpha_1 \int_{z\in\delta} \psi_2(z) = 2\pi i \alpha_1 h_2(z_{01}^+)\end{aligned}$$

since $h_2(z_0^+) = 0$. In order to calculate the integral over the circle $\sigma_1 = \sigma_2$ introduce the parametrization $z = z_1 + re^{i\theta}$ with the parameter $\theta \in \mathbb{R}$, assuming for convenience that the point $z_{01} = z_{02}$ corresponds to the parameter values $\theta \in 2\pi\mathbb{Z}$; and also introduce the Laurent expansions of the differentials $\psi_j$ about the point $z_1 = z_2$ and integrate them term by term to obtain expansions of the form

$$h_j(z) = \alpha_j \log(z - z_j) + \sum_v a_v^j (z - z_j)^v,$$

where $h_j^*(z) = \sum_v a_v^j (z - z_j)^v$ are meromorphic functions near $z_1 = z_2$ and the branches of the logarithm are chosen so that $\log(z_{0j}^+ - z_j) = \log r$ and hence

$$h_j(z_{0j}^+) = \alpha_j \log r + \sum_v a_v^j r^v.$$

The integral over $\sigma_1 = \sigma_2$ is then given by

$$\begin{aligned}\int_{\sigma_1} h_1(z)\psi_2(z) &= \int_{\sigma_1}[\alpha_1 \log(z - z_1) + h_1^*(z)]\,\psi_2(z) \\ &= \int_{\sigma_1} \alpha_1 \log(z - z_1)\psi_2(z) + 2\pi i\, \text{residue}_{z=z_1} h_1^*(z)\psi_2(z),\end{aligned}$$

where

$$\begin{aligned}\int_{\sigma_1} \alpha_1 \log(z - z_1)\psi_2(z) &= \int_{\theta=0}^{2\pi} \alpha_1 \log(re^{i\theta})\psi_2(z_2 + re^{i\theta}) \\ &= \alpha_1 \log r \int_{\sigma_1} \psi_2(z) + i\alpha_1 \int_{\theta=0}^{2\pi} \theta[\alpha_2 (re^{i\theta})^{-1} + \sum_v a_v^2 v (re^{i\theta})^{v-1}]\, i re^{i\theta} d\theta \\ &= 2\pi i \alpha_1 \alpha_2 \log r - 2\pi^2 \alpha_1 \alpha_2 + 2\pi i \alpha_1 \sum_{v \neq 0} a_v^2 r^v \\ &= 2\pi i \alpha_1 [h_2(z_{02}^+) - a_0^2] - 2\pi^2 \alpha_1 \alpha_2.\end{aligned}$$

Thus if $z_1 = z_2$ then

$$\int_{\delta_1^- - \sigma_1 - \delta_1^+} h_1(z)\psi_2(z) = 2\pi^2 \alpha_1 \alpha_2 + 2\pi i \alpha_1 a_0^2 - 2\pi i\, \text{residue}_{z=z_1} h_1^*(z)\psi_2(z).$$

The constants $a_0^j$ depend on the explicit choice of the branch of the logarithm; but it is quite easy to obtain a more intrinsic form for this evaluation of the integral. Note that since the same branch of $\log(z - z_1)$ is chosen in the expansions of the functions $h_1(z)$ and $h_2(z)$ then $\alpha_2 h_1(z) - \alpha_1 h_2(z) = \alpha_2 h_1^*(z) - \alpha_1 h_2^*(z)$; here

$\alpha_2 h_1(z)-\alpha_1 h_2(z)$ is of course a well defined meromorphic function throughout $\tilde{\Delta}$, and its constant term is $\alpha_2 a_0^1-\alpha_1 a_0^2$. Now

$$\begin{aligned}&\operatorname{residue}_{z=z_1} h_1^*(z)\psi_2(z)\\ &\quad=\operatorname{residue}_{z=z_1}\left(\sum_\mu a_\mu^1(z-z_1)^\mu\right)\left(\alpha_2(z-z_1)^{-1}+\sum_\nu \nu a_\nu^2(z-z_1)^{\nu-1}\right)\\ &\quad=\alpha_2 a_0^1+\sum_\nu \nu a_{-\nu}^1 a_\nu^2\end{aligned}$$

and therefore

$$\begin{aligned}\int_{\delta_1^- -\sigma_1-\delta_1^+} h_1(z)\psi_2(z)&=2\pi^2\alpha_1\alpha_2+2\pi i\left(\alpha_1 a_0^2-\alpha_2 a_0^1-\sum_\nu \nu a_{-\nu}^1 a_\nu^2\right)\\ &=2\pi^2\alpha_1\alpha_2+2\pi i\operatorname{residue}_{z=z_1}(z-z_1)^{-1}(\alpha_1 h_2(z)-\alpha_2 h_1(z))+2\pi i\sum_\nu \nu a_\nu^1 a_{-\nu}^2\\ &=2\pi^2 c_{-1}^1 c_{-1}^2-2\pi i\operatorname{residue}_{z=z_1}(z-z_1)^{-1}h(z)-2\pi i\sum_{\nu\neq 0}\frac{1}{\nu}c_{\nu-1}^1 c_{-\nu-1}^2\end{aligned}$$

in terms of the coefficients in the Laurent expansion of the differentials $\psi_j(z)$, where $h(z)=\alpha_2 h_1(z)-\alpha_1 h_2(z)=c_{-1}^2 h_1(z)-c_{-1}^1 h_2(z)$. Upon combining the preceding observations the proof of the theorem is readily completed.

**Corollary 1 to Theorem 17.** *If the Prym differentials $\psi_1(z), \psi_2(z)$ have at most simple poles as singularities and if $z_1\neq z_2$ then the right hand side of the equality of Theorem 17 is equal to*

$$\sum_{k=1}^g[\psi_1(\tilde{\delta}_2)\psi_2(\gamma_k)-\psi_2(\tilde{\delta}_1)\psi_1(\gamma_k)]+\sum_{k,l=1}^g\psi_1(\gamma_k)\psi_2(\gamma_l).$$

*Proof.* Recalling the expression for the total residue of a meromorphic Prym differential as given in Theorem 16b and using the notation (93) note that the right hand side of the equality of Theorem 17 can be rewritten as

$$\sum_{k=1}^g[h_1(z_2)\psi_2(\gamma_k)-h_2(z_1)\psi_1(\gamma_k)].$$

Since the Prym integrals $h_j(z)$ were normalized so that $h_j(z)$ approaches zero as $z$ approaches $z_0$ between the paths $\tilde{\delta}_j$ and $\tilde{\alpha}_1$ in $\tilde{\Delta}$ then referring to Fig. 5 note also that

$$\begin{aligned}h_1(z_2)&=\int_{\delta_1^+ +\sigma_1-\delta_1^- +\tilde{\delta}_2}\psi_1(z)=2\pi i\operatorname{residue}_{z=z_1}\psi_1(z)+\int_{\tilde{\delta}_2}\psi_1(z)\\ &=\sum_{k=1}^g\psi_1(\gamma_k)+\psi_1(\tilde{\delta}_2)\end{aligned}$$

and

$$h_2(z_1)=\int_{\tilde{\delta}_1}\psi_2(z)=\psi_2(\tilde{\delta}_1).$$

The desired result follows immediately upon combining these observations.

The functional equations of Theorem 16a completely determine the Prym periods $\psi(z_1,0,t;\tilde{\alpha}_k)$, as noted in the discussion following the proof of that theorem; hence those periods can be viewed as another class of known functions. It is but natural to ask whether the periods $\psi(z_1,0,t;\tilde{\beta}_k)$ are correspondingly determined.

The simplest direct analogue of the argument of Theorem 16a fails, since the periods $\psi(z_1, 0, t; \tilde{\beta}_k)$ generally have logarithmic singularities at the points $\Gamma z_0$ when viewed as functions of the variable $z_1$. Of course for suitable linear combinations of such periods the logarithmic singularities cancel, and the argument used in the proof of Theorem 16a shows that these linear combinations are completely determined by the appropriate functional equations and hence can also be viewed as known functions; and a similar procedure can be used to handle the appropriate linear combinations of the periods $\psi(z_1, s, t; \tilde{\alpha}_k)$ and $\psi(z_1, s, t; \tilde{\beta}_k)$ when $s \neq 0$. However it is perhaps more interesting to observe that Theorem 17 can be used to express some combinations of the periods $\psi(z_1, 0, t; \tilde{\beta}_k)$ in terms of the already known periods $\psi(z_1, 0, t; \tilde{\alpha}_k)$.

Consider again the alternative canonical meromorphic Prym differentials $\psi_i^*(z_1, 0, t; z)$, which are described in terms of the usual meromorphic Prym differentials $\psi_i(z_1, 0, t; z)$ and their periods $\psi_i(z_1, 0, t; \tilde{\alpha}_k)$ and which therefore can be used in place of the differentials $\psi_i(z_1, 0, t; z)$ for present purposes; and to simplify the notation set

$$\psi_i^0(z_1, 0, t; z) = \alpha(z_1, 0, t)^{-1} \psi_i^*(z_1, 0, t; z).$$

Thus $\psi_i^0(z_1, 0, t; z)$ are meromorphic in all variables, form a basis for the space of meromorphic Prym differentials associated to the representation $\rho_t$ and having as singularities at most simple poles at the points $\Gamma z_1$ whenever $\alpha(z_1, 0, t) \neq 0$, have periods $\psi_i^0(z_1, 0, t; \tilde{\alpha}_k) = \delta_k^i$ whenever $\alpha(z_1, 0, t) \neq 0$, and can be considered as known; and the problem is that of determining the periods $\psi_i^0(z_1, 0, t; \tilde{\beta}_k)$.

**Corollary 2 to Theorem 17.** *If $z_1$ and $z_2$ are distinct points in $\tilde{\Delta}$, $\tilde{\delta}_1$ and $\tilde{\delta}_2$ are paths from $z_0$ to $z_1$ and $z_2$ respectively as before, and $t_1$ and $t_2$ are points of $\mathbb{C}^g$ such that the representations $\rho_1 = \rho_{t_1}$ and $\rho_2 = \rho_{t_2}$ are inverse to one another, then for any indices $1 \leq i, j \leq g$*

$$\begin{aligned} \rho_1(B_i)\psi_j^0(z_2, 0, t_2; \tilde{\beta}_i) + (1 - \rho_1(B_i))\psi_j^0(z_2, 0, t_2; \tilde{\delta}_1) \\ = \rho_2(B_j)\psi_i^0(z_1, 0, t_1; \tilde{\beta}_j) + (1 - \rho_2(B_j))\psi_i^0(z_1, 0, t_1; \tilde{\delta}_2) \\ + \delta^{i>j}(1 - \rho_1(B_i))(1 - \rho_2(B_j)) + \delta_j^i(1 - \rho_1(B_i)) \end{aligned}$$

*whenever $\alpha(z_1, 0, t_1) \cdot \alpha(z_2, 0, t_2) \neq 0$, where $\delta_j^i$ is the usual Kronecker symbol and*

$$\delta^{i>j} = \begin{cases} 1 & \text{if } i > j \\ 0 & \text{otherwise.} \end{cases}$$

*Proof.* The asserted result follows immediately from Theorem 17 and its Corollary 1, taking $\psi_1(z) = \psi_i^0(z_1, 0, t_1; z)$ and $\psi_2(z) = \psi_j^0(z_2, 0, t_2; z)$ and noting that $\rho_1(A_k) = \rho_2(A_k) = 1$ and

$$\psi_1(\tilde{\alpha}_k) = \delta_k^i, \quad \psi_1(\tilde{\beta}_k) = \psi_i^0(z_1, 0, t_1; \tilde{\beta}_k), \quad \psi_1(\gamma_k) = \delta_k^i(1 - \rho_1(B_i)),$$
$$\psi_2(\tilde{\alpha}_k) = \delta_k^j, \quad \psi_2(\tilde{\beta}_k) = \psi_j^0(z_2, 0, t_2; \tilde{\beta}_k), \quad \psi_2(\gamma_k) = \delta_k^j(1 - \rho_2(B_k)).$$

Upon taking the derivative of the identity given in the preceding Corollary 2 with respect to the variable $z_1$ it follows that

$$(94) \qquad (1-\rho_1(B_i))\psi_j^0(z_2, 0, t_2; z_1) = \frac{\partial}{\partial z_1}[\rho_2(B_j)\psi_i^0(z_1, 0, t_1; \tilde{\beta}_j) + (1-\rho_2(B_j))\psi_i^0(z_1, 0, t_1; \tilde{\delta}_2)]\,dz_1;$$

thus at least the derivative with respect to the variable $z_1$ of the linear combination of periods

$$\rho_2(B_j)\psi_i^0(z_1, 0, t_1; \tilde{\beta}_j) + (1-\rho_2(B_j))\psi_i^0(z_1, 0, t_1; \tilde{\delta}_2)$$

is expressed in terms of previously known functions. This can also be viewed as determining at least the derivative with respect to the variable $z_1$ of any period $\psi_i^0(z_1, 0, t_1; \tilde{\delta}_2)$ in terms of the derivatives of the basic Prym periods and other known functions. Upon differentiating the identity (94) with respect to the variable $z_2$ there further follows the interesting dual relation

$$(95) \qquad (1-\rho_1(B_i))\frac{\partial}{\partial z_2}\psi_j^0(z_2, 0, t; z_1)\,dz_2 = (1-\rho_2(B_j))\frac{\partial}{\partial z_1}\psi_i^0(z_1, 0, t_1; z_2)\,dz_1,$$

which it least implicitly involves the periods along the paths $\tilde{\alpha}_k$.

As an example of another sort of application of the preceding theorem, decompose the $g \times 2g$ Prym period matrix into square blocks

$$\Psi(z_1, s, t) = (\Psi_1(z_1, s, t), \Psi_2(z_1, s, t))$$

as before, where $\Psi_1(z_1, s, t) = (\psi_i(z_1, s, t; \tilde{\alpha}_j))$.

**Corollary 3 to Theorem 17.** *If $t_1, t_2 \in \mathbb{C}^g$ are not contained in $\mathscr{L}$ and the representations $\rho_1 = \rho_{t_1}, \rho_2 = \rho_{t_2}$ are inverse to one another then for any point $z_1 \in \tilde{M}$*

$$\operatorname{rank} \Psi_1(z_1, 0, t_1) \geq \max_{z_2 \in \tilde{M}} \operatorname{rank} \Psi_1(z_2, 0, t_2) - 1.$$

*Proof.* If $\det \psi_1(z_1, 0, t_1) \neq 0$ the assertion is of course trivial. If $\det \Psi_1(z_1, 0, t_1) = 0$ it is possible to choose Prym differentials $\psi_1(z) = \sum_j c_j \psi_j(z_1, 0, t_1; z)$ as in Theorem 17 so that $\psi_1(\tilde{\alpha}_k) = 0$ for all $k$; indeed the number of linearly independent such differentials is equal to $g - \operatorname{rank} \Psi_1(z_1, 0, t)$. For any of these differentials $\psi_1(z)$ note that $\psi_1(\gamma_k) = 0$ for all $k$ also; hence by Theorem 16b the differential $\psi_1(z)$ is everywhere regular, so that $\psi_1(z) = dh_1(z)$ for a holomorphic function $h_1(z)$ on $\tilde{M}$ with $h_1(z_0) = 0$. It then follows from Theorem 17 and Eq. (74), taking $\psi_2(z) = \psi_j(z_2, 0, t_2; z)$, that

$$\sum_k \psi_j(z_2, 0, t_2; \tilde{\alpha}_k)\rho_2(B_k)\psi_1(\tilde{\beta}_k) = -2\pi i\, h_1(z_2) q(z_2)^{-1} \theta_j(t_2 + k + \tilde{\phi}(z_2 - z_0), z_2)$$

for any fixed $z_2 \in \tilde{M}$. Now the right hand side of this equality represents a vector in a fixed one-dimensional subspace of $\mathbb{C}^g$, independent of the choice of the differential $\psi_1(z)$; hence for any choice of $\psi_1(z)$ the vector having components $\rho_2(B_k)\psi_1(\tilde{\beta}_k)$ must lie in a fixed subspace of dimension $g+1-\operatorname{rank}\Psi_1(z_2,0,t_2)$ in $\mathbb{C}^g$. Since $t_1\notin\mathscr{L}$ then linearly independent differentials $\psi_1(z)$ necessarily have linearly independent modified period vectors $\{\rho_2(B_k)\psi_1(\tilde{\beta}_k)\}$, so that

$$g-\operatorname{rank}\Psi_1(z_1,0,t_1)\leq g+1-\operatorname{rank}\Psi_1(z_2,0,t_2);$$

and since this is true for any $z_2\in\tilde{M}$ the desired result has been demonstrated.

As a consequence of this observation, when $t_1, t_2$ are fixed in $\mathbb{C}^g-\mathscr{L}$ then either $\det\Psi_1(z,0,t_2)=0$ for all $z\in\tilde{M}$ or $\operatorname{rank}\Psi_1(z,0,t_1)\geq g-1$ for all $z\in\tilde{M}$; and for any fixed $z_1\in\tilde{M}$ and $t_1\in\mathbb{C}^g-\mathscr{L}$,

$$\operatorname{rank}\Psi_1(z_1,0,t_1)\geq\max_{z\in\tilde{M}}\operatorname{rank}\Psi_1(z,0,t_1)-2.$$

## § 15. Regular Prym Differentials

In conclusion a few comments about the everywhere regular Prym differentials and their periods are perhaps in order here. The complex analytic Prym differentials associated to the representations $\rho_{s,t}$ do not form vector spaces of a constant dimension as $s$ and $t$ vary throughout $\mathbb{C}^g$, so cannot be described directly by global functions on $\mathbb{C}^g\times\mathbb{C}^g\times\tilde{M}$. However paralleling the description of meromorphic Prym differentials by means of generalized theta functions as given in Theorem 15 it is possible to describe the complex analytic Prym differentials in terms of generalized theta functions of rank $g-1$ in the variable $t$, viewed as cross-sections of a complex analytic vector bundle defined over the complement of the origin in the Jacobi variety $J(M)$ of the Riemann surface $M$. It is directer and certainly somewhat easier though to begin with the meromorphic Prym differentials $\psi_i(z_1,s,t;z)$ for some fixed point $z_1\in\tilde{\Delta}$, recalling that the residues $R_1\psi_i(z_1,s,t)$ of these differentials are given quite explicitly in terms of the generalized theta functions of rank $g$ as in Eq. (74), and to observe that the complex analytic Prym differentials associated to the representation $\rho_{s,t}$ can be written in the form $\sum_{i=1}^g c_i\psi_i(z_1,s,t;z)$ where $c\in\mathbb{C}^g$ are vectors such that $\sum_{i=1}^g c_iR_1\psi_i(z_1,s,t)=0$, hence such that

$$\sum_{i=1}^g c_i\theta_i(t+k+\tilde{\phi}(z_1-z_0),z_1)=0. \tag{96}$$

The problem of describing the complex analytic Prym differentials as local functions of the parameters $s$, $t$ is thus reduced to the problem of finding bases for the subspaces consisting of those vectors $c\in\mathbb{C}^g$ satisfying (96) so that these bases are local complex analytic functions of the parameter $t$, noting that the parameter $s$ does not appear in (96); and since $\theta(t+k+\tilde{\phi}(z_1-z_0),z_1)=0$ precisely when $t\in\mathscr{L}$ there exist such bases in an open neighborhood of any point $t\in\mathbb{C}^g-\mathscr{L}$.

The behavior of these bases near points $t \in \mathscr{L}$ can also be read off from Eq. (96). Furthermore this description of the complex analytic Prym differentials in terms of the meromorphic Prym differentials immediately yields a corresponding description of the periods of the complex analytic Prym differentials in terms of the periods of the meromorphic Prym differentials; thus the basic complex analytic Prym periods are ${}^t c \cdot \Psi(z_1, s, t)$, where $c \in \mathbb{C}^g$ is any vector satisfying (96) and $\Psi(z_1, s, t)$ is the $g \times 2g$ Prym period matrix for the meromorphic Prym differentials. This should be compared with the discussion at the end of §.12.

As an alternative to (96), which is particularly interesting when considering the periods of the complex analytic Prym differentials, the residue

$$R_1 \psi_i(z_1, s, t) = \text{residue}_{z \in \tilde{\Delta}} \psi_i(z_1, s, t; z)$$

can be described in terms of the basic periods of the meromorphic Prym differential $\psi_i(z_1, s, t; z)$ by using Theorem 16b. Thus the complex analytic Prym differentials associated to the representation $\rho_{s,t}$ can be written in the form $\sum_{i=1}^g c_i \psi_i(z_1, s, t; z)$ where $c \in \mathbb{C}^g$ are vectors such that

$$\textstyle\sum_{i,j=1}^g c_i [\psi_i(z_1, s, t; \tilde{\alpha}_j)(1 - \rho_{s,t}(B_j)) - \psi_i(z_1, s, t; \tilde{\beta}_j)(1 - \rho_{s,t}(B_j))] = 0\,;$$

and this can be rewritten more succinctly as the condition that

$$(97) \qquad {}^t c \cdot \Psi(z_1, s, t) \cdot \sigma(s, t) = 0\,,$$

where $\sigma(s, t) \in \mathbb{C}^{2g}$ is the vector having components

$$\sigma_j(s, t) = \begin{cases} 1 - \rho_{s,t}(B_j) & \text{for} \quad 1 \le j \le g\,, \\ -1 + \rho_{s,t}(A_j) & \text{for} \quad g+1 \le j \le 2g\,. \end{cases}$$

Conditions (96) and (97) are precisely equivalent, so either can be used in describing the complex analytic Prym differentials and their periods.

The basic periods for complex analytic Prym differentials can be described somewhat more invariantly than those for meromorphic Prym differentials. If $\psi(z)$ is a complex analytic Prym differential associated to a representation $\rho \in \operatorname{Hom}(\Gamma, \mathbb{C}^*)$ then there exist complex analytic functions $h(z)$ on $\tilde{M}$ such that $dh(z) = \psi(z)$; these functions will be called *integrals* of the Prym differential $\psi(z)$. Any two integrals of the same Prym differential of course differ by a complex constant. If $h(z)$ is an integral of the Prym differential $\psi(z)$ then clearly

$$h(Tz) = \rho(T)\, h(z) + \psi(T)$$

for some complex constant $\psi(T)$ whenever $T \in \Gamma$; the mapping $\psi: \Gamma \longrightarrow \mathbb{C}$ which associates to each element $T \in \Gamma$ the constant $\psi(T)$ is called the *period mapping* for the Prym differential $\psi(z)$ with respect to the integral $h(z)$. The terminology is suggested by the observation that if $h(z_0) = 0$ then $\psi(T) = \int_{z_0}^{Tz_0} \psi(z)$. It is readily verified that any such period mapping has the property that

$$(98) \qquad \psi(ST) = \psi(S) + \rho(S)\, \psi(T)$$

for any elements $S, T \in \Gamma$. In general a mapping $\psi: \Gamma \longrightarrow \mathbb{C}$ satisfying (98) is called a one-cocycle of the group $\Gamma$ with coefficients in the $\Gamma$-module defined by the representation $\rho$; and the set of all such one-cocycles form a complex vector space which is denoted by $Z^1(\Gamma,\rho)$. The terminology is that of the cohomology theory of abstract groups, [7]. Thus the period mapping for the Prym differential $\psi(z)$ with respect to any integral $h(z)$ is an element $\psi \in Z^1(\Gamma,\rho)$. The period mapping for the Prym differential $\psi(z)$ with respect to another integral $h(z)+c$ associates to an element $T \in \Gamma$ the value $\psi(T)+c(1-\rho(T))$. In general a mapping $\psi_0: \Gamma \longrightarrow \mathbb{C}$ of the form

$$\psi_0(T) = c(1-\rho(T)) \tag{99}$$

for some complex constant $c$ is called a one-coboundary of the group $\Gamma$ with coefficients in the $\Gamma$-module defined by the representation $\rho$; and the set of all such one-coboundaries form a complex vector space which is denoted by $B^1(\Gamma,\rho)$. Clearly $B^1(\Gamma,\rho) \subseteq Z^1(\Gamma,\rho)$; the quotient space $H^1(\Gamma,\rho) = Z^1(\Gamma,\rho)/B^1(\Gamma,\rho)$ is called the first cohomology group of the group $\Gamma$ with coefficients in the $\Gamma$-module defined by the representation $\rho$. The period mappings for the Prym differential $\psi(z)$ with respect to all integrals of $\psi(z)$ form a coset $\psi + B^1(\Gamma,\rho)$ of $B^1(\Gamma,\rho)$ in $Z^1(\Gamma,\rho)$, and hence determine a unique element $\psi \in H^1(\Gamma,\rho)$; the latter element is called the *period class* for the Prym differential $\psi(z)$.

Since the group $\Gamma$ is generated by the elements $A_1, \ldots, A_g, B_1, \ldots, B_g$ and these generators are subject to the single relation $C_1 \ldots C_g = 1$ where $C_i = A_i B_i A_i^{-1} B_i^{-1}$, as noted in § 4, any one-cocycle $\psi \in Z^1(\Gamma,\rho)$ is fully determined by the values $\psi(A_i), \psi(B_i)$ alone; and it is a straightforward consequence of (98) that these values determine a one-cocycle precisely when

$$0 = \textstyle\sum_{i=1}^{g} \left[(1-\rho(B_i))\psi(A_i) - (1-\rho(A_i))\psi(B_i)\right]. \tag{100}$$

It is therefore evident that

$$\dim_{\mathbb{C}} Z^1(\Gamma,\rho) = \begin{cases} 2g & \text{if } \rho(T)=1 \text{ for all } T \in \Gamma, \\ 2g-1 & \text{otherwise.} \end{cases}$$

Recalling (99) it is equally evident that

$$\dim_{\mathbb{C}} B^1(\Gamma,\rho) = \begin{cases} 0 & \text{if } \rho(T)=1 \text{ for all } T \in \Gamma, \\ 1 & \text{otherwise;} \end{cases}$$

and consequently

$$\dim_{\mathbb{C}} H^1(\Gamma,\rho) = \begin{cases} 2g & \text{if } \rho(T)=1 \text{ for all } T \in \Gamma, \\ 2g-2 & \text{otherwise.} \end{cases}$$

Of course if $\rho = 1$, that is to say, if $\rho(T) = 1$ for all $T \in \Gamma$, then $H^1(\Gamma,\rho) = Z^1(\Gamma,\rho) = \mathrm{Hom}(\Gamma,\mathbb{C})$. Now if $\rho = 1$ then the complex analytic Prym differentials associated to the representation $\rho$ are just the Abelian differentials on $M$, and the period classes of these Prym differentials in $H^1(\Gamma,\rho)$ are just the period classes of the Abelian differentials in $\mathrm{Hom}(\Gamma,\mathbb{C})$; these analytic period classes thus form a $g$-dimensional linear subspace of the $2g$-dimensional vector space $H^1(\Gamma,\rho)$. If $\rho\neq 1$ but $\rho$ is analytically trivial then there are $g$ linearly independent complex analytic Prym differentials associated to the representation $\rho$. If one of these differentials $\psi(z)$ has a zero period class in $H^1(\Gamma,\rho)$ then there is an integral $h(z)$ of $\psi(z)$ such that $h(Tz)=\rho(T)h(z)$ for all $T\in\Gamma$; and since the space of complex analytic functions $h(z)$ satisfying this condition is one-dimensional it follows that the period classes of the complex analytic Prym differentials form a $(g-1)$-dimensional linear subspace of the $2(g-1)$-dimensional vector space $H^1(\Gamma,\rho)$. If $\rho\neq 1$ and $\rho$ is not analytically trivial then there are $g-1$ linearly independent complex analytic Prym differentials associated to the representation $\rho$ and their period classes form a $(g-1)$-dimensional linear subspace of the $2(g-1)$-dimensional vector space $H^1(\Gamma,\rho)$. Thus for all representations $\rho$ the analytic period classes form a linear subspace of $H^1(\Gamma,\rho)$ having dimension half that of $H^1(\Gamma,\rho)$.

The basic periods for a complex analytic Prym differential $\psi(z)$ are just the values $\psi(A_i), \psi(B_i)$ of the period mapping for that Prym differential with respect to the integral $h(z)$ for which $h(z_0)=0$. The period class $\psi\in H^1(\Gamma,\rho)$ is the natural invariant form in which to describe these periods, removing the dependence on a particular choice of base point or of generators for the group $\Gamma$; and the basic periods can almost entirely be recovered from the period class.

The basic periods for complex analytic Prym differentials are particularly interesting. For example, if $M_1 \longrightarrow M$ is a finite covering space of $M$ with an abelian group of covering translations then the Abelian differentials for $M_1$ can be viewed as complex analytic Prym differentials for $M$ associated to some representation $\rho\in\mathrm{Hom}(\Gamma,\mathbb{C}^*)$; and the problem of determining the period matrix for the covering space $M_1$ can thus be reduced to the problem of determining the periods of the Prym differentials on $M$.

# Appendix. Some Topics in the Classical Theory of Theta Functions

## § 16. Classification of Scalar Factors of Automorphy for Complex Tori

The classical theory of theta functions treats of complex analytic relatively automorphic functions for scalar factors of automorphy for compact complex tori. The first step in this study is the classification of all such factors of automorphy, or equivalently the classification of complex analytic line bundles over such tori. This will be carried out here in a manner parallel to the classification of factors of automorphy for compact Riemann surfaces in § 4, following roughly and in abbreviated form the treatment in [5]. Of course this classification can be accomplished more quickly if one is willing to make more extensive use of sheaf theory and other machinery; but an explicit description of canonical forms for factors of automorphy is still an essential part of the theory, and the classical approach yields this quite directly.

Consider a complex torus $J=\mathbb{C}^g/\mathscr{L}$ defined by a lattice subgroup $\mathscr{L}\subseteq\mathbb{C}^g$ of rank $2g$; and choose a set of $2g$ generators $\omega_1,\ldots,\omega_{2g}$ for the lattice $\mathscr{L}$, so that $\mathscr{L}=\Omega\cdot\mathbb{Z}^{2g}$ where $\Omega$ is the $g\times 2g$ period matrix $\Omega=(\omega_1,\ldots,\omega_{2g})$. This choice of a set of generators for $\mathscr{L}$ will be called a *marking* of the torus $J$, since it is evidently the analogue of a marking of a Riemann surface; the origin $0\in\mathbb{C}^g$ plays the role of the base point of $J$, and the paths along the vectors $\omega_i$ represent generators for the fundamental group of the torus $J$ or a basis for the first homology group of $J$. A scalar factor of automorphy $\xi\colon\mathscr{L}\times\mathbb{C}^g\longrightarrow\mathbb{C}^*$ for this torus $J$ is completely determined by the $2g$ functions $\xi(\omega_i,t)$ associated to the generators $\omega_i$ of $\mathscr{L}$; and these can be arbitrary complex analytic nowhere vanishing functions on $\mathbb{C}^g$ subject only to the conditions

$$(101)\qquad \xi(\omega_i,t+\omega_j)\,\xi(\omega_j,t)=\xi(\omega_j,t+\omega_i)\,\xi(\omega_i,t),\quad 1\leq i,j\leq 2g,$$

corresponding to the basic relations $\omega_i+\omega_j=\omega_j+\omega_i$ among the generators of $\mathscr{L}$, since $\mathscr{L}$ is the free abelian group on these generators. Choose arbitrary branches of the logarithms to define $\sigma(\omega_i,t)=\frac{1}{2\pi i}\log\xi(\omega_i,t)$; and note that these functions $\sigma(\omega_i,t)$ can be extended to a summand of automorphy for the action of the group

$\mathscr{L}$ on $\mathbb{C}^g$ precisely when $\chi_{ij}=0$ for $1\leq i,j\leq 2g$, where

$$\chi_{ij}=\sigma(\omega_i, t+\omega_j)+\sigma(\omega_j, t)-\sigma(\omega_j, t+\omega_i)-\sigma(\omega_i, t). \tag{102}$$

It is an evident consequence of (101) that $\chi_{ij}$ is always an integer; and it is evident from the definition (102) that this integer is independent of the choices of the branches of the logarithms, and that $\chi_{ij}=-\chi_{ji}$. These integers can be viewed as the entries of a $2g\times 2g$ skew-symmetric integral matrix $\chi=(\chi_{ij})$ uniquely associated to the factor of automorphy $\xi$; and the matrix $\chi=\chi(\xi)$ will be called the *characteristic matrix* of the factor of automorphy $\xi$. The preceding observation can then be summarized as the assertion that *there exists a summand of automorphy $\sigma$ such that $\xi=\exp 2\pi i\sigma$ precisely when $\chi(\xi)=0$.* The set of scalar factors of automorphy for $\mathscr{L}$ form a group under multiplication, and the mapping which associates to each factor of automorphy its characteristic matrix is evidently a homomorphism from this group into the additive group of all $2g\times 2g$ skew-symmetric integral matrices.

In order to determine all possible characteristic matrices it is convenient first to consider a more geometrical alternative interpretation of these matrices. It is quite easy to see that there exists a $C^\infty$ real-valued function $g$ such that

$$g(t)>0 \quad\text{and}\quad g(t+\omega)=|\xi(\omega, t)|^2 g(t) \quad\text{for all}\quad \omega\in\mathscr{L},\ t\in\mathbb{C}^g; \tag{103}$$

indeed the real logarithms of $|\xi(\omega, t)|^2$ form a real-valued summand of automorphy for the action of the group $\mathscr{L}$ on $\mathbb{C}^g$ and the real logarithm of $g(t)$ can then be constructed as in the proof of Theorem 2b. For any branch of $\log\xi(\omega, t)$, which must of course be a complex analytic function of $t$, and for the real branch of $\log g(t)$ it follows from (103) that

$$\log g(t+\omega)=\log\xi(\omega, t)+\overline{\log\xi(\omega, t)}+\log g(t)$$

and hence that $\bar\partial\partial\log g(t+\omega)=\bar\partial\partial\log g(t)$ for all $\omega\in\mathscr{L}$; therefore

$$\gamma(t)=\frac{1}{2\pi i}\bar\partial\partial\log g(t)$$

is a closed $C^\infty$ differential form of type (1, 1) on $\mathbb{C}^g$ which is invariant under the action of $\mathscr{L}$ and which consequently defines a closed $C^\infty$ differential form of type (1, 1) on $J=\mathbb{C}^g/\mathscr{L}$. This is evidently a real differential form, since $\overline{\gamma(t)}=\gamma(t)$. Moreover this form is determined uniquely up to the exterior derivative of a $C^\infty$ real-valued function on $J=\mathbb{C}^g/\mathscr{L}$; for if $g'$ is any other function satisfying (103) then $g'(t)=g(t)h(t)$ where $h(t)>0$ and $h(t+\omega)=h(t)$ for all $\omega\in\mathscr{L}$ and $\gamma'(t)=\gamma(t)+\frac{1}{2\pi i}\bar\partial\partial\log h(t)$. Thus to each factor of automorphy $\xi$ there is associated a unique real deRham class of type (1, 1) on $J$, represented by the differential form $\gamma(t)$, hence of course also a cohomology class in $H^2(J,\mathbb{R})$; and the characteristic matrix $\chi(\xi)$ is the matrix representing the values of this cohomology class on the canonical basis for the homology group $H_2(J,\mathbb{Z})$ of the marked torus $J$, in the following sense. For any index $1\leq i\leq 2g$ consider the singular one-chain in $\mathbb{C}^g$ represented by the mapping $\sigma_i\colon[0,1]\longrightarrow\mathbb{C}^g$ defined by $\sigma_i(x)=x\cdot\omega_i$. Since $\sigma_i(1)=\sigma_i(0)+\omega_i$ this chain represents a singular one-cycle in $J=\mathbb{C}^g/\mathscr{L}$; and these $2g$ cycles represent a basis for $H_1(J,\mathbb{Z})$. Similarly for any distinct indices $1\leq i,j\leq 2g$ consider the singular two-chain in $\mathbb{C}^g$ represented by the mapping $\sigma_{ij}\colon[0,1]^2\longrightarrow\mathbb{C}^g$ defined by $\sigma_{ij}(x_1,x_2)=x_1\cdot\omega_i+x_2\cdot\omega_j$; and note that $\sigma_{ji}=-\sigma_{ij}$, since these chains differ only in orientation. Since

$$\partial\sigma_{ij}=\sigma_i-(\sigma_i+\omega_j)+(\sigma_j+\omega_i)-\sigma_j,$$

where $(\sigma_i+\omega_j)(x)=\sigma_i(x)+\omega_j$ the chain $\sigma_{ij}$ represents a singular two-cycle in $J=\mathbb{C}^g/\mathscr{L}$; and these $\binom{2g}{2}$ cycles for $1\leq i<j\leq 2g$ represent a basis for $H_2(J,\mathbb{Z})$. Now recalling all the definitions and applying Stokes' theorem note that

$$\begin{aligned}
\int_{t\in\sigma_{ij}}\gamma(t)&=\frac{1}{2\pi i}\int_{t\in\sigma_{ij}}\bar\partial\partial\log g(t)=\frac{1}{2\pi i}\int_{t\in\sigma_{ij}}d\partial\log g(t)\\
&=\frac{1}{2\pi i}\int_{t\in\partial\sigma_{ij}}\partial\log g(t)\\
&=\frac{1}{2\pi i}\int_{t\in\sigma_i}[\partial\log g(t)-\partial\log g(t+\omega_j)]\\
&\quad+\frac{1}{2\pi i}\int_{t\in\sigma_j}[\partial\log g(t+\omega_i)-\partial\log g(t)]\\
&=-\frac{1}{2\pi i}\int_{t\in\sigma_i}\partial\log\xi(\omega_j,t)+\frac{1}{2\pi i}\int_{t\in\sigma_j}\partial\log\xi(\omega_i,t)\\
&=-\frac{1}{2\pi i}\int_{t\in\sigma_i}d\log\xi(\omega_j,t)+\frac{1}{2\pi i}\int_{t\in\sigma_j}d\log\xi(\omega_i,t)\\
&=\frac{1}{2\pi i}[\log\xi(\omega_j,0)-\log\xi(\omega_j,\omega_i)+\log\xi(\omega_i,\omega_j)-\log\xi(\omega_i,0)]\\
&=\chi_{ij}(\xi),
\end{aligned}$$

so that the matrix $\chi(\xi)$ represents the period classes of the differential form $\gamma(t)$ on the canonical basis for the homology group $H_2(J,\mathbb{Z})$ of the marked torus $J$.

Of course it follows from deRham's theorem that an arbitrary skew-symmetric matrix $\chi$ represents the period classes of some closed $C^\infty$ differential form of total degree 2 on $J$; the point of the preceding observation is that the characteristic matrix $\chi(\xi)$ of a factor of automorphy represents the period classes of a closed $C^\infty$ differential form of total degree 2 *and of type* (1, 1) on $J$. This condition can be rewritten much more simply and explicitly as follows. The cycle $\sigma_{ij}$ is evidently homologous to any translate $\sigma_{ij}+s$ by an element $s$ of the group $J$, so that

$$\chi_{ij}(\xi)=\int_{t\in\sigma_{ij}+s}\gamma(t)=\int_{t\in\sigma_{ij}}\gamma(t-s)$$

for any $s\in J$; and consequently

$$\chi_{ij}(\xi)=\int_{s\in J}\chi_{ij}(\xi)d\mu(s)=\int_{t\in\sigma_{ij}}\int_{s\in J}\gamma(t-s)d\mu(s)$$

where $d\mu(s)$ is normalized Haar measure on the group $J$. If $\gamma(t)=\sum_{kl}\gamma_{kl}(t)dt_k\wedge d\bar{t}_l$ for some $\mathscr{L}$-invariant $C^\infty$ functions $\gamma_{kl}(t)$ on $\mathbb{C}^g$ then

$$\begin{aligned}\int_{s\in J}\gamma(t-s)d\mu(s)&=\sum_{kl}\Big(\int_{s\in J}\gamma_{kl}(t-s)d\mu(s)\Big)dt_k\wedge d\bar{t}_l\\&=\sum_{kl}g_{kl}dt_k\wedge d\bar{t}_l,\end{aligned}$$

where $g_{kl}$ are constants; thus the characteristic matrix actually represents the period classes of a differential form $\gamma(t)=\sum_{kl}g_{kl}dt_k\wedge d\bar{t}_l$ with constant coefficients. These coefficients can be viewed as forming a $g\times g$ complex matrix $G=(g_{kl})$; and since $\overline{\gamma(t)}=\gamma(t)$ it follows that $\bar{G}=-{}^tG$, so that this matrix is skew-Hermitian. Now

$$\begin{aligned}\chi_{ij}&=\int_{t\in\sigma_{ij}}\gamma(t)=\sum_{kl}g_{kl}\int_{t\in\sigma_{ij}}dt_k\wedge d\bar{t}_l\\&=\sum_{kl}g_{kl}\int_{(x_1,x_2)\in[0,1]^2}(dx_1\omega_{ki}+dx_2\omega_{kj})\wedge(dx_1\bar{\omega}_{li}+dx_2\bar{\omega}_{lj})\\&=\sum_{kl}g_{kl}(\omega_{ki}\bar{\omega}_{lj}-\omega_{kj}\bar{\omega}_{li})\int_{(x_1,x_2)\in[0,1]^2}dx_1\wedge dx_2\\&=\sum_{kl}g_{kl}(\omega_{ki}\bar{\omega}_{lj}-\bar{\omega}_{li}\omega_{kj}),\end{aligned}$$

or in matrix terms

$$\chi={}^t\Omega G\bar{\Omega}-{}^t\bar{\Omega}{}^tG\Omega. \tag{104}$$

Thus the condition that the characteristic matrix $\chi(\xi)$ of a factor of automorphy $\xi$ represents the period classes of a closed $C^\infty$ differential form of type (1, 1) on $J$ is equivalent to the condition that the matrix $\chi(\xi)$ satisfies (104) for some $g\times g$ skew-Hermitian matrix $G$.

Now it is quite easy to see conversely that every $2g\times 2g$ skew-symmetric integral matrix $\chi$ satisfying (104) for some $g\times g$ skew-Hermitian matrix $G$ is actually the period matrix of a factor of automorphy; indeed this can be accomplished by writing down the requisite factor of automorphy quite explicitly, taking that factor of automorphy in the particularly simple form

$$\xi(\omega_i,t)=\exp 2\pi i\sum_{k=1}^{g}t_k\lambda_{ki},\quad 1\leq i\leq 2g, \tag{105}$$

for some complex constants $\lambda_{ki}$. These constants can be viewed as forming a $g\times 2g$ matrix $\Lambda=(\lambda_{ki})$, which will be called the *factor matrix*. The necessary and sufficient condition that the functions (105) described by the factor matrix $\Lambda$ form a factor of automorphy is that they satisfy (101); or rather more precisely, the necessary and sufficient condition that the functions (105) described by the factor matrix $\Lambda$ form a factor of automorphy with characteristic matrix $\chi$ is that the functions $\sigma(\omega_i, t)=\sum_k t_k\lambda_{ki}$ satisfy (102). Writing the latter condition out in detail, it has the form

$$\begin{aligned}\chi_{ij}&=\textstyle\sum_k\left[(t_k+\omega_{kj})\lambda_{ki}+t_k\lambda_{kj}-(t_k+\omega_{ki})\lambda_{kj}-t_k\lambda_{ki}\right]\\&=\textstyle\sum_k(\omega_{kj}\lambda_{ki}-\omega_{ki}\lambda_{kj}),\end{aligned}$$

or in matrix terms

$$\chi={}^t\Lambda\Omega-{}^t\Omega\Lambda. \tag{106}$$

Thus the necessary and sufficient condition that the functions (105) form a factor of automorphy with characteristic matrix $\chi$ is that the factor matrix $\Lambda$ satisfies (106). Now any matrix $\chi$ which satisfies (104) also satisfies (106) with $\Lambda=-G\bar{\Omega}$; and consequently any $2g\times 2g$ skew-symmetric integral matrix $\chi$ which satisfies (104) for some $g\times g$ skew-Hermitian matrix $G$ is the characteristic matrix for the factor of automorphy (105) with $\Lambda=-G\bar{\Omega}$.

It is in some ways more convenient, and it is more traditional, to rephrase the preceding discussion of the characteristic matrix of a factor of automorphy in terms of the inverse period matrix. Recall from Theorem 1a that the period matrix $\Omega$ of a complex torus has the property that the $2g\times 2g$ matrix $({}^t\Omega\,{}^t\bar{\Omega})$ is nonsingular. Writing

$$({}^t\Omega\,{}^t\bar{\Omega})^{-1}=\begin{pmatrix}P\\P_1\end{pmatrix}$$

where $P$ and $P_1$ are $g\times 2g$ matrices, note that

$$\begin{aligned}&P\,{}^t\Omega=P_1{}^t\bar{\Omega}=I,\quad \text{the } g\times g \text{ identity matrix, and}\\&P\,{}^t\bar{\Omega}=P_1{}^t\Omega=0,\quad \text{the } g\times g \text{ zero matrix;}\end{aligned}$$

thus $(P_1-\bar{P})({}^t\Omega\,{}^t\bar{\Omega})=0$, so that $P_1=\bar{P}$. Therefore

$$({}^t\Omega\,{}^t\bar{\Omega})^{-1}=\begin{pmatrix}P\\\bar{P}\end{pmatrix};$$

or equivalently, there is a unique $g\times 2g$ complex matrix $P$ such that

$$\begin{aligned}&P\,{}^t\Omega=\bar{P}\,{}^t\bar{\Omega}=I,\\&P\,{}^t\bar{\Omega}=\bar{P}\,{}^t\Omega=0,\\&{}^t\Omega P+{}^t\bar{\Omega}\bar{P}=I.\end{aligned}\tag{107}$$

Now if the $2g \times 2g$ matrix $\chi$ satisfies (104) for some $g \times g$ skew-Hermitian matrix $G$ then $\chi$ is real and skew-symmetric, and using (107) it follows that $P\chi{}^tP=0$ and $P\chi{}^t\bar{P}=G$; and conversely if the $2g \times 2g$ matrix $\chi$ is real and skew-symmetric and $P\chi{}^tP=0$ then the $g \times g$ matrix $G=P\chi{}^t\bar{P}$ is skew-Hermitian, and using (107) it follows that

$$\begin{aligned}{}^t\Omega G\bar{\Omega}-{}^t\bar{\Omega}{}^tG\Omega &= {}^t\Omega P\chi{}^t\bar{P}\bar{\Omega}-{}^t\bar{\Omega}\bar{P}{}^t\chi{}^tP\Omega\\ &=({}^t\Omega P\chi{}^t\bar{P}\bar{\Omega}+{}^t\bar{\Omega}\bar{P}\chi{}^tP\Omega)+({}^t\Omega P\chi{}^tP\Omega+{}^t\bar{\Omega}\bar{P}\chi{}^t\bar{P}\bar{\Omega})\\ &=({}^t\Omega P+{}^t\bar{\Omega}\bar{P})\chi({}^tP\Omega+{}^t\bar{P}\bar{\Omega})\\ &=\chi\,.\end{aligned}$$

Therefore the necessary and sufficient condition that a $2g \times 2g$ skew-symmetric integral matrix $\chi$ be the characteristic matrix of a factor of automorphy is that $P\chi{}^tP=0$. If this condition is fulfilled then $\chi$ is the characteristic matrix of a factor of automorphy of the form (105) where $\Lambda$ is any $g \times 2g$ matrix satisfying (106); in particular it is always possible to take $\Lambda=-G\bar{\Omega}=-P\chi{}^t\overline{P\Omega}$, but there are other possibilities as well. If $\Lambda$ satisfies (106) for the characteristic matrix $\chi$ then using (107) and recalling the properties of $\chi$ it follows firstly that $0=P\chi{}^tP=P{}^t\Lambda-\Lambda{}^tP$, so that the $g \times g$ matrix $F=\Lambda{}^tP$ is symmetric, and secondly that $G=P\chi{}^t\bar{P}=-\Lambda{}^t\bar{P}$; and therefore

$$\Lambda=\Lambda({}^tP,{}^t\bar{P})\begin{pmatrix}\Omega\\ \bar{\Omega}\end{pmatrix}=(F,\,-G)\begin{pmatrix}\Omega\\ \bar{\Omega}\end{pmatrix}=F\Omega-G\bar{\Omega}\,.$$

Conversely if $\Lambda=F\Omega-G\bar{\Omega}$ where $F$ is any symmetric matrix and $G=P\chi{}^t\bar{P}$ then

$$\begin{aligned}{}^t\Lambda\Omega-{}^t\Omega\Lambda &=({}^t\Omega{}^tF-{}^t\bar{\Omega}{}^tG)\Omega-{}^t\Omega(F\Omega-G\bar{\Omega})\\ &=-{}^t\bar{\Omega}{}^tG\Omega+{}^t\Omega G\bar{\Omega}=\chi\end{aligned}$$

as above. Therefore the most general possible factor matrix is $\Lambda=F\Omega-G\bar{\Omega}$ where $G=P\chi{}^t\bar{P}$ and $F$ is any $g \times g$ symmetric matrix.

In summary of the results so far, *a $2g \times 2g$ skew-symmetric integral matrix $\chi$ is the characteristic matrix of a factor of automorphy precisely when $P\chi{}^tP=0$, where $({}^tP\,{}^t\bar{P})$ is the inverse of the extended period matrix; indeed if this condition is satisfied then $\chi$ is the characteristic matrix of a factor of automorphy of the form $\xi(\omega_i,t)=\exp 2\pi i\sum_k t_k\lambda_{ki}$ precisely when the factor matrix $\Lambda=(\lambda_{ki})$ is of the form $\Lambda=F\Omega-G\bar{\Omega}$, where $G=P\chi{}^t\bar{P}$ and $F$ is any $g \times g$ symmetric matrix.* Moreover the characteristic matrix can be viewed as the period matrix of the exterior differential form $\gamma(t)=\sum_{kl} g_{kl}dt_k d\bar{t}_l$ where $G=(g_{kl})$.

The characteristic matrix of a factor of automorphy for a complex torus also satisfies the analogues of Theorems 2a and 2b. Indeed the proof of Theorem 2b carries over immediately to the case of a complex torus, so that any two factors of automorphy having the same characteristic matrix are continuously equivalent; in particular therefore any factor of automorphy for a complex torus is continuously equivalent to a factor of automorphy of the form (105). The analogue of Theorem 2a is somewhat more sophisticated, since it requires a knowledge of some notions and elementary properties of the divisors of complex analytic functions of several complex variables; so this result will only be discussed quite briefly here as a bit of a digression, assuming the requisite additional knowledge. If $f$ is any meromorphic relatively automorphic function for a factor of automorphy $\xi$ then the function $g(t)=|f(t)|^2$ is $C^\infty$ and satisfies (103) for all points $t$ outside the divisor of $f$; and $g$ can be modified in an arbitrarily small open neighborhood of the divisor of $f$ to yield a $C^\infty$ function $g_0$ which satisfies (103) throughout $\mathbb{C}^g$. It then follows as before that the characteristic matrix $\chi(\xi)$ is given by

$$\chi_{ij}(\xi)=\frac{1}{2\pi i}\int_{t\in\sigma_{ij}}\bar{\partial}\partial\log g_0(t)=\frac{1}{2\pi i}\int_{t\in\partial\sigma_{ij}}\partial\log g_0(t)\,;$$

and assuming that $\partial\sigma_{ij}$ is disjoint from the divisor of $f$, and hence that $g_0(t)=g(t)$ for $t\in\partial\sigma_{ij}$, it further follows that

$$\chi_{ij}(\xi)=\frac{1}{2\pi i}\int_{t\in\partial\sigma_{ij}}\partial\log g(t)=\frac{1}{2\pi i}\int_{t\in\partial\sigma_{ij}}\partial(\log f(t)+\log\bar{f}(t))$$
$$=\frac{1}{2\pi i}\int_{t\in\partial\sigma_{ij}}d\log f(t)\,.$$

This last integral can be interpreted as the intersection number or Kronecker index of the divisor of $f$ and the two-cycle $\sigma_{ij}$ on the torus, so that

$$\chi_{ij}(\xi)=\mathrm{K.I.}(\sigma_{ij},\mathfrak{d}(f))\,;$$

and this is the analogue of Theorem 2a.

The analogue of the weak form of Abel's theorem also holds for complex tori; *a scalar factor of automorphy is analytically equivalent to a flat factor of automorphy precisely when it has zero characteristic class.* As in the case of Riemann surfaces there are various proofs of this result, one rather function-theoretic proof going back to Frobenius and Appell and discussed in detail in [5], and another based on potential-theoretic methods on Kaehler manifolds which can be found in [27] or [28] for the case of complex tori or in [9] for the case of compact Kaehler manifolds of the appropriate general structure. These proofs will not be given here, but some consequences should be pointed out. The principal consequence is of course the assertion that any scalar factor of automorphy is analytically equivalent to a factor of automorphy of the form

(108) $$\xi(\omega_i,t)=\exp 2\pi i(\sigma_i+\textstyle\sum_{k=1}^{g}t_k\lambda_{ki})\,,\quad 1\le i\le 2g\,,$$

for some constants $\Lambda=(\lambda_{ki})$ and $\sigma=(\sigma_i)$; the $g\times 2g$ matrix $\Lambda$ is a factor matrix as discussed above, and the vector $\sigma\in\mathbb{C}^{2g}$ is called the *parameter.* This is indeed an immediate consequence of the weak form of Abel's theorem and the observation that any scalar factor of automorphy can be written as the product of a factor of automorphy of the form (105) and a factor of automorphy having zero characteristic matrix.

The complete classification of scalar factors of automorphy then requires an analysis of the analytic equivalences among factors of automorphy of the form (108); but, noting that the set of factors of automorphy of this form is closed under multiplication, it is sufficient merely to determine when any such factor of automorphy is analytically trivial. Clearly the factor of automorphy (108) is analytically trivial precisely when there is a complex analytic function $f$ in $\mathbb{C}^g$ such that

$$(109) \qquad f(t+\omega_i)=f(t)+n_i+\sigma_i+\textstyle\sum_k t_k\lambda_{ki}, \quad 1\le i\le 2g,$$

for some integers $n_i$. The partial derivatives $\partial^2 f/\partial t_i\partial t_j$ of any such function are necessarily $\mathscr{L}$-invariant complex analytic functions, hence constants; so that the only possible such functions are quadratic polynomials

$$f(t)=\textstyle\sum_{k,l=1}^g a_{kl}t_kt_l+\sum_{k=1}^g b_kt_k+c$$

for some complex constants $a_{kl}, b_k, c,$ where it can of course be assumed that the $g\times g$ matrix $A=(a_{kl})$ is symmetric. A function of this form is easily seen to satisfy (109) precisely when $\Lambda=2A\Omega$ and $n_i+\sigma_i=\sum_{kl}a_{kl}\omega_{ki}\omega_{li}+\sum_k b_k\omega_{ki}$; consequently the factor of automorphy (108) is analytically trivial precisely when

$$(110) \qquad \Lambda=2A\Omega \quad \text{and} \quad \sigma_i=m_i+\textstyle\sum_{kl}a_{kl}\omega_{ki}\omega_{li}+\sum_k b_k\omega_{ki}$$

for some integers $m_i$ and complex constants $a_{kl}, b_k$ such that the matrix $A=(a_{kl})$ is symmetric, or equivalently two factors of automorphy $\xi'$ and $\xi''$ of the form (108) are analytically equivalent precisely when the differences $\Lambda=\Lambda'-\Lambda''$ of their factor matrices and $\sigma=\sigma'-\sigma''$ of their parameters satisfy (110). As noted earlier though this restriction on the factor matrix is equivalent to the requirement that the two factors of automorphy have the same characteristic matrix. Furthermore then two factors of automorphy $\xi'$ and $\xi''$ of the form (108) and having the same factor matrix are analytically equivalent precisely when their parameters $\sigma'$ and $\sigma''$ have the property that $\sigma'-\sigma''\in\mathbb{Z}^{2g}+{}^t\Omega\cdot\mathbb{C}^g\subseteq\mathbb{C}^{2g}$. Therefore *the set of complex analytic equivalence classes of scalar factors of automorphy for the torus* $J=\mathbb{C}^g/\Omega\mathbb{Z}^{2g}$ *can be identified with the product*

$$X\times\mathbb{C}^{2g}/(\mathbb{Z}^{2g}+{}^t\Omega\mathbb{C}^g),$$

*where*

$$X=\{\chi\in\mathbb{Z}^{2g\times 2g}\,|\,{}^t\chi=-\chi \quad \text{and} \quad P\chi\,{}^tP=0\}.$$

Indeed a standard if not canonical representative of any equivalence class is a factor of automorphy of the form (108) where $\Lambda$ is any factor matrix with characteristic matrix $\chi\in X$. Considering the nonsingular linear isomorphism $p:\mathbb{C}^{2g}\longrightarrow\mathbb{C}^g\times\mathbb{C}^g$ defined by $p(\sigma)=(P\sigma,\overline{P\sigma})$ and recalling (107) note that $p({}^t\Omega\mathbb{C}^g)=\mathbb{C}^g\times 0$; hence $p$ induces an isomorphism

$$p:\mathbb{C}^{2g}/(\mathbb{Z}^{2g}+{}^t\Omega\mathbb{C}^g)\longrightarrow\mathbb{C}^g/\overline{P\mathbb{Z}^{2g}},$$

which can be viewed as identifying $\mathbb{C}^{2g}/(\mathbb{Z}^{2g}+{}^t\Omega\mathbb{C}^g)$ with the complex torus dual to $J=\mathbb{C}^g/\Omega\mathbb{Z}^{2g}$.

## § 17. Relatively Automorphic Functions: the Theta Series

Consider next the problem of determining the complex analytic relatively automorphic functions for a fixed factor of automorphy of the canonical form (108). As convenient preliminary observations note first that Eq. (108) can be rewritten in the form

$$\text{(111)} \qquad \xi(\omega_i, t)=\exp 2\pi i({}^t\sigma+{}^t tA)\cdot\delta_i,$$

where $\delta_i\in\mathbb{R}^{2g}$ is the column vector having the entry 1 in row $i$ and entries 0 elsewhere. Note also that the real linear transformation

$$\Omega\colon \mathbb{R}^{2g}\longrightarrow\mathbb{C}^g$$

defined by the period matrix $\Omega$ is an isomorphism, as proved in Theorem 1a, and transforms the lattice subgroup $\mathbb{Z}^{2g}\subset\mathbb{R}^{2g}$ into the lattice subgroup $\mathscr{L}\subset\mathbb{C}^g$, since $\Omega\delta_i=\omega_i$; and if $t=\Omega x$ then evidently $x={}^tPt+{}^t\overline{Pt}$.

If $f(t)$ is a complex analytic relatively automorphic function for the factor of automorphy (111) and is not identically zero then introduce the auxiliary function

$$F(x)=f(\Omega x)\exp 2\pi i({}^t xAx+{}^t ax)$$

of the variable $x\in\mathbb{R}^{2g}$, for some $2g\times 2g$ complex symmetric matrix $A$ and some point $a\in\mathbb{C}^{2g}$, and note that this function is also nontrivial and satisfies

$$\begin{aligned}F(x+\delta_i)&=f(\Omega x+\omega_i)\exp 2\pi i[{}^t(x+\delta_i)A(x+\delta_i)+{}^t a(x+\delta_i)]\\&=F(x)\exp 2\pi i[{}^t x({}^t\Omega\Lambda+2A)\delta_i+({}^t\sigma+{}^t\delta_i A+{}^t a)\delta_i]\end{aligned}$$

for any index in the range $1\leq i\leq 2g$. In particular choosing $A=-\frac{1}{4}({}^t\Omega\Lambda+{}^t\Lambda\Omega)$ and $a_i=-\sigma_i-A_{ii}$ note that ${}^t\Omega\Lambda+2A=-\frac{1}{2}\chi$, where $\chi$ is the characteristic matrix of the factor of automorphy (111), and that

$$F(x+\delta_i)=F(x)\exp(-\pi i\,{}^t x\chi\delta_i).$$

Since ${}^t x\chi\delta_i\in\mathbb{R}$ it follows that $|F(x+\delta_i)|=|F(x)|$; the continuous function $|F(x)|$ is thus invariant under the lattice subgroup $\mathbb{Z}^{2g}\subset\mathbb{R}^{2g}$, and since $\mathbb{R}^{2g}/\mathbb{Z}^{2g}$ is compact this function is consequently uniformly bounded in $\mathbb{R}^{2g}$. For the constant

$$M=\sup_{\chi\in\mathbb{R}^{2g}}|F(x)|<\infty$$

it then follows that

$$|f(\Omega x)| \le M |\exp - 2\pi i({}^t x A x + {}^t a x)|$$

for all points $x \in \mathbb{R}^{2g}$. Note that if $t = \Omega x$ then

$$\begin{aligned} {}^t x A x &= -\tfrac{1}{4}({}^t t P + {}^t \overline{tP})({}^t \Omega \Lambda + {}^t \Lambda \Omega)({}^t P t + {}^t \overline{Pt}) \\ &= -\tfrac{1}{2}{}^t t F t + \tfrac{1}{2}{}^t t G \bar{t}, \end{aligned}$$

where $F = \Lambda\,{}^t P$ is symmetric and $G = -\Lambda\,{}^t\bar{P} = P\chi\,{}^t\bar{P}$, and

$$\begin{aligned} {}^t a x &= {}^t a({}^t P t + {}^t \overline{Pt}) \\ &= ({}^t a - {}^t \bar{a}){}^t P t + (\text{real number}); \end{aligned}$$

so the inequality can be rewritten

$$|f(t) \exp h(t)| \le M |\exp - \pi i\,{}^t t G \bar{t}|,$$

where $h(t) = -\pi i\,{}^t t F t + 2\pi i({}^t a - {}^t \bar{a}){}^t P t$ is a complex analytic function of $t$. Note that the matrix $G$ is skew-Hermitian, so the matrix $-\pi i G = H$ is Hermitian. By a nonsingular complex linear change of coordinates $t = Cs$ in $\mathbb{C}^g$ the Hermitian form ${}^t t H \bar{t}$ can be reduced to the diagonal form $\sum_{i=1}^g h_i |s_i|^2$ where $h_i = 0$ or $\pm 1$; and in terms of these new coordinates

$$|f(Cs) \exp h(Cs)| \le M \cdot \prod_{i=1}^g \exp h_i |s_i|^2$$

for all points $s = (s_1, \ldots, s_g) \in \mathbb{C}^g$. Now if $h_i \le 0$ for some index $i$ then the function $f(Cs) \exp h(Cs)$ is a bounded entire function of $s_i$ for any fixed values of the remaining variables, hence must of course be constant in $s_i$; that is, if $h_i \le 0$ the function $f(Cs) \exp h(Cs)$ is independent of the variable $s_i$. The function $f(Cs)$ is a complex analytic relatively automorphic function for the given factor of automorphy, in terms of the new variable $s$, and the function $f(Cs) \exp h(Cs)$ is a complex analytic relatively automorphic function for an analytically equivalent factor of automorphy, since $\exp h(Cs)$ is nowhere vanishing; hence there is a factor of automorphy which is analytically equivalent to the given factor of automorphy and which is independent of the variable $s_i$.

To summarize these observations, a factor of automorphy for the action of the lattice subgroup $\mathscr{L}$ on the space $\mathbb{C}^g$ will be called *degenerate* if there is an analytically equivalent factor of automorphy which can be made independent of some of the variables in $\mathbb{C}^g$ after a nonsingular linear change of coordinates. *If a nondegenerate factor of automorphy with characteristic matrix $\chi$ admits a nontrivial complex analytic relatively automorphic function then the matrix $-\pi i P\chi\,{}^t\bar{P}$ is positive definite Hermitian.*

This necessary condition that a nondegenerate factor of automorphy admit a nontrivial complex analytic relatively automorphic function is also sufficient, as can be seen from a closer analysis of the form of such functions. For this analysis it is convenient to simplify the factor of automorphy (111) further by suitable changes of coordinates in $\mathbb{C}^g$ and of generators for the lattice subgroup $\mathscr{L}\subset\mathbb{C}^g$. First under a nonsingular linear change of coordinates $t=Cs$ in $\mathbb{C}^g$ effected by a matrix $C\in GL(g,\mathbb{C})$ the lattice vector $\omega_i$ is transformed into the lattice vector $\omega_i'=C^{-1}\omega_i$, hence the period matrix $\Omega$ is transformed into $\Omega'=C^{-1}\Omega$; and the inverse period matrix $P$ is easily seen to be transformed into $P'={}^tCP$. Correspondingly the factor of automorphy (111) becomes

$$\xi'(\omega_i', s)=\xi(\omega_i, Cs)=\exp 2\pi i({}^t\sigma+{}^ts\,{}^tC\Lambda)\cdot\delta_i\,,$$

and that has the effect merely of transforming the factor matrix $\Lambda$ into $\Lambda'={}^tC\Lambda$. It is apparent upon substituting these results into (106) that the characteristic matrix $\chi$ is unchanged under such a coordinate transformation; but of course that is even more apparent as a consequence of the geometric interpretation of the characteristic matrix. On the other hand the Hermitian matrix $H=-\pi iP\chi{}^t\bar{P}$ is transformed into $H'={}^tCH\bar{C}$; hence that matrix can be reduced to standard normal form by such a change of coordinates. Next under a change of basis for the lattice subgroup $\mathscr{L}\subset\mathbb{C}^g$ of the form $\omega_i'=\sum_{j=1}^{g} m_{ij}\omega_j$ where $M=(m_{ij})\in SL(2g,\mathbb{Z})$ the period matrix $\Omega$ is of course transformed into $\Omega'=\Omega{}^tM$; and the inverse period matrix $P$ is easily seen to be transformed into $P'=PM^{-1}$. As for the factor of automorphy (111) note first as a consequence of the basic defining relations for a factor of automorphy that for any integer $m>0$

$$\begin{aligned}\xi(m\omega_i, t)&=\prod_{j=1}^{m}\xi(\omega_i, t+(j-1)\omega_i)\\&=\exp 2\pi i\sum_{j=1}^{m}({}^t\sigma+{}^tt\Lambda+(j-1){}^t\omega_i\Lambda)\cdot\delta_i\\&=\exp 2\pi i(m{}^t\sigma+\tfrac{1}{2}m(m-1){}^t\omega_i\Lambda+m{}^tt\Lambda)\cdot\delta_i\,;\end{aligned}$$

and for any integer $m<0$

$$\begin{aligned}\xi(m\omega_i, t)&=\xi(-m\omega_i, t+m\omega_i)^{-1}\\&=\exp -2\pi i(-m{}^t\sigma+\tfrac{1}{2}m(m+1){}^t\omega_i\Lambda-m{}^t(t+m\omega_i)\Lambda)\cdot\delta_i\\&=\exp 2\pi i(m{}^t\sigma+\tfrac{1}{2}m(m-1){}^t\omega_i\Lambda+n{}^tt\Lambda)\cdot\delta_i\,,\end{aligned}$$

so the same formula holds for all integers $m\in\mathbb{Z}$. Then using this formula and the basic defining relations for a factor of automorphy note that

$$\begin{aligned}\xi(\omega_i', t)&=\xi(\textstyle\sum_{j=1}^{g} m_{ij}\omega_j, t)\\&=\prod_{j=1}^{g}\xi(m_{ij}\omega_j, t+\textstyle\sum_{k=j+1}^{g} m_{ik}\omega_k)\\&=\exp 2\pi i\sum_{j=1}^{g}(m_{ij}{}^t\sigma+\tfrac{1}{2}m_{ij}(m_{ij}-1){}^t\omega_j\Lambda+m_{ij}{}^t(t+\textstyle\sum_{k=j+1}^{g} m_{ik}\omega_k)\Lambda)\cdot\delta_j\\&=\exp 2\pi i(\sigma_i'+\textstyle\sum_{j=1}^{g} m_{ij}{}^tt\Lambda\delta_j)\quad\text{for some constant}\quad\sigma_i'\\&=\exp 2\pi i({}^t\sigma'+{}^tt\Lambda')\cdot\delta_i\,,\end{aligned}$$

where ${}^t\sigma'=(\sigma_1',\ldots,\sigma_g')$ and $\Lambda'=\Lambda{}^tM$. It is apparent upon substituting these results into (106) that the characteristic matrix $\chi$ is transformed into $\chi'=M\chi{}^tM$; hence that matrix can be reduced to standard normal form by such a change of coordinates. On the other hand the Hermitian matrix $H=-\pi i P\chi {}^t\bar{P}$ is unchanged under such a transformation; and that too is evident from the interpretation of the matrix $G=\frac{i}{\pi}H$ as a differential form on the quotient manifold $J=\mathbb{C}^g/\mathcal{L}$.

Now consider a factor of automorphy of the form (111) with characteristic matrix $\chi$ such that $H=-\pi i G=-\pi i P\chi {}^t\bar{P}$ is positive definite Hermitian. Since $P\chi {}^tP=0$ the matrix

$$\begin{pmatrix} P \\ \bar{P} \end{pmatrix}\chi({}^t\bar{P}\,{}^tP)=\begin{pmatrix} G & 0 \\ 0 & \bar{G} \end{pmatrix} \tag{112}$$

is necessarily nonsingular, hence $\chi$ must also be nonsingular. It is a theorem of Frobenius [5] that for any nonsingular $2g\times 2g$ skew-symmetric integer matrix $\chi$ there exists a matrix $M\in SL(2g,\mathbb{Z})$ such that

$$M\chi {}^tM=\begin{pmatrix} 0 & D \\ -D & 0 \end{pmatrix},$$

where $D=\operatorname{diag}(d_1,\dots,d_g)$ is a diagonal matrix with positive integers $d_i$ along the diagonal; these integers can be so chosen that $d_i$ is a divisor of $d_{i+1}$, and are then uniquely determined by the matrix $\chi$. Thus after a suitable change of basis for the lattice subgroup $\mathcal{L}\subset\mathbb{C}^g$ the characteristic matrix $\chi$ can be reduced to such a normal form

$$\chi=\begin{pmatrix} 0 & D \\ -D & 0 \end{pmatrix};$$

the factor of automorphy (111) must be changed correspondingly, but remains of the same general form. Upon taking inverses on both sides of Eq. (112) that equation becomes

$$\begin{pmatrix} \bar{\Omega} \\ \Omega \end{pmatrix}\chi^{-1}({}^t\Omega\,{}^t\bar{\Omega})=\begin{pmatrix} G^{-1} & 0 \\ 0 & \bar{G}^{-1} \end{pmatrix},$$

so that $\bar{\Omega}\chi^{-1}{}^t\Omega=G^{-1}$ and $\Omega\chi^{-1}{}^t\Omega=0$; and upon writing $\Omega=(\Omega_1,\Omega_2)$ and using the above normal form for the characteristic matrix $\chi$ the latter two equations reduce to

$$\begin{aligned} &\bar{\Omega}_2D^{-1}{}^t\Omega_1-\bar{\Omega}_1D^{-1}{}^t\Omega_2=G^{-1}, \\ &\Omega_2D^{-1}{}^t\Omega_1-\Omega_1D^{-1}{}^t\Omega_2=0. \end{aligned} \tag{113}$$

It follows readily from the first of these equations that the matrix $\Omega_1$ is necessarily nonsingular; for if $\Omega_1$ were singular there would exist a nonzero vector $c\in\mathbb{C}^g$ such that ${}^t\Omega_1c=0$, and then ${}^t\bar{c}(-\pi i G)c=0$ in contradiction to the assumption that $-\pi i G$ is positive definite Hermitian. Then applying the nonsingular linear change of coordinates in $\mathbb{C}^g$ determined by the matrix $\Omega_1$ reduces the period matrix to the form $\Omega=(I, \Omega_2)$ without changing the characteristic matrix $\chi$; again the factor of automorphy (111) must be changed correspondingly, but remains of the same general form. After this further change Eqs. (113) reduce to $\bar{\Omega}_2 D^{-1} - D^{-1}{}^t\Omega_2 = G^{-1}$ and $\Omega_2 D^{-1} - D^{-1}{}^t\Omega_2 = 0$. The matrix $\Omega_2 D^{-1}$ is thus symmetric, hence so is $\bar{\Omega}_2 D^{-1}$; and the first equation can be rewritten as $G^{-1} = (\bar{\Omega}_2 - \Omega_2)D^{-1}$ so that the matrix $G^{-1}$ is also symmetric. Finally the factor matrix has the form $\Lambda = F\Omega - G\bar{\Omega}$ where $F$ is some $g\times g$ symmetric complex matrix; indeed $F$ can be taken to be an arbitrary $g\times g$ symmetric matrix by passing to an analytically equivalent factor of automorphy of the same form. In particular taking $F=G$, which is permissible since $G$ was shown to be symmetric, it follows that $\Lambda = G(\Omega-\bar{\Omega}) = (0, G(\Omega_2 - \bar{\Omega}_2))$; and recalling the above formula for $G^{-1}$ it further follows that $\Lambda = (0, -D)$.

In summary, *for any factor of automorphy with characteristic matrix* $\chi$ *such that* $H = -\pi i P\chi{}^t\bar{P}$ *is positive definite Hermitian there exist a choice of basis for the lattice subgroup* $\mathscr{L}\subset\mathbb{C}^g$ *and a choice of coordinates in* $\mathbb{C}^g$ *such that* $\Omega=(I, \Omega_2)$, *that*

$$\chi = \begin{pmatrix} 0 & D \\ -D & 0 \end{pmatrix}$$

*where* $D = \operatorname{diag}(d_1, \dots, d_g)$ *for some positive integers* $d_i$, *and that the factor of automorphy is analytically equivalent to a factor of automorphy of the form* (111) *with the factor matrix* $\Lambda = (0, -D)$. Furthermore the $g\times g$ complex matrix $Z = D\Omega_2 = D(\Omega_2 D^{-1})D$ is symmetric and its imaginary part $\operatorname{Im} Z = \frac{1}{2i}(Z-\bar{Z}) = \frac{\pi}{2} D H^{-1} D$ is a $g\times g$ real positive definite symmetric matrix.

After having reduced the characteristic matrix $\chi$ and the factor matrix $\Lambda$ to the preceding normal form the factor of automorphy (111) becomes

$$\xi(\omega_i, t) = \begin{cases} \exp 2\pi i \sigma_i & \text{for} \quad 1\le i\le g, \\ \exp 2\pi i(\sigma_i - d_{i-g} t_{i-g}) & \text{for} \quad g+1 \le i \le 2g\,. \end{cases}$$

By passing to an analytically equivalent factor of automorphy the parameter $\sigma\in\mathbb{C}^{2g}$ can be replaced by $\sigma + n + {}^t\Omega c$ for any $n\in\mathbb{Z}^{2g}$ and $c\in\mathbb{C}^g$, as noted previously; hence it can be assumed that $\sigma_i = 0$ for $1\le i\le g$. Then upon writing $\tau_i = \sigma_{i+g}$ for $1\le i\le g$ the factor of automorphy (111) finally becomes

$$\begin{aligned} &\xi(\omega_i, t) = 1, \\ &\xi(\omega_{i+g}, t) = \exp 2\pi i(\tau_i - d_i t_i) = \exp 2\pi i({}^t\tau - {}^t t D)\cdot\delta_i \end{aligned} \tag{114}$$

for $1\le i\le g$.

Now if $f(t)$ is any complex analytic relatively automorphic function for the factor of automorphy (114) then $f(t)$ is invariant under the lattice subgroup $\mathbb{Z}^g\subset\mathbb{C}^g$ generated by the vectors $\omega_1, \dots, \omega_g$ and hence $f(t)$ has an analytic Fourier expansion of the form

$$f(t) = \textstyle\sum_{n\in\mathbb{Z}^g} a_n \exp 2\pi i\, {}^t n t\,.$$

It follows by simple substitution that

$$f(t+\omega_{g+i})=\sum_n a_n \exp 2\pi i\,{}^t n(t+\omega_{g+i})$$

for $1\leq i\leq g$, while if $f(t)$ is relatively automorphic then

$$\begin{aligned} f(t+\omega_{g+i}) &= \xi(\omega_{g+i}, t) f(t) \\ &= \sum_n a_n \exp 2\pi i[{}^t\tau\delta_i + {}^t(n-D\delta_i)t] \end{aligned}$$

for $1\leq i\leq g$; hence upon comparison,

$$(115)\qquad a_{n+D\delta_i}=a_n \exp 2\pi i({}^t n\Omega_2-{}^t\tau)\cdot\delta_i\,.$$

Conversely it is clear that any convergent analytic Fourier expansion in which the coefficients satisfy (115) is relatively automorphic for the factor of automorphy (114). Introducing the rational vectors $v\in\mathbb{R}^g$ with components satisfying $0\leq v_i<1$ and $d_i v_i\in\mathbb{Z}$, note that the vectors $D(v+n)$ vary simply over the lattice $\mathbb{Z}^g$ as $n$ varies over $\mathbb{Z}^g$ and $v$ varies over its $d=\det D=d_1\ldots d_g$ possible values. It is evident that (115) can only be satisfied when the coefficients $a_{D(v+n)}$ are given by a formula of the form

$$a_{D(v+n)}=a_{Dv}\exp 2\pi i(\tfrac{1}{2}{}^t nZn+{}^t zn)$$

for some $g\times g$ complex symmetric matrix $Z$ and some vector $z\in\mathbb{C}^g$. To determine these parameters explicitly note that by substitution

$$\begin{aligned} a_{D(v+n+\delta_i)} &= a_{Dv}\exp 2\pi i[\tfrac{1}{2}{}^t(n+\delta_i)Z(n+\delta_i)+{}^t z(n+\delta_i)] \\ &= a_{D(v+n)}\exp 2\pi i({}^t nZ+\tfrac{1}{2}{}^t\delta_i Z+{}^t z)\cdot\delta_i\,, \end{aligned}$$

while by (115)

$$a_{D(v+n+\delta_i)}=a_{D(v+n)}\exp 2\pi i({}^t nD\Omega_2+{}^t vD\Omega_2-{}^t\tau)\cdot\delta_i\,;$$

therefore $Z=D\Omega_2$, which is symmetric as noted earlier, and $z=D\Omega_2 v-\tau-\varepsilon$ where $\varepsilon_i=\frac{1}{2}d_i\omega_{i,i+g}$. Thus the complex analytic relatively automorphic functions for the factor of automorphy (114) are precisely the convergent analytic Fourier expansions of the form

$$\begin{aligned} f(t) &= \sum_v\sum_n a_{D(v+n)}\exp 2\pi i\,{}^t(v+n)Dt \\ &= \sum_v\sum_n a_{Dv}\exp 2\pi i(\tfrac{1}{2}{}^t nZn+{}^t zn+{}^t(v+n)Dt) \\ &= \sum_v\sum_n a_{Dv}\exp 2\pi i[\tfrac{1}{2}{}^t nD\Omega_2 n+({}^t vD\Omega_2-{}^t\tau-{}^t\varepsilon)n+{}^t(v+n)Dt] \\ &= \sum_v a_{Dv}\exp 2\pi i[{}^t v(\tau+\varepsilon)-\tfrac{1}{2}{}^t vD\Omega_2 v]\,\theta[v\,|-\tau-\varepsilon](t)\,, \end{aligned}$$

where

$$(116)\qquad \theta[v\,|-\tau-\varepsilon](t)=\sum_n \exp 2\pi i[\tfrac{1}{2}{}^t(v+n)D\Omega_2(v+n)+{}^t(v+n)(Dt-\tau-\varepsilon)]\,.$$

The latter series, called the *theta series with characteristic* $[v|-\tau-\varepsilon]$, is easily seen to be absolutely uniformly convergent on any compact subset of $\mathbb{C}^g$, since the matrix $\operatorname{Im} D\Omega_2$ is positive definite; and the $d$ coefficients $a_{Dv}$ can be quite arbitrary complex numbers. Thus *the space of complex analytic relatively automorphic functions for the factor of automorphy* (114) *is the d-dimensional vector space spanned by the theta series* (116). Consequently of course *a nondegenerate factor of automorphy with characteristic matrix* $\chi$ *admits a nontrivial complex analytic relatively automorphic function precisely when the matrix* $H=-\pi i P\chi{}^t\bar{P}$ *is positive definite Hermitian.*

## § 18. Jacobi Varieties: Abelian and Riemannian Theta Functions

Considering now the special tori which occur as Jacobi varieties of compact Riemann surfaces, recall from the discussion of Riemann's equality and inequality (Theorem 3) that the period matrix $\Omega$ of the Jacobi variety $J(M)$ of a compact Riemann surface $M$ can always be put into the canonical form $\Omega=(I,\Omega_2)$ where $\Omega_2$ is symmetric and has positive definite imaginary part. In view of the preceding discussion that means that the matrix

$$\chi=\begin{pmatrix} 0 & I \\ -I & 0 \end{pmatrix}$$

is the characteristic matrix of a factor of automorphy of the canonical form (114) with the factor matrix $\Lambda=(0,-I)$; that factor of automorphy can then be written

$$(117)\qquad \begin{cases} \xi_\tau(\omega_i,t)=1\,, \\ \xi_\tau(\omega_{i+g},t)=\exp 2\pi i(\tau_i-t_i)\,, \end{cases}$$

for $1\le i\le g$, where $\tau\in\mathbb{C}^g$ is the appropriate parameter. Furthermore the space of complex analytic relatively automorphic functions for this factors of automorphy is the one-dimensional vector space spanned by the theta series

$$\theta[0|-\tau-\varepsilon](t)=\theta(t-\tau-\varepsilon)\,,$$

where

$$(118)\qquad \theta(t)=\textstyle\sum_n \exp 2\pi i[\tfrac{1}{2}{}^t n\Omega_2 n+{}^t nt]$$

and $\varepsilon\in\mathbb{C}^g$ is the vector with components $\varepsilon_i=\frac{1}{2}\omega_{i,i+g}$. The theta series (118) represents the *Abelian theta function* for the Jacobi variety $J(M)$.

The restriction of any translate $\theta(t-\tau-\varepsilon)$ of the Abelian theta function to the image of the normalized Jacobi imbedding $\tilde{\phi}_1\colon \tilde{M}\longrightarrow\mathbb{C}^g$ yields a complex analytic function $\theta(\tilde{\phi}(z)-\tilde{\phi}(z_0)-\tau-\varepsilon)$ on $\tilde{M}$; and since the mapping $\tilde{\phi}_1$ commutes with the actions of the group $\Gamma$ on $\tilde{M}$ and of the group $\mathscr{L}$ on $\mathbb{C}^g$ it follows that $\theta(\tilde{\phi}(z)-$

$\tilde{\phi}(z_0)-\tau-\varepsilon)$ is a complex analytic relatively automorphic function for the factor of automorphy defined by

$$(119)\qquad \begin{cases} \xi_\tau(A_i, z)=1\,, \\ \xi_\tau(B_i, z)=\exp 2\pi i(\tau_i+w_i(z_0)-w_i(z)) \end{cases}$$

for $1\leq i\leq g$. The function $\theta(\tilde{\phi}(z)-\tilde{\phi}(z_0)-\tau-\varepsilon)$ is called the *Riemannian theta function* for the surface $M$ associated to the parameter $\tau$; and the corresponding factor of automorphy (114) is called the *Riemannian theta factor of automorphy* associated to the parameter $\tau$.

The factor of automorphy (117) is of course merely the translate $T_\tau\xi$ of the basic factor of automorphy $\xi$ for $J(M)$ defined by

$$\xi(\omega_i, t)=1\,,\qquad \xi(\omega_{i+g}, t)=e^{-2\pi i t_i}$$

for $1\leq i\leq g$; thus $\xi_\tau=T_\tau\xi$. This factor of automorphy $\xi$ represents a complex analytic line bundle over the Jacobi variety $J(M)$; and the Riemannian theta factor of automorphy (119) represents the restriction of the translated line bundle $T_\tau\xi$ to the submanifold $W_1\subseteq J(M)$ when $W_1$ is identified with the Riemann surface $M$ by the normalized Jacobi mapping, or equivalently the restriction of the line bundle $\xi$ to the translated submanifold $W_1-\tau$ when that too is identified with the Riemann surface $M$. It is an interesting exercise to identify this restricted line bundle over $M$, as follows.

Choosing as logarithms of the functions (119) the functions

$$\sigma(A_i, z)=0\,,\qquad \sigma(B_i, z)=\tau_i+w_i(z_0)-w_i(z)\,,$$

the characteristic class of the factor of automorphy (119) is the integer

$$\begin{aligned} c(\xi)&=\textstyle\sum_{i=1}^{g}[\sigma(A_i, B_iA_i^{-1}B_i^{-1}C_{i+1}\dots C_gz)+\sigma(B_i, A_i^{-1}B_i^{-1}C_{i+1}\dots C_gz) \\ &\qquad -\sigma(A_i, A_i^{-1}B_i^{-1}C_{i+1}\dots C_gz)-\sigma(B_i, B_i^{-1}C_{i+1}\dots C_gz)] \\ &=\textstyle\sum_{i=1}^{g}[0+\tau_i+w_i(z_0)-w_i(A_i^{-1}B_i^{-1}C_{i+1}\dots C_gz) \\ &\qquad -0-\tau_i-w_i(z_0)+w_i(B_i^{-1}C_{i+1}\dots C_gz)] \\ &=\textstyle\sum_{i=1}^{g}[-\omega_i(A_i^{-1}B_i^{-1}C_{i+1}\dots C_g)+\omega_i(B_i^{-1}C_{i+1}\dots C_g)] \\ &=\textstyle\sum_{i=1}^{g}\omega_i(A_i) \\ &=g\,, \end{aligned}$$

recalling that $\omega_i(A_j)=\delta^i_j$. Thus the factor of automorphy (119) is analytically equivalent to a factor of automorphy of the form $\rho_t\zeta^g$ for some point $t\in\mathbb{C}^g$. To determine this point $t$ more explicitly, choose a nontrivial meromorphic relatively automorphic function $f$ for the factor of automorphy (119) and write $\mathfrak{d}(f)=\sum_j \nu_j\cdot\Gamma z_j$ where $z_j$ are distinct points lying inside the canonical fundamental polygon $\Delta$; of course the marking can be so chosen that $\mathfrak{d}(f)$ is disjoint from $\partial\Delta$.

The flat factor of automorphy $\rho_t$ then represents the complex analytic line bundle associated to the divisor $\sum_j \nu_j \cdot z_j - g \cdot z_0$ on $M$, and it follows from Theorem 5 that

$$t_i = \sum_j \nu_j w_i(z_j) - g w_i(z_0) = \sum_j \nu_j (w_i(z_j) - w_i(z_0));$$

and by the residue theorem therefore

$$t_i = \frac{1}{2\pi i} \int_{\partial \Delta} (w_i(z) - w_i(z_0)) df(z)/f(z).$$

Note that $df(Tz)/f(Tz) = df(z)/f(z) + d\log \xi(T, z)$ for any $T \in \Gamma$. Note further as a consequence of (119) that $d\log \xi(A_k, z) = 0$ and $d\log \xi(B_k, z) = -2\pi i \omega_k(z)$; and either as the result of a simple calculation using these two formulas, or better as an immediate consequence of the observation that the factor of automorphy (119) is induced by a factor of automorphy for the abelianized group $\Gamma/[\Gamma, \Gamma] \cong \mathscr{L}$, that $d\log \xi(C_k, z) = 0$. Then recalling the explicit form for the boundary of the canonical fundamental polygon $\Delta$ as given in § 4 it follows easily that

$$\begin{aligned} t_i = \frac{1}{2\pi i} \sum_{k=1}^{g} \{ & \int_{\tilde{\alpha}_k} [w_i(z) - w_i(z_0)] \cdot [df(z)/f(z) + d\log \xi(C_1 \dots C_{k-1}, z)] \\ & + \int_{\tilde{\beta}_k} [w_i(z) - w_i(z_0) + \delta_k^i][df(z)/f(z) + d\log \xi(C_1 \dots C_{k-1} A_k, z)] \\ & - \int_{\tilde{\alpha}_k} [w_i(z) - w_i(z_0) + \omega_{ik}][df(z)/f(z) + d\log \xi(C_1 \dots C_k B_k, z)] \\ & - \int_{\tilde{\beta}_k} [w_i(z) - w_i(z_0)][df(z)/f(z) + d\log \xi(C_1 \dots C_k, z)] \} \\ = \frac{1}{2\pi i} \sum_{k=1}^{g} \{ & -\omega_{ik} \int_{\tilde{\alpha}_k} df(z)/f(z) + 2\pi i \int_{\tilde{\alpha}_k} [w_i(z) - w_i(z_0) + \omega_{ik}] \omega_k(z) \\ & + \delta_k^i \int_{\tilde{\beta}_k} df(z)/f(z) \}. \end{aligned}$$

Choosing any branch of $\log f(z)$ along the arc $\tilde{\alpha}_k$ note that

$$\int_{\tilde{\alpha}_k} df(z)/f(z) = \int_{\tilde{\alpha}_k} d\log f(z) = \log f(A_k z_0) - \log f(z_0) = 2\pi i\, m_k$$

for some integer $m_k$, since $f(A_k z_0) = f(z_0)$; and choosing any branch of $\log f(z)$ along the arc $\tilde{\beta}_k$ note similarly that

$$\int_{\tilde{\beta}_k} df(z)/f(z) = \log f(B_k z_0) - \log f(z_0) = 2\pi i(\tau_k + n_k)$$

for some integer $n_k$, since $f(B_k z_0) = \xi(B_k, z_0) f(z_0) = e^{2\pi i \tau_k} f(z_0)$. The constants

$$r_i = \sum_{k=1}^{g} \int_{\tilde{\alpha}_k} [w_i(z) - w_i(z_0)] \omega_k(z) \tag{120}$$

are uniquely determined by the marking of the surface, and are called the *Riemann constants*; the vector ${}^t r = (r_1, \dots, r_g)$ represents a point $r \in J(M)$ called the *Riemann point* in the Jacobi variety of the marked Riemann surface $M$. (This differs slightly from the traditional definition of the Riemann vector $r \in \mathbb{C}^g$.) Combining these observations it follows that

$$\begin{aligned} t_i &= r_i + \sum_{k=1}^{g} \{-\omega_{ik} m_k + \omega_{ik} + \delta_k^i(\tau_k + n_k)\} \\ &= r_i + \tau_i + \lambda_i , \end{aligned}$$

where $\lambda_i = \sum_{k=1}^{g} (\delta_k^i n_k + \omega_{ik}(1 - m_k))$ and hence where

$$\lambda = {}^t(\lambda_1, \dots, \lambda_g) \in \mathscr{L} = \Omega \mathbb{Z}^{2g} .$$

Since the analytic equivalence class of the flat factor of automorphy $\rho_t$ is unchanged when $t$ is replaced by $t+\lambda$ for any lattice vector $\lambda \in \mathscr{L}$, the term $\lambda$ can be ignored altogether. In summary then *the Riemannian theta factor of automorphy* (119) *associated to the parameter* $\tau \in \mathbb{C}^g$ *has characteristic class g and is analytically equivalent to the factor of automorphy* $\rho_{\tau+r}\zeta^g$, where $r \in J(M)$ is the Riemann point represented by the vector with components (120). Consequently *whenever the Riemannian theta function* $\theta(\tilde{\phi}(z) - \tilde{\phi}(z_0) - \tau - \varepsilon)$ *is not identically zero it is a relatively automorphic function for a factor of automorphy analytically equivalent to* $\rho_{\tau+r}\zeta^g$; *so that if*

$$\mathfrak{d}(\theta(\tilde{\phi}(z) - \tilde{\phi}(z_0) - \tau - \varepsilon)) = p_1 + \cdots + p_g$$

*for some points* $p_1, \dots, p_g \in M$ *then*

$$\phi(p_1 + \cdots + p_g - g p_0) = r + \tau .$$

These observations can now be used to determine the zero locus of the Abelian theta function $\theta(t)$ in $J(M)$. First consider a point $t_0 \in \mathbb{C}^g$ such that $\theta(t_0) = 0$. If the function $f(z) = \theta(\tilde{\phi}(z) - \tilde{\phi}(z_0) + t_0)$ does not vanish identically then it is a nontrivial Riemannian theta function associated to the parameter $\tau = -\varepsilon - t_0$. This function vanishes at $g$ points of the Riemann surface $M$, including the base point $p_0 \in M$ since $\theta(t_0) = 0$ by assumption; and if $\mathfrak{d}(f) = p_0 + p_1 + \cdots + p_{g-1}$ then $r + \tau = r - \varepsilon - t_0 = \phi(p_0 + p_1 + \cdots + p_{g-1} - g p_0) = \phi(p_1 + \cdots + p_{g-1} - (g-1) p_0)$, so that $r - \varepsilon - t_0 \in W_{g-1}$ or equivalently $t_0 \in -W_{g-1} + r - \varepsilon$. If the function $f(z)$ does vanish identically let $n$ be the least integer such that there are points $z_1, \dots, z_n \in \tilde{M}$ for which the function $f_n(z) = \theta(\tilde{\phi}(z) - \tilde{\phi}(z_0) + t_0 - \tilde{\phi}(z_1 + \cdots + z_n - n z_0))$ does not vanish identically; there evidently exists such an integer, since $f(t) \not\equiv 0$ and any point of $J(M)$ can be represented by $\tilde{\phi}(z_1 + \cdots + z_g - g z_0)$ for some points $z_i \in \tilde{M}$ as a consequence of the Jacobi inversion theorem. The function $f_n(z)$ is then a nontrivial Riemannian theta function associated to the parameter $\tau = \tilde{\phi}(z_1 + \cdots + z_n - n z_0) - \varepsilon - t_0$. This function vanishes at $g$ points of the Riemann surface $M$, including the points $p_1, \dots, p_n$ represented by $z_1, \dots, z_n$ since $f_n(z_i) = \theta(t_0 - \tilde{\phi}(z_1 + \cdots + z_{i-1} + z_{i+1} + \cdots + z_n - (n-1) z_0) = f_{n-1}(z_0)$ and $f_{n-1}(z) \equiv 0$ by assumption; and if $\mathfrak{d}(f_n) = p_1 + \cdots + p_n + p_{n+1} + \cdots + p_g$ then

$$r + \tau = r + \tilde{\phi}(z_1 + \cdots + z_n - n z_0) - \varepsilon - t_0 \equiv \phi(p_1 + \cdots + p_g - g p_0) (\operatorname{mod} \mathscr{L}) ,$$

hence

$$r-\varepsilon-t_0=\phi(p_{n+1}+\cdots+p_g-(g-n)p_0)\in W_{g-n}\subseteq W_{g-1}$$

so that even in this case $t_0\in -W_{g-1}+r-\varepsilon$. Thus the zero locus of the function $\theta(t)$ must be contained in the irreducible complex analytic subvariety $-W_{g-1}+r-\varepsilon\subset J(M)$ of dimension $g-1$; but the zero locus of $\theta(t)$ is nontrivial, since $\theta(t)$ is relatively automorphic with respect to a factor of automorphy having a nontrivial characteristic matrix, so it must be a $(g-1)$-dimensional complex analytic subvariety of $-W_{g-1}+r-\varepsilon$ and hence as a consequence of irreducibility must coincide with $-W_{g-1}+r-\varepsilon$. Therefore it has been proved that

$$(121)\qquad -W_{g-1}+r-\varepsilon=\{t\in\mathbb{C}^g|\theta(t)=0\}/\mathscr{L}.$$

It is quite easy to see further that the function $\theta(t)$ vanishes to first order along the subvariety $-W_{g-1}+r-\varepsilon$, or equivalently that *the function $\theta(t)$ generates the proper ideal of the subvariety $-W_{g-1}+r-\varepsilon$ at each point of $J(M)$*. For this purpose consider the Riemannian theta function $\theta(\tilde{\phi}(z)-\tilde{\phi}(z_0)-\tau-\varepsilon)$, and recalling (121) note first as a useful preliminary observation that this function vanishes identically as a function of $z$ precisely when $r+\tau\in W_g^2\subset W_{g-1}$. Indeed recalling (121) this function vanishes identically precisely when $\tilde{\phi}(z)-\tilde{\phi}(z_0)-\tau-\varepsilon\in -W_{g-1}+r-\varepsilon$ for all points $z\in\tilde{M}$, hence when $r+\tau-\tilde{\phi}(z-z_0)\in W_{g-1}$ for all points $z\in\tilde{M}$; and that is equivalent to the condition that

$$\gamma(\rho_{r+\tau}\zeta_z^{-1}\zeta^g)=\gamma(\rho_{r+\tau-\tilde{\phi}(z-z_0)}\zeta^{g-1})\geq 1$$

for all points $z\in\tilde{M}$. This last condition really means that for any point $z\in\tilde{M}$ there must exist a nontrivial complex analytic relatively automorphic function for the factor of automorphy $\rho_{r+\tau}\zeta^g$ vanishing at $z$; and this in turn is evidently equivalent to the condition that $\gamma(\rho_{r+\tau}\zeta^g)\geq 2$, hence that $r+\tau\in W_g^2$ as desired. It follows readily from the Riemann-Roch theorem that $W_g^2=k-W_{g-2}$, hence that $W_g^2$ is a subvariety of dimension $g-2$ in $J(M)$. Now, returning to the proof of the assertion at the beginning of this paragraph, suppose conversely that the function $\theta(t)$ does not vanish to first order on the subvariety $-W_{g-1}+r-\varepsilon$; since that subvariety is irreducible the function $\theta(t)$ will vanish to some fixed order $n>1$ at all points of the subvariety, and consequently

$$\mathfrak{d}(\theta(\tilde{\phi}(z)-\tilde{\phi}(z_0)-\tau-\varepsilon))=n(q_1+\cdots+q_{g/n})$$

for some points $q_i\in M$ whenever the function $\theta(\tilde{\phi}(z)-\tilde{\phi}(z_0)-\tau-\varepsilon)$ does not vanish identically. Thus whenever $r+\tau\notin W_g^2$ the factors of automorphy $\rho_{r+\tau}\zeta^g$ and $(\zeta_{q_1}\cdots\zeta_{q_{g/n}})^n$ are analytically equivalent, for some points $q_i\in M$. Choosing any points $p_i\in M$ such that $\phi(p_1+\cdots+p_g-gp_0)\notin W_g^2$, and taking $r+\tau=\phi(p_1+\cdots+p_g-gp_0)$, it follows that $\zeta_{p_1}\cdots\zeta_{p_g}$ and $(\zeta_{q_1}\cdots\zeta_{q_{g/n}})^n$ are analytically equivalent for some points $q_i\in M$, and also that $\gamma(\zeta_{p_1}\cdots\zeta_{p_g})=1$; but that implies that $p_1\ldots+p_g=n(q_1+\cdots+q_{g/n})$, an obvious contradiction in general if $n>1$, and that suffices to prove the desired result.

Upon replacing $n$ by $-n$ as the index of summation in (118) it is clear that $\theta(t)=\theta(-t)$, hence that the Abelian theta function $\theta(t)$ is an even function. The zero locus of $\theta(t)$ is thus symmetric about the origin, so that $-W_{g-1}+r-\varepsilon=W_{g-1}-r+\varepsilon$ or equivalently

(122) $$W_{g-1}=-W_{g-1}+2(r-\varepsilon);$$

and that has an interesting consequence. Recall first that $t\in W_{g-1}$ precisely when $\gamma(\rho_t\zeta^{g-1})\geq 1$, and note that from the Riemann-Roch theorem $\gamma(\rho_t\zeta^{g-1})=\gamma(\kappa\rho_t^{-1}\zeta^{1-g})=\gamma(\rho_{k-t}\zeta^{g-1})$; hence $t\in W_{g-1}$ precisely when $k-t\in W_{g-1}$, or equivalently $W_{g-1}=-W_{g-1}+k$. On the other hand suppose that $W_{g-1}=W_{g-1}+x$ for some point $x\in J(M)$. Then for any points $p_1,\ldots,p_{g-1}\in M$ necessarily

$$x-\phi(p_1+\cdots+p_{g-1}-(g-1)p_0)\in W_{g-1},$$

so that

$$\gamma(\rho_x\zeta_{p_1}^{-1}\ldots\zeta_{p_g}^{-1}\zeta^{2g-2})=\gamma(\rho_{x-\phi(p_1+\cdots+p_{g-1}-(g-1)p_0)}\zeta^{g-1})\geq 1;$$

thus there is at least one nontrivial complex analytic relatively automorphic function for the factor of automorphy $\rho_x\zeta^{2g-2}$ vanishing at any $g-1$ specified points of $M$, hence evidently $\gamma(\rho_x\zeta^{2g-2})\geq g$. Applying the Riemann-Roch theorem, it follows that $\gamma(\rho_{k-x})=\gamma(\kappa\rho_x^{-1}\zeta^{2-2g})\geq 1$, hence that $\rho_{k-x}$ is analytically equivalent to the identity bundle and $x=k$. Therefore $W_{g-1}=-W_{g-1}+x$ if and only if $x=k$; and it then follows from (122) that

(123) $$2(r-\varepsilon)=k.$$

Thus the Riemann point $r\in J(M)$ is determined up to elements of order 2 in the group $J(M)$ by the canonical point $k\in J(M)$. Of course the elements of order 2 in $J(M)$ are the $2^{2g}$ half-periods, the representatives of the points of $\frac{1}{2}\mathscr{L}$ modulo $\mathscr{L}$.

Since the function $\theta(t)$ vanishes on the subvariety $-W_{g-1}+r-\varepsilon=W_{g-1}-r+\varepsilon$ it is clear that the translated function

(124) $$\theta_0(t)=\theta(t-r+\varepsilon)=\theta(t-(k-r)-\varepsilon)$$

vanishes on the subvariety $W_{g-1}$ itself; indeed, recalling the finer point established above, the function $\theta_0(t)$ generates the proper ideal of the subvariety $W_{g-1}$ at each of its points. Riemann observed and sketched a proof of an even more refined result in this direction, usually called **Riemann's vanishing theorem** [20, pp. 212–224]. To establish a convenient notation for stating this theorem, for any integer $n\geq 0$ let $\mathscr{S}^n(\theta_0)$ be the set of those points in $\mathbb{C}^g$ at which the function $\theta_0$ and its partial derivatives of all orders $\leq n$ simultaneously vanish. Thus $\mathscr{S}^n(\theta_0)$ are complex analytic subvarieties of $\mathbb{C}^g$, and $\mathscr{S}^0(\theta_0)\supseteq\mathscr{S}^1(\theta_0)\supseteq\mathscr{S}^2(\theta_0)\supseteq\cdots$. Since $\theta_0$ is a relatively automorphic function for a factor of automorphy for the lattice subgroup $\mathscr{L}\subseteq\mathbb{C}^g$ it is evident that the subvarieties $\mathscr{S}^n(\theta_0)$ are invariant under that lattice subgroup, hence can be considered as subvarieties of the Jacobi variety $J(M)=\mathbb{C}^g/\mathscr{L}$. Now as already noted $\mathscr{S}^0(\theta_0)=W_{g-1}$. Furthermore, since $\theta_0$ generates the proper ideal of the subvariety $W_{g-1}$, it follows that $\mathscr{S}^1(\theta_0)$ is precisely the singular set of that subvariety; and using the observation, stated without proof at the end of § 8, that $W^2_{g-1}$ is precisely the singular set of the subvariety $W_{g-1}$, this can be rewritten as the assertion that $\mathscr{S}^1(\theta_0)=W^2_{g-1}$. More generally, *the Riemann vanishing theorem states that* $\mathscr{S}^n(\theta_0)=W^{n+1}_{g-1}$ *for any integer* $n\geq 0$. Complete proofs of this assertion have only recently been given; the following elegant proof is that of Henrik H. Martens [18], in whose paper references to the other proofs can be found.

As some useful preliminaries note first of all that $t\in W^{n+1}_{g-1}$ *for some integer* $n\geq 1$ *if and only if* $t+\phi(p_1-p_2)\in W^n_{g-1}$ *for all points* $p_i\in M$. Indeed by definition $t\in W^{n+1}_{g-1}$ precisely when $\gamma(\rho_t\zeta^{g-1})\geq n+1$; and as noted a few pages previously that is easily seen to be equivalent to the condition that $\gamma(\rho_t\zeta^{g-1}\zeta^{-1}_{p_2})\geq n$ for all points $p_2\in M$. Now by the Riemann-Roch theorem $\gamma(\kappa\rho_t^{-1}\zeta^{1-g}\zeta_{p_2})=\gamma(\rho_t\zeta^{g-1}\zeta^{-1}_{p_2})+1\geq n+1$; as before this inequality is equivalent to $\gamma(\kappa\rho_t^{-1}\zeta^{1-g}\zeta_{p_2}\zeta^{-1}_{p_1})\geq n$ for all points $p_i\in M$, and after yet another application of the Riemann-Roch theorem this is in turn equivalent to the desired condition that $\gamma(\rho_t\zeta^{g-1}\zeta_{p_1}\zeta^{-1}_{p_2})\geq n$. Note next that since the Riemann vanishing theorem involved various derivatives of the theta function it might be expected that some detailed properties of these derivatives would be of use in the proof; actually only a few remarkably simple observations about these derivatives are required. First recall that the function $\theta_0(t)=\theta(t-(k-r)-\varepsilon)$ is relatively automorphic for the factor of automorphy (117) with the parameter value $\tau=k-r$, so that $\theta_0(t+\lambda)=\xi_{k-r}(\lambda,t)\theta_0(t)$ whenever $\lambda\in\mathscr{L}$. Now if $D$ is any constant coefficient linear partial differential operator of order $\leq n$ then

$$\begin{aligned}(D\theta_0)(t+\lambda)&=D[\xi_{k-r}(\lambda,t)\theta_0(t)]\\&=\xi_{k-r}(\lambda,t)(D\theta_0)(t)\\&\quad+(\text{terms involving derivatives of }\theta_0\text{ of orders }\leq n-1),\end{aligned}$$

and consequently the restriction $D\theta_0|\mathscr{S}^{n-1}(\theta_0)$ of the function $D\theta_0$ to the subvariety $\mathscr{S}^{n-1}(\theta_0)\subseteq\mathbb{C}^g$ is a relatively automorphic function for the same factor of automorphy on that subvariety. The same observation of course holds for any translate of the function $\theta_0$. The partial differential operators of particular interest here are those most closely related to the Jacobi imbedding of the Riemann surface $M$ in $J(M)$. In terms of any given local coordinate $z$ at a point $p_1\in M$ introduce the first order constant coefficient linear partial differential operator $\partial_{p_1}$ defined by

$$\begin{aligned}(\partial_{p_1}\theta_0)(t)&=\textstyle\sum_{j=1}^g w_j'(p_1)\,\dfrac{\partial\theta_0(t)}{\partial t_j}\\&=\dfrac{d}{dz}\,\theta_0(t+\phi(z)-\phi(p_1))\Big|_{z=p_1},\end{aligned}$$

where $w_j'(p_1)=dw_j(z)/dz|_{z=p_1}$. Since the Abelian differentials $\omega_j=dw_j$ are linearly independent these differential operators span the space of all first order constant coefficient linear partial differential operators, for various points $p_1$; thus any constant coefficient linear partial differential operator of order $n$ can be written as a linear combination of the operators $\partial_{p_1}\ldots\partial_{p_n}$ for various points $p_i\in M$.

Using these observations the Riemann vanishing theorem will be proved by induction on the index $n$, the case $n=0$ having already been established. Assume therefore that $\mathscr{S}^n(\theta_0)=W^{n+1}_{g-1}$ for some index $n\geq 0$. If $t\in W^{n+2}_{g-1}$ then $t+\phi(p_1-p_2)\in W^{n+1}_{g-1}=\mathscr{S}^n(\theta_0)$ for all points $p_i\in M$, and consequently $(D\theta_0)(t+\phi(p_1-p_2))=0$ for any constant coefficient linear partial differential operator $D$ of order $\leq n$ and any points $p_i\in M$. Then

$$(\partial_{p_1}D\theta_0)(t)=\frac{d}{dz}D\theta_0(t+\phi(z-p_1))\bigg|_{z=p_1}=\frac{d}{dz}0\bigg|_{z=p_1}=0$$

for any point $p_1\in M$; and since the differential operators $\partial_{p_1}D$ span all constant coefficient linear partial differential operators of orders $\leq n+1$ it follows that $t\in\mathscr{S}^{n+1}(\theta_0)$. Thus it has been demonstrated that $W^{n+2}_{g-1}\subseteq\mathscr{S}^{n+1}(\theta_0)$. To demonstrate the other containment it is convenient to consider separately the cases $n=0$ and $n\geq 1$. First suppose that $t\in\mathscr{S}^1(\theta_0)$, and for any point $z_1\in\tilde{M}$ consider the function $f(z)=\theta_0(t+\tilde{\phi}(z-z_1))$ as a complex analytic function of $z\in\tilde{M}$. If this function is identically zero for any choice of $z_1\in\tilde{M}$ then of course $t+\tilde{\phi}(z-z_1)\in W_{g-1}$ for all points $z,z_1\in\tilde{M}$ and consequently $t\in W^2_{g-1}$. Otherwise for any point $z_1$ outside an analytic subvariety of $\tilde{M}$ the function $f(z)$ is not identically zero; and writing $f(z)=\theta(\tilde{\phi}(z-z_0)-(\tilde{\phi}(z_1-z_0)-t+k-r)-\varepsilon)$ note that this function is then a nontrivial relatively automorphic function for the Riemannian theta factor of automorphy associated to the parameter $\tau=\tilde{\phi}(z_1-z_0)-t+k-r$. Note further that $f(z_1)=\theta_0(t)=0$ and that for any local coordinate $z$ at $z_1$

$$f'(z_1)=\frac{d}{dz}f(z)\bigg|_{z=z_1}=\frac{d}{dz}\theta_0(t+\tilde{\phi}(z-z_1))=\partial_{p_1}\theta_0(t)=0$$

since $t\in\mathscr{S}^1(\theta_0)$, where $z_1\in\tilde{M}$ represents $p_1\in M$; thus $f$ has a double zero as the point $p_1$, and hence $\mathfrak{d}(f)=2p_1+p_3+\cdots+p_g$ for some points $p_i\in M$. Then upon recalling the properties of the Riemannian theta factor of automorphy it follows that

$$\phi(2p_1+p_3+\cdots+p_g-gp_0)=r+\tau=\phi(p_1-p_0)-t+k$$

hence that

$$k-t-\phi(p_1-p_0)=\phi(p_3+\cdots+p_g-(g-2)p_0)\in W_{g-2}$$

or equivalently that

$$\gamma(\rho_{k-t}\zeta_{p_1}^{-1}\zeta^{g-1})=\gamma(\rho_{k-t-\phi(p_1-p_0)}\zeta^{g-2})\geq 1\,;$$

but since that holds for all but finitely many points $p_1\in M$ it follows as before that $\gamma(\rho_{k-t}\zeta^{g-1})\geq 2$, hence applying the Riemann-Roch theorem that $\gamma(\rho_t\zeta^{g-1})\geq 2$ and $t\in W^2_{g-1}$. Thus in either case $t\in W^2_{g-1}$, and therefore $\mathscr{S}^1(\theta_0)\subseteq W^2_{g-1}$. Next suppose that $t\in\mathscr{S}^{n+1}(\theta_0)$ for some index $n\geq 1$. Then of course $t\in\mathscr{S}^n(\theta_0)=W^{n+1}_{g-1}$, so that $t+\tilde{\phi}(z_1-z_2)\in W^n_{g-1}=\mathscr{S}^{n-1}(\theta_0)$ for all points $z_i\in\tilde{M}$. If $D_n\theta_0(t+\tilde{\phi}(z_1-z_2))=0$

for all constant coefficient linear partial differential operators $D_n$ of order $\leq n$ and all points $z_i \in \tilde{M}$ then $t+\tilde{\phi}(z_1-z_2) \in \mathcal{S}^n(\theta_0)=W_{g-1}^{n+1}$ for all points $z_i \in \tilde{M}$ and consequently $t \in W_{g-1}^{n+2}$. Otherwise there exist some constant coefficient linear partial differential operator $D_n$ of order $n$ and some points $z_i \in \tilde{M}$ such that $D_n \theta_0(t+\tilde{\phi}(z_1-z_2)) \neq 0$; indeed there exist points $z_1, \ldots, z_n, z' \in \tilde{M}$ representing points $p_1, \ldots, p_n, p' \in M$ such that the function $f(z)=\partial_{p_1} \ldots \partial_{p_n} \theta_0(t+\tilde{\phi}(z-z'))$ is not identically zero, and moreover that will be the case for all points $p_1, \ldots, p_n, p'$ outside a proper complex analytic subvariety of $M^{n+1}$. As before $\theta_0(t+\tilde{\phi}(z-z'))$ is a nontrivial relatively automorphic function for the Riemannian theta factor of automorphy associated to the parameter $\tau=\tilde{\phi}(z'-z_0)-t+k-r$; and since $t+\tilde{\phi}(z'-z_1) \in \mathcal{S}^{n-1}(\theta_0)$ for all $z \in \tilde{M}$ it then follows that $f(z)$ is also a nontrivial relatively automorphic function for the same factor of automorphy. Note that for the local coordinate $z$ at $z_1$

$$\begin{aligned}
f(z_1) &= \partial_{p_1} \ldots \partial_{p_n} \theta_0(t+\tilde{\phi}(z_1-z')) \\
&= \frac{d}{dz} \partial_{p_2} \ldots \partial_{p_n} \theta_0(t+\tilde{\phi}(z_1-z')+\tilde{\phi}(z-z_1)) \Big|_{z=z_1} \\
&= \frac{d}{dz} \partial_{p_2} \ldots \partial_{p_n} \theta_0(t+\tilde{\phi}(z-z')) \Big|_{z=z_1} = \frac{d}{dz} 0 \Big|_{z=z_1} \\
&= 0,
\end{aligned}$$

since $t+\tilde{\phi}(z-z') \in \mathcal{S}^{n-1}(\theta_0)$; thus $f$ vanishes at the point $p_1$, and by symmetry also at the points $p_2, \ldots, p_n$. Note further that $f(z')=\partial_{p_1} \ldots \partial_{p_n} \theta_0(t)=0$ and that for any local coordinate $z$ at $z'$

$$\begin{aligned}
f'(z') &= \frac{d}{dz} f(z) \Big|_{z=z'} = \frac{d}{dz} \partial_{p_1} \ldots \partial_{p_n} \theta_0(t+\tilde{\phi}(z-z')) \Big|_{z=z'} \\
&= \partial_{p'} \partial_{p_1} \ldots \partial_{p_n} \theta_0(t)=0,
\end{aligned}$$

since $t \in \mathcal{S}^{n+1}(\theta_0)$; thus $f(z)$ has a double zero at $p'$. Altogether then

$$\mathfrak{d}(f)=2p'+p_1+\cdots+p_n+q_{n+3}+\cdots q_g$$

for some points $q_i \in M$; and recalling again the properties of the Riemannian theta factor of automorphy it follows that

$$\phi(2p'+p_1+\cdots+p_n+q_{n+3}+\cdots+q_g-gp_0)=r+\tau=\phi(p'-p_0)-t+k,$$

hence that

$$k-t-\phi(p'+p_1+\cdots+p_n-(n+1)p_0)=\phi(q_{n+3}+\cdots+q_{g-n-2}-(g-n-2)p_0) \in W_{g-n-2}.$$

Therefore

$$\gamma(\rho_{k-t}\zeta_{p'}^{-1}\zeta_{p_1}^{-1}\ldots\zeta_{p_n}^{-1}\zeta^{g-1})=\gamma(\rho_{k-t-\phi(p'+p_1+\cdots+p_n-(n+1)p_0}\zeta^{g-n-2}) \geq 1;$$

but this holds for all points $(p', p_1, \ldots, p_n)$ outside a proper complex analytic subvariety of $M^{n+1}$, and as before it then follows that $\gamma(\rho_{k-t}\zeta^{g-1}) \geq n+2$, or after applying the Riemann-Roch theorem, that $\gamma(\rho_t\zeta^{g-1}) \geq n+2$. Thus again $t \in W_{g-1}^{n+2}$, and as a result it has been shown that $\mathscr{S}^{n+1}(\theta_0) \subseteq W_{g-1}^{n+2}$; and that is enough to conclude the proof of the Riemann vanishing theorem.

Finally a few further properties of the Riemannian theta function which follow from quite similar lines of reasoning should also be mentioned. For this purpose consider a fixed point $t \in W_{g-1}$ such that $t \notin W_{g-1}^2$; this point can thus be represented as $t = \phi(p_1 + \cdots + p_{g-1} - (g-1)p_0)$ for a unique divisor $\mathfrak{d} = p_1 + \cdots + p_{g-1}$ on $M$. Viewing $t$ as lying in $\mathbb{C}^g$ rather than in $J(M)$, consider then the Riemannian theta function

$$f(z) = \theta_0(\tilde{\phi}(z) - \tilde{\phi}(z') + t) = \theta(\tilde{\phi}(z - z_0) - (\tilde{\phi}(z' - z_0) + k - r - t) - \varepsilon)$$

associated to the parameter value $\tau = \tilde{\phi}(z' - z_0) + k - r - t$, for some fixed point $z' \in \tilde{M}$. Note first that this function vanishes identically in $z$ precisely when $z' \in \tilde{M}$ represents one of the points $p_1, \ldots, p_{g-1}$ of $M$. Indeed since $W_{g-1}$ is the zero locus of $\theta_0$ the function $f(z)$ vanishes identically precisely when $t + \tilde{\phi}(z - z') = t' + \tilde{\phi}(z - z_0) \in W_{g-1}$ for all points $z \in \tilde{M}$, where $t' = t - \tilde{\phi}(z' - z_0)$, hence precisely when

$$\gamma(\rho_{t'}\zeta_z\zeta^{g-2}) = \gamma(\rho_{t' + \tilde{\phi}(z - z_0)}\zeta^{g-1}) \geq 1$$

for all points $z \in \tilde{M}$. By the Riemann-Roch theorem that is equivalent to the condition that $\gamma(\kappa\rho_{t'}^{-1}\zeta_z^{-1}\zeta^{2-g}) \geq 1$ for all points $z \in \tilde{M}$, which as usual reduces to the condition that $\gamma(\kappa\rho_{t'}^{-1}\zeta^{2-g}) \geq 2$; and by another application of the Riemann-Roch theorem that is in turn equivalent to the condition that $\gamma(\rho_{t'}\zeta^{g-2}) \geq 1$, hence that $t - \tilde{\phi}(z' - z_0) = t' \in W_{g-2}$. This means however that $t = \phi(p' + q_1 + \cdots + q_{g-2} - (g-1)p_0)$ for some points $q_i \in M$, where $z' \in \tilde{M}$ represents $p' \in M$; but since $t$ is uniquely expressible in such a form it follows that $p'$ (and the points $q_i$) must belong to the set of points $p_1, \ldots, p_{g-1}$, as desired. Thus if $z_i \in \tilde{M}$ represents $p_i \in M$ then $f(z) = \theta_0(\tilde{\phi}(z) - \tilde{\phi}(z_i) + t)$ vanishes identically; so viewing $z$ as a local coordinate near $z_i$,

$$0 = f'(z_i) = \frac{d}{dz} f(z)\bigg|_{z = z_i} = \sum_{j=1}^{g} \frac{\partial \theta_0}{\partial t_j}(t) \cdot w_j'(z_i).$$

This can of course be written more invariantly in the form

$$\sum_{j=1}^{g} \frac{\partial \theta_0}{\partial t_j}(t) \cdot \omega_j(z_i) = 0;$$

thus the Abelian differential

$$(125) \qquad \omega_t(z) = \sum_{j=1}^{g} \frac{\partial \theta_0}{\partial t_j}(t) \cdot \omega_j(z),$$

which is nontrivial since as a consequence of the Riemann vanishing theorem not all the first partial derivatives of $\theta_0$ vanish at the point $t$, vanishes at the points $p_1,\dots,p_{g-1}$. Noting by an application of the Riemann-Roch theorem that $\gamma(\kappa\zeta_{p_1}^{-1}\dots\zeta_{p_{g-1}}^{-1})=\gamma(\zeta_{p_1}\dots\zeta_{p_{g-1}})=\gamma(\rho_t\zeta^{g-1})=1$, it follows that there is, up to a nonzero constant factor, a unique Abelian differential vanishing at the points $p_1,\dots,p_{g-1}$; (125) gives the expression of that differential in terms of the canonical Abelian differentials.

On the other hand if $p'$ is not one of the points $p_1,\dots,p_{g-1}$ then $f(z)$ is a nontrivial Riemannian theta function and $f(z')=\theta_0(t)=0$; thus, recalling again the properties of the Riemannian theta function,

$$\mathfrak{d}(f)=p'+q_1+\dots+q_{g-1}$$

for some points $q_i\in M$ and

$$\phi(p'+q_1+\dots+q_{g-1}-gp_0)=r+\tau=\phi(p'-p_0)+k-t\,.$$

This last formula can be rewritten

$$\begin{aligned}k&=t+\phi(q_1+\dots+q_{g-1}-(g-1)p_0)\\&=\phi(p_1+\dots+p_{g-1}+q_1+\dots+q_{g-1}-2(g-1)p_0)\,;\end{aligned}$$

the divisor $p_1+\dots+p_{g-1}+q_1+\dots+q_{g-1}$ is therefore the divisor of an Abelian differential on $M$, recalling the definition of the canonical point $k\in J(M)$. Indeed this divisor must be the divisor of the Abelian differential $\omega_t$ vanishing at the points $p_1,\dots,p_{g-1}$, and is consequently uniquely determined and quite independent of the choice of the point $p'$. That is to say, as $p'$ varies over the complement of the points $p_1,\dots,p_{g-1}$ in $M$ the divisors of the corresponding nontrivial Riemannian theta functions $f(z)$ consist of the variable point $p'$ and the $g-1$ fixed points $q_1,\dots,q_{g-1}$. Now if $z_+,z_-$ are two points of $\tilde{M}$ representing points $p_+,p_-$ of $M$, neither of which is one of the points $p_1,\dots,p_{g-1}$, then the quotient function

$$f_{z_+z_-}(z)=\frac{\theta_0(\tilde{\phi}(z)-\tilde{\phi}(z_+)+t)}{\theta_0(\tilde{\phi}(z)-\tilde{\phi}(z_-)+t)}$$

will be a meromorphic relatively automorphic function on $\tilde{M}$ with divisor

$$\mathfrak{d}(f_{z_+z_-})=p_+-p_-$$

on $M$. This function is of course a relatively automorphic function for the factor of automorphy $\xi_{\tau_+}/\xi_{\tau_-}$, where $\xi_\tau$ is the Riemannian theta factor of automorphy (119) and $\tau_+=\tilde{\phi}(z_+-z_0)+k-r-t$ and $\tau_-=\tilde{\phi}(z_--z_0)+k-r-t$; and since from (119)

$$\begin{aligned}&\xi_{\tau_+}(A_i,z)/\xi_{\tau_-}(A_i,z)=1\,,\\&\xi_{\tau_+}(B_i,z)/\xi_{\tau_-}(B_i,z)=\exp 2\pi i(\tau_i^+-\tau_i^-)=\exp 2\pi i(w_i(z_+)-w_i(z_-))\end{aligned}$$

it follows that $\xi_{\tau_+}/\xi_{\tau_-}$ coincides with the flat factor of automorphy $\rho_{\tilde{\phi}(z_+-z_-)}$. Recall from § 6 though that the prime function $p(z, z_j, z_+, z_-)$ is also a well defined meromorphic relatively automorphic function on $\tilde{M}$, since $z_j$ represents a point $p_j \in M$ distinct from $p_+$ and $p_-$, and that it has the same divisor and the same factor of automorphy as the function $f_{z_+z_-}$; consequently $f_{z_+z_-}(z) = C_j p(z, z_j, z_+, z_-)$ for some nontrivial complex constant $C_j$, indeed for $C_j = f_{z_+z_-}(z_j)$. Therefore *the prime function for the marked Riemann surface M can be expressed in terms of the Riemannian theta functions in the form*

$$p(z, z_j, z_+, z_-) = \frac{f_{z_+z_-}(z)}{f_{z_+z_-}(z_j)} = \frac{\theta_0(\tilde{\phi}(z) - \tilde{\phi}(z_+) + t)\,\theta_0(\tilde{\phi}(z_j) - \tilde{\phi}(z_-) + t)}{\theta_0(\tilde{\phi}(z) - \tilde{\phi}(z_-) + t)\,\theta_0(\tilde{\phi}(z_j) - \tilde{\phi}(z_+) + t)}$$

*whenever* $t = \tilde{\phi}(z_1 + \cdots + z_{g-1} - (g-1)z_0)$ *represents a point of* $W_{g-1}$ *not contained in the singular locus* $W^2_{g-1} \subset W_{g-1}$.

## § 19. Some Analytic Cohomology Groups for Complex Tori

Consider again a general complex torus $J$ described by a period matrix $\Omega$, so that $J = \mathbb{C}^g/\mathscr{L}$ where $\mathscr{L} = \Omega \cdot \mathbb{Z}^{2g}$; and let $\xi$ be a scalar factor of automorphy for the action of the lattice subgroup $\mathscr{L}$ on $\mathbb{C}^g$. The factor of automorphy $\xi$ can be used to impose on the space of all complex analytic functions on $\mathbb{C}^g$ the structure of an $\mathscr{L}$-module, by associating to each function $f$ and each element $\lambda \in \mathscr{L}$ the function $f^\lambda$ defined by

$$(126) \qquad f^\lambda(t) = \xi(\lambda, t)^{-1} f(t + \lambda).$$

Indeed the mapping $f \longrightarrow f^\lambda$ is evidently linear, and it is readily verified that $(f^{\lambda_1})^{\lambda_2} = f^{\lambda_1 + \lambda_2}$. Now to any $\mathscr{L}$-module there are associated in a well known manner the formal cohomology groups of the group $\mathscr{L}$ with coefficients in that $\mathscr{L}$-module, [7]; in this particular example these cohomology groups will be denoted by $H^p(\mathscr{L}, \xi)$. The zero-dimensional cohomology group is by definition just the space of $\mathscr{L}$-invariant elements in the $\mathscr{L}$-module, the space consisting of those elements $f$ such that $f^\lambda = f$ for all $\lambda \in \mathscr{L}$; thus $H^0(\mathscr{L}, \xi)$ is the space of complex analytic relatively automorphic functions for the factor of automorphy $\xi$, and is of an abvious analytical interest. The one-dimensional cohomology group is of somewhat less obvious interest, but is nonetheless occasionally both useful and relatively easy to calculate; so it will be considered next, at least for the simplest factors of automorphy.

The space $Z^1(\mathscr{L}, \xi)$ of one-cocycles is defined as the space of those mappings $\sigma: \mathscr{L} \times \mathbb{C}^g \longrightarrow \mathbb{C}$ such that $\sigma(\lambda, t)$ is a complex analytic function of $t \in \mathbb{C}^g$ for each $\lambda \in \mathscr{L}$ and that

$$(127) \qquad \sigma(\lambda_1 + \lambda_2, t) = \xi(\lambda_2, t)^{-1} \sigma(\lambda_1, t + \lambda_2) + \sigma(\lambda_2, t)$$

for each $\lambda_1, \lambda_2 \in \mathscr{L}$; or equivalently $Z^1(\mathscr{L}, \xi)$ consists of those mappings $\sigma$ from $\mathscr{L}$ into the $\mathscr{L}$-module of complex analytic functions on $\mathbb{C}^g$ such that $\sigma(\lambda_1+\lambda_2)=\sigma(\lambda_1)^{\lambda_2}+\sigma(\lambda_2)$ for each $\lambda_1, \lambda_2 \in \mathscr{L}$. The subspace $B^1(\mathscr{L}, \xi) \subseteq Z^1(\mathscr{L}, \xi)$ of one-coboundaries is defined as the space of one-cocycles of the form

$$\sigma(\lambda, t)=\xi(\lambda, t)^{-1} f(t+\lambda)-f(t) \tag{128}$$

for some complex analytic function $f$ on $\mathbb{C}^g$; or equivalently $B^1(\mathscr{L}, \xi)$ consists of those one-cocycles $\sigma$ of the form $\sigma(\lambda)=f^\lambda-f$ for some element $f$ of the $\mathscr{L}$-module of complex analytic functions on $\mathbb{C}^g$. It is easily verified that any expression of the form (128) automatically satisfies the cocycle condition. The one-dimensional cohomology group is then defined to be the quotient space

$$H^1(\mathscr{L}, \xi)=Z^1(\mathscr{L}, \xi)/B^1(\mathscr{L}, \xi).$$

Note that this cohomology group really depends only on the analytic equivalence class of the factor of automorphy $\xi$, up to isomorphism. Indeed suppose that $\xi$ and $\xi_1$ are analytically equivalent factors of automorphy, hence that $\xi_1(\lambda, t)=\xi(\lambda, t)h(t+\lambda)/h(t)$ for some nowhere vanishing complex analytic function $h$ on $\mathbb{C}^g$. Then for any cocycle $\sigma \in Z^1(\mathscr{L}, \xi)$ the product $h(t)\sigma(\lambda, t)$ is easily seen to be a cocycle in $Z^1(\mathscr{L}, \xi_1)$; and if $\sigma$ is a coboundary, so that $\sigma(\lambda, t)=\xi(\lambda, t)^{-1} f(t+\lambda)-f(t)$ for some complex analytic function $f$ on $\mathbb{C}^g$, then $h(t)\sigma(\lambda, t)=\xi_1(\lambda, t)^{-1}h(t+\lambda)f(t+\lambda)-h(t)f(t)$ is also a coboundary. Since $1/h(t)$ is also complex analytic this correspondence clearly establishes a linear isomorphism between $H^1(\mathscr{L}, \xi)$ and $H^1(\mathscr{L}, \xi_1)$, as desired.

When $\xi$ is the trivial factor of automorphy $\xi(\lambda, t)\equiv 1$ then the one-cocycles in $Z^1(\mathscr{L}, 1)$ are precisely the summands of automorphy for the action of the lattice group $\mathscr{L}$ on $\mathbb{C}^g$, the additive analogues of factors of automorphy; and the one-coboundaries in $B^1(\mathscr{L}, 1)$ are those summands of automorphy which are analytically trivial, when that is defined as the additive analogue of analytic triviality for factors of automorphy. The weak form of Abel's theorem for complex tori, as discussed in § 1 of this appendix, then merely amounts to the assertion that any one-cocycle in $Z^1(\mathscr{L}, 1)$ is cohomologous to a one-cocycle which is independent of $t$; and the one-cocycles which are independent of $t$ of course form the subspace $\operatorname{Hom}(\mathscr{L}, \mathbb{C}) \subseteq Z^1(\mathscr{L}, 1)$, so that

$$H^1(\mathscr{L}, 1) \cong \operatorname{Hom}(\mathscr{L}, \mathbb{C})/\operatorname{Hom}(\mathscr{L}, \mathbb{C}) \cap B^1(\mathscr{L}, 1).$$

Now a cocycle $\sigma \in \operatorname{Hom}(\mathscr{L}, \mathbb{C})$ is determined uniquely by its values on the $2g$ generators $\omega_i$ of the lattice $\mathscr{L}$, where $\omega_i$ are the column vectors of the period matrix $\Omega$; and these values $\sigma(\omega_i)$ can be prescribed quite arbitrarily, so that $\operatorname{Hom}(\mathscr{L}, \mathbb{C}) \cong \mathbb{C}^{2g}$. On the other hand such a cocycle is a coboundary precisely when $\sigma(\lambda)=f(t+\lambda)-f(t)$ for some complex analytic function $f$, which must of course be a linear function; and hence the coboundaries of this form are described precisely by the homogeneous linear functions on $\mathbb{C}^g$, so that $\operatorname{Hom}(\mathscr{L}, \mathbb{C}) \cap B^1(\mathscr{L}, 1) \cong \mathbb{C}^g$ and consequently $H^1(\mathscr{L}, 1) \cong \mathbb{C}^g$. In view of the observations at the end of the preceding paragraph it therefore follows that $H^1(\mathscr{L}, \xi) \cong \mathbb{C}^g$ *for any analytically trivial factor of automorphy* $\xi$.

Next consider topologically trivial factors of automorphy, but only for complex tori with period matrices satisfying the Riemann conditions, hence for Jacobi varieties of compact Riemann surfaces among other complex tori. The period matrix $\Omega$ can be reduced to the normal form $\Omega = (I, \Omega_2)$, where $\Omega_2$ is symmetric and has positive definite imaginary part. The generators of the lattice $\mathscr{L}$ are the $2g$ column vectors $\omega_i$ of the period matrix $\Omega$; and the first $g$ of these are the standard generators $\omega_i = \delta_i$ of the lattice subgroup $\mathbb{Z}^g \subset \mathscr{L} \subset \mathbb{C}^g$. If $\xi$ is a topologically trivial factor of automorphy then as noted before it is analytically equivalent to a flat factor of automorphy $\rho \in \mathrm{Hom}(\mathscr{L}, \mathbb{C}^*)$, indeed to a flat factor of automorphy for which $\rho(\omega_i) = 1$ for $1 \leq i \leq g$. If $\sigma \in Z^1(\mathscr{L}, \rho)$ is any associated one-cocycle then the restriction of $\sigma$ to the lattice subgroup $\mathbb{Z}^g \subset \mathscr{L}$ is evidently a one-cocycle in $Z^1(\mathbb{Z}^g, 1)$. Now an auxiliary result in the function-theoretic proof of the weak form of Abel's theorem for complex tori, as discussed in [5], is equivalent to the assertion that $H^1(\mathbb{Z}^g, 1) = 0$; or alternatively the cohomology group $H^1(\mathbb{Z}^g, 1)$ can be identified with the first cohomology group of the quotient space $\mathbb{C}^g/\mathbb{Z}^g$ with coefficients in the sheaf of germs of complex analytic functions, and the latter cohomology group is zero by Cartan's Theorem B since $\mathbb{C}^g/\mathbb{Z}^g \cong (\mathbb{C}^*)^g$ is a Stein manifold [22]. At any rate there exists a complex analytic function $f$ on $\mathbb{C}^g$ such that $\sigma(\lambda, t) = f(t+\lambda) - f(t)$ whenever $\lambda \in \mathbb{Z}^g$; and the cocycle $\sigma_1 \in Z^1(\mathscr{L}, \rho)$ given by

$$\sigma_1(\lambda, t) = \sigma(\lambda, t) - \rho(\lambda)^{-1} f(t+\lambda) - f(t)$$

is cohomologous to $\sigma$ and has the property that $\sigma_1(\lambda, t) = 0$ whenever $\lambda \in \mathbb{Z}^g$. Thus any cocycle in $Z^1(\mathscr{L}, \rho)$ is cohomologous to one which is trivial on the sublattice $\mathbb{Z}^g \subset \mathscr{L}$.

Having made these preliminary simplifications the calculation of the cohomology group $H^1(\mathscr{L}, \rho)$ is a relatively straightforward matter. A cocycle $\sigma \in Z^1(\mathscr{L}, \rho)$ is determined uniquely by the functions $\sigma(\omega_i, t)$ for $1 \leq i \leq 2g$; and since the only relations among the generators $\omega_i$ of the lattice subgroup $\mathscr{L}$ are those of commutativity it is an evident consequence of the cocycle condition (127) that the functions $\sigma(\omega_i, t)$ can be quite arbitrary, subject only to the restrictions that

$$(129) \qquad \rho(\omega_j)^{-1} \sigma(\omega_i, t+\omega_j) + \sigma(\omega_j, t) = \rho(\omega_i)^{-1} \sigma(\omega_j, t+\omega_i) + \sigma(\omega_i, t)$$

for $1 \leq i, j \leq 2g$. Of course it can also be assumed that $\rho(\omega_i) = 1$ and $\sigma(\omega_i, t) = 0$ for $1 \leq i \leq g$, as observed above. Then for $g+1 \leq i \leq 2g$, $1 \leq j \leq g$, condition (129) reduces to $\sigma(\omega_i, t+\omega_j) = \sigma(\omega_i, t)$, which implies that the functions $\sigma(\omega_i, t)$ are invariant under the lattice subgroup $\mathbb{Z}^g$ and hence have complex analytic Fourier expansions

$$\sigma(\omega_i, t) = \sum_{\nu \in \mathbb{Z}^g} \sigma_\nu(\omega_i) \exp 2\pi i\, {}^t\nu \cdot t$$

for $g+1 \leq i \leq 2g$. Furthermore condition (129) is then equivalent to the condition that these Fourier coefficients satisfy

$$\sigma_\nu(\omega_i)(1-\rho(\omega_j)^{-1}\exp 2\pi i\,{}^t\nu\cdot\omega_j)=\sigma_\nu(\omega_j)(1-\rho(\omega_i)^{-1}\exp 2\pi i\,{}^t\nu\cdot\omega_i) \tag{130}$$

for $g+1 \leq i, j \leq 2g$. This cocycle is a coboundary precisely when there exists a complex analytic function $f$ on $\mathbb{C}^g$ such that

$$\sigma(\omega_i, t)=\rho(\omega_i)^{-1} f(t+\omega_i)-f(t) \tag{131}$$

for $1 \leq i \leq 2g$. For $1 \leq i \leq g$ condition (131) reduces to $f(t+\omega_i)=f(t)$, which implies that the function $f$ is also invariant under the lattice subgroup $\mathbb{Z}^g$ and hence has a complex analytic Fourier expansion

$$f(t)=\sum_{\nu\in\mathbb{Z}^g} f_\nu \exp 2\pi i\,{}^t\nu\cdot t\,.$$

Then condition (131) is equivalent to the condition that these Fourier coefficients satisfy

$$-\sigma_\nu(\omega_i)=f_\nu(1-\rho(\omega_i)^{-1}\exp 2\pi i\,{}^t\nu\cdot\omega_i) \tag{132}$$

for $g+1 \leq i \leq 2g$. Now if $\rho(\omega_i)=\exp 2\pi i\,{}^t\nu\cdot\omega_i$ for some index $\nu\in\mathbb{Z}^g$ and for all indices $g+1 \leq i \leq 2g$ then the nowhere vanishing complex analytic function $h(t)=\exp 2\pi i\,{}^t\nu\cdot t$ has the property that $h(t+\omega_i)=\rho(\omega_i)h(t)$ for $1 \leq i \leq 2g$, and consequently the flat factor of automorphy $\rho$ is analytically trivial. Excluding the previously considered case that $\rho$ is analytically trivial, it therefore follows that for each index $\nu\in\mathbb{Z}^g$ there exists an index $g+1 \leq i \leq 2g$ such that $(1-\rho(\omega_i)^{-1}\exp 2\pi i\,{}^t\nu\cdot\omega_i)\neq 0$. For this index $i$ Eq. (132) determines the Fourier coefficient $f_\nu$ explicitly in terms of the coefficient $\sigma_\nu(\omega_i)$; and it further follows immediately from (130) that condition (132) holds for all indices $g+1 \leq i \leq 2g$. The cocycle $\sigma$ is consequently always the coboundary of a formal complex analytic Fourier series $f$; and if it can be proved that that Fourier series converges it will follow that $H^1(\mathscr{L}, \rho)=0$. Actually the convergence is an easy consequence of an estimate of the form.

$$\min_{\nu\in\mathbb{Z}^g} \max_{g+1\leq i\leq 2g}|1-\rho(\omega_i)^{-1}\exp 2\pi i\,{}^t\nu\cdot\omega_i| \geq \varepsilon > 0\,,$$

since the Fourier series for $\sigma(\omega_i, t)$ are convergent; and this estimate is in turn an evident consequence of the assertion that for some $\varepsilon > 0$ the set of those indices $\nu\in\mathbb{Z}^g$ such that

$$\max_{g+1\leq i\leq 2g}|1-\rho(\omega_i)^{-1}\exp 2\pi i\,{}^t\nu\cdot\omega_i| < \varepsilon \tag{133}$$

is finite, since the assumption that $\rho$ is not analytically trivial implies that for each index $\nu\in\mathbb{Z}^g$ there exists an index $g+1 \leq i \leq 2g$ such that

$(1-\rho(\omega_i)^{-1}\exp 2\pi i\,{}^t v\cdot\omega_i)\neq 0$. To demonstrate this finiteness write $\rho(\omega_i)=\exp 2\pi i\, s_i$ and note that if $\varepsilon$ is sufficiently small then (133) implies that for some $\delta>0$

$$\max_i|({}^t v\cdot\omega_i - s_i) - (\text{nearest integer})|<\delta\,,$$

hence that

$$\max_i|{}^t v\cdot \operatorname{Im}\omega_i - \operatorname{Im} s_i|<\delta\,;$$

and therefore there is a positive constant $C$ such that

$$\max_i|{}^t v\cdot \operatorname{Im}\omega_i|<C\,,$$

hence such that

$$|{}^t v\cdot \operatorname{Im}\Omega_2\cdot v| = |\textstyle\sum_i {}^t v\cdot \operatorname{Im}\omega_i\cdot v_i|\leq gC\max_i|v_i|\,.$$

On the other hand since the matrix $\operatorname{Im}\Omega_2$ is positive definite there is a positive constant $c>0$ such that

$${}^t v\cdot\operatorname{Im}\Omega_2\cdot v\geq c\textstyle\sum_i v_i^2\geq c\max_i|v_i|^2\,;$$

and upon combining this with the preceding inequality it follows that

$$\max_i|v_i|\leq c^{-1}gC\,.$$

There are of course only finitely many indices $v\in\mathbb{Z}^g$ satisfying such an inequality. In summary then, it has been demonstrated that *if $\xi$ is a topologically trivial factor of automorphy for a complex torus with period matrix satisfying the Riemann conditions then*

$$H^1(\mathscr{L},\xi)\cong\begin{cases}\mathbb{C}^g & \textit{if } \xi \textit{ is analytically trivial,}\\ 0 & \textit{otherwise.}\end{cases}$$

# References


1. Atiyah, M. F.: Vector bundles over an elliptic curve. Proc. London Math. Soc. **7**, 414—452 (1957).
2. Behnke, H., Sommer, F.: Theorie der analytischen Funktionen einer komplexen Veränderlichen. 2d ed. Berlin-Heidelberg-New York: Springer 1962.
3. Behnke, H., Thullen, P.: Theorie der Funktionen mehrerer komplexer Veränderlichen. 2d ed. Berlin-Heidelberg-New York: Springer 1970.
4. Cartan, H.: Espaces fibrés analytiques. In: Symposium Internacional de Topologia Algebraica, pp. 97—121. México: Univ. Nacional de México 1958.
5. Conforto, F.: Abelsche Funktionen und algebraische Geometrie. Berlin-Heidelberg-New York: Springer 1956.
6. Cornalba, M., Griffiths, P.: Analytic cycles and vector bundles on non-compact algebraic varieties. Inventiones Math. **28**, 1—106 (1975).
7. Eilenberg, S., MacLane, S.: Cohomology theory in abstract groups. Annals of Math. **48**, 51—78 and 326—341 (1947).
8. Grauert, H.: Analytische Faserungen über holomorph-vollständigen Räumen. Math. Annalen **135**, 266—273 (1958).
9. Gunning, R. C.: The structure of factors of automorphy. Amer. J. Math. **78**, 357—382 (1956).
10. Gunning, R. C.: Lectures on Riemann Surfaces. Mathematical Notes, vol. 2. Princeton: Princeton University Press 1966.
11. Gunning, R. C.: Lectures on Vector Bundles over Riemann Surfaces. Mathematical Notes, vol. 6. Princeton: Princeton University Press 1967.
12. Gunning, R. C.: Lectures on Riemann Surfaces: Jacobi Varieties. Mathematical Notes, vol. 12. Princeton: Princeton University Press 1972.
13. Gunning, R. C.: Some special complex vector bundles over Jacobi varieties. Inventiones Math. **22**, 187—210 (1973).
14. Gunning, R. C.: Lectures on Complex Analytic Varieties: Finite Analytic Mappings. Mathematical Notes, vol. 14. Princeton: Princeton University Press 1974.
15. Husemoller, D.: Fibre Bundles. New York: McGraw-Hill 1966.
16. Kempf, G.: A property of the periods of Prym differentials. Proc. Amer. Math. Soc. **54**, 181—184 (1976).
17. Krazer, A.: Lehrbuch der Thetafunktionen. Leipzig: Teubner 1903.
18. Martens, H. H.: On the partial derivatives of thetafunctions. Commentarii Math. Helvetici **48**, 394—408 (1973).
19. Porteous, I. R.: Simple singularities of maps. In: Proceedings of Liverpool Singularities Symposium I, pp. 286—307. Lecture Notes in Mathematics, vol. 192. Berlin-Heidelberg-New York: Springer 1971.
20. Riemann, B.: Collected Works. New York: Dover Publications, 1953.
21. Röhrl, H.: Holomorphic fiber bundles over Riemann surfaces. Bulletin Amer. Math. Soc. **68**, 125—160 (1962).
22. Serre, J. P.: Quelques problèmes globaux relatifs aux variétés de Stein. In: Coll. sur les fonctions de plusiers variables, Bruxelles 1953, pp. 57—68.
23. Siegel, C. L.: Topics in Complex Function Theory, vols. I, II, III. New York: Wiley-Interscience 1969—1973.
24. Springer, G.: Introduction to Riemann Surfaces. Reading, Mass.: Addison-Wesley 1957.

25. Titchmarsh, E. C.: The Theory of Functions. London: Oxford Univ. Press 1939.
26. Weil, A.: Généralisation des fonctions abéliennes. J. math. pures appl. **17**, 47—87 (1938).
27. Weil, A.: Théorèmes fondamentaux de la théorie des fonctions thêta. Séminaire Bourbaki, 1949.
28. Weil, A.: Introduction à l'étude des variétés kähleriennes. Paris: Hermann 1958.

# Index of Theorems

# Index of Notation

# Index

# Ergebnisse der Mathematik und ihrer Grenzgebiete

62. Robinson: Finiteness Conditions and Generalized Soluble Groups, Part 1
63. Robinson: Finiteness Conditions and Generalized Soluble Groups, Part 2
64. Hakim: Topos annelés et schémas relatifs
65. Browder: Surgery on Simply-Connected Manifolds
66. Pietsch: Nuclear Locally Convex Spaces
67. Dellacherie: Capacités et processus stochastiques
68. Raghunathan: Discrete Subgroups of Lie Groups
69. Rourke/Sanderson: Introduction to Piecewise-Linear Topology
70. Kobayashi: Transformation Groups in Differential Geometry
71. Tougeron: Idéaux de fonctions différentiables
72. Gihman/Skorohod: Stochastic Differential Equations
73. Milnor/Husemoller: Symmetric Bilinear Forms
74. Fǫssum: The Divisor Class Group of a Krull Domain
75. Springer: Jordan Algebras and Algebraic Groups
76. Wehrfritz: Infinite Linear Groups
77. Radjavi/Rosenthal: Invariant Subspaces
78. Bognár: Indefinite Inner Product Spaces
79. Skorohod: Integration in Hilbert Space
80. Bonsall/Duncan: Complete Normed Algebras
81. Crossley/Nerode: Combinatorial Functors
82. Petrov: Sums of Independent Random Variables
83. Walker: The Stone-Čech Compactification
84. Wells/Williams: Embeddings and Extensions in Analysis
85. Hsiang: Cohomology Theory of Topological Transformation Groups
86. Olevskiĭ: Fourier Series with Respect to General Orthogonal Systems
87. Berg/Forst: Potential Theory on Locally Compact Abelian Groups
88. Weil: Elliptic Functions according to Eisenstein and Kronecker
89. Lyndon/Schupp: Combinatorial Group Theory
90. Edwards/Gaudry: Littlewood-Paley and Multiplier Theory